Surface Science and Advanced Functional Applications of Cellulose Composites

Interface Science and Technology

Surface Science and Advanced Functional Applications of Cellulose Composites

Volume 38

Edited by

Mingguo Ma
MOE Engineering Research Center of Forestry Biomass Materials and Bioenergy, Research Center of Biomass Clean Utilization, Beijing Key Laboratory of Lignocellulosic Chemistry, College of Materials Science and Technology, Beijing Forestry University, Beijing, P.R. China

Xing-Xiang Ji
State Key Laboratory of Biobased Material and Green Papermaking, Qilu University of Technology (Shandong Academy of Sciences), Jinan, P.R. China

ELSEVIER

ACADEMIC PRESS
An imprint of Elsevier

Academic Press is an imprint of Elsevier
125 London Wall, London EC2Y 5AS, United Kingdom
525 B Street, Suite 1650, San Diego, CA 92101, United States
50 Hampshire Street, 5th Floor, Cambridge, MA 02139, United States

ISBN: 978-0-443-13675-7
ISSN: 1573-4285

For information on all Academic Press publications visit our website at https://www.elsevier.com/books-and-journals

Publisher: Peter Llewelyn
Acquisition Editor: Charles Bath
Editorial Project Manager: Emerald Li
Production Project Manager: Kumar Anbazhagan
Cover Designer: Greg Harris

Typeset by TNQ Technologies

Interface Science and Technology

Series Editor: Arthur T. Hubbard

In this Series:

Vol. 1 Clay Surfaces: Fundamentals and Applications
Edited by F. Wypych and K.G. Satyanarayana

Vol. 2 Electrokinetics in Microfluids
By Dongqing Li

Vol. 3 Radiotracer Studies of Interfaces
Edited By G. Horanyi

Vol. 4 Emulsions: Structure Stability and Interactions
Edited By D.N. Petsev

Vol. 5 Inhaled Particles
By Chiu-sen Wang

Vol. 6 Heavy Metals in the Environment: Origin, Interaction and Remediation
Edited By H.B. Bradl

Vol. 7 Activated Carbon Surfaces in Environmental Remediation
Edited By Teresa J. Bandosz

Vol. 8 Tissue Engineering: Fundamentals and Applications
By Yoshito Ikada

Vol. 9 Particles at Interfaces: Interactions, Deposition, Structure
By Zbigniew Adamczyk

Vol. 10 Interface Science in Drinking Water Treatment: Theory and Applications
Edited By G. Newcombe and D. Dixon

Vol. 11 Surface Complexation Modelling
Edited By Johannes Lützenkirchen

Vol. 12 Theory of Colloid and Interfacial Electric Phenomena
By Hiroyüki Ohshima

Vol. 13 Sorbent Deformation
By A.V. Tvardovskiy

Vol. 14 Advanced Chemistry of Monolayers at Interfaces: Trends in Methodology and Technology
Edited By Toyoko Imae

Vol. 15 Macromolecüles in Solütion and Brownian Relativity
By Stefano A. Mezzasalma

Vol. 16 The Properties of Water and their Role in Colloidal and Biological Systems
By Carel Jan van Oss

Vol. 17 Chemistry of Modified Oxide and Phosphate Sürfaces: Fundamentals and Applications
By Robson Fernandes de Farias

Vol. 18 Interface Science and Composites
By Soo-Jin Park and Min-Kang Seo

Vol. 19 Nanoparticle Technologies: From Lab to Market
By Farid Bensebaa

Vol. 20 Particles at Interfaces
By Zbigniew Adamczyk

Vol. 21 Self-Assembly Processes at Interfaces: Multiscale Phenomena
By Vincent Ball

Vol. 22 Rheology of Emulsions: Electrohydrodynamics Principles
By Aleksandar M. Spasic

Vol. 23 Tailored Thin Coatings for Corrosion Inhibition using Molecular Approach
By Simo Olavi Pehkonen and Shaojun Yuan

Vol. 24 Charge and Energy Storage in Electrical Double Layers
By Silvia Ahualli and Angel V. Delgado

Vol. 25 Stimuli Responsive Polymeric Membranes
By Mihir Kumar Purkait, Manish Kumar Sinha, Piyal Mondal, and Randeep Singh

Vol. 26 Theory of Electrophoresis and Diffusiophoresis of Highly Charged Colloidal Particles
By Eric Lee

Vol. 27 Graphene Surfaces: Particles and Catalysts
By Karim Kakaei, Mehdi Esrafili and Ali Ehsani

Vol. 28 An Introduction to Green Nanotechnology
By Mahmoud Nasrollahzadeh, Mohammad Sajadi, Monireh Atarod, Mohadeseh Sajjadi and Zahra Isaabadi

Vol. 29 Emerging Natural and Tailored Nanomaterials for Radioactive Waste Treatment and Environmental Remediation
Edited by Changlun Chen

Vol. 30 Advanced Low-Cost Separation Techniques in Interface Science
Edited by George Z. Kyzas and Athanasios C. Mitropoulos

Vol. 31 Surface Science of Photocatalysis
Edited by Jiaguo Yu, Mietek Jaroniec and Chuanjia Jiang

Vol. 32 Photocatalysis: Fundamental Processes and Applications
Edited by Jiaguo Yu, Mietek Jaroniec and Chuanjia Jiang

Vol. 33 Adsorption: Fundamental Processes and Applications
Edited by Jiaguo Yu, Mietek Jaroniec and Chuanjia Jiang

Vol. 34 Surface Science of Adsorbents and Nanoadsorbents
Edited by Tawfik A. Saleh

Vol. 35 S-scheme Heterojunction Photocatalysts: Fundamentals and Applications
By Jiaguo Yu, Liuyang Zhang, Linxi Wang and Bicheng Zhu

Vol. 36 Chemistry of Modified Oxide and Phosphate Surfaces: Fundamentals and Applications
Edited by Robson Fernandes de Farias

Vol. 37 Fundamentals of Soft Interfaces in Colloid and Surface Chemistry
By Hiroyuki Ohshima

Contents

6. Advanced cellulose-based electrodes and electrolytes for supercapacitor energy storage devices

Dandan Li, Tian Mai and Mingguo Ma

7. Cellulose composites for optical application

Ting Xu, Chuanling Si and Mingguo Ma

8. Conductive nanocomposite hydrogels for flexible wearable sensor

Wenyan Guo and Mingguo Ma

Contributors

Wentao Cao, Department of Prosthodontics Shanghai, Stomatological Hospital & School of Stomatology, Fudan University, Shanghai, PR China

Yanyan Dong, Institute of Environment and Sustainable Development in Agriculture, Chinese Academy of Agricultural Sciences, Beijing, PR China

Lianhua Fu, School of Biomedical Engineering, Shenzhen University Medical School, Shenzhen University, Shenzhen, Guangdong, PR China

Wenyan Guo, MOE Engineering Research Center of Forestry Biomass Materials and Bioenergy, Research Center of Biomass Clean Utilization, Beijing Key Laboratory of Lignocellulosic Chemistry, College of Materials Science and Technology, Beijing Forestry University, Beijing, PR China

Xing-Xiang Ji, State Key Laboratory of Biobased Material and Green Papermaking, Qilu University of Technology (Shandong Academy of Sciences), Jinan, PR China

Dandan Li, MOE Engineering Research Center of Forestry Biomass Materials and Bioenergy, Research Center of Biomass Clean Utilization, Beijing Key Laboratory of Lignocellulosic Chemistry, College of Materials Science and Technology, Beijing Forestry University, Beijing, PR China

Qi Liu, MOE Engineering Research Center of Forestry Biomass Materials and Bioenergy, Research Center of Biomass Clean Utilization, Beijing Key Laboratory of Lignocellulosic Chemistry, College of Materials Science and Technology, Beijing Forestry University, Beijing, PR China

Chang Ma, College of Mechanical Engineering, Tianjin University of Commerce, Tianjin, PR China

Mingguo Ma, MOE Engineering Research Center of Forestry Biomass Materials and Bioenergy, Research Center of Biomass Clean Utilization, Beijing Key Laboratory of Lignocellulosic Chemistry, College of Materials Science and Technology, Beijing Forestry University, Beijing, PR China; State Silica-based Materials Laboratory of Anhui Province, Bengbu, PR China

Tian Mai, MOE Engineering Research Center of Forestry Biomass Materials and Bioenergy, Research Center of Biomass Clean Utilization, Beijing Key Laboratory of Lignocellulosic Chemistry, College of Materials Science and Technology, Beijing Forestry University, Beijing, PR China

Chuanling Si, State Key Laboratory of Biobased Fiber Manufacturing Technology, Tianjin Key Laboratory of Pulp and Paper, College of Light Industry and Engineering, Tianjin University of Science and Technology, Tianjin, PR China

Peilin Wang, MOE Engineering Research Center of Forestry Biomass Materials and Bioenergy, Research Center of Biomass Clean Utilization, Beijing Key Laboratory of Lignocellulosic Chemistry, College of Materials Science and Technology, Beijing Forestry University, Beijing, PR China

Ting Xu, State Key Laboratory of Biobased Fiber Manufacturing Technology, Tianjin Key Laboratory of Pulp and Paper, College of Light Industry and Engineering, Tianjin University of Science and Technology, Tianjin, PR China

Chapter 1

Introduction

Mingguo Ma[1,*] **and Xing-Xiang Ji**[2,*]
[1]*MOE Engineering Research Center of Forestry Biomass Materials and Bioenergy, Research Center of Biomass Clean Utilization, Beijing Key Laboratory of Lignocellulosic Chemistry, College of Materials Science and Technology, Beijing Forestry University, Beijing, PR China;* [2]*State Key Laboratory of Biobased Material and Green Papermaking, Qilu University of Technology (Shandong Academy of Sciences), Jinan, PR China*
[*]*Corresponding authors E-mail: mg_ma@bjfu.edu.cn and xxjt78@163.com*

1.1 Process and production of cellulose and regenerated cellulose

Cellulose, as one of the most extensive biomass resources with an annual output of $10^{11} \sim 10^{12}$ tons on earth, has the characteristics of cheap, easily available, biodegradability, biocompatibility, and environmentally friendly (Klemm et al., 2005). The chemical structure of cellulose is connected by β-linear polymer composed of 1,4-dehydrated-D-glucopyranosyl unit. Cellulose has strong intramolecular and intermolecular hydrogen bonding, high degree of orientation, and crystallinity, which makes it insoluble in general organic and inorganic solvents. The aggregation structure of cellulose includes crystalline region and amorphous region. So far, there are five crystalline polymorphs of cellulose, having different cell structures, chain conformation, and stacking form. Under certain conditions, crystalline polymorphs can be transformed into each other.

In general, regenerated cellulose takes natural cellulose (cotton, hemp, bamboo, trees, and shrubs) as raw materials, which does not change its chemical structure, only change the physical structure of natural cellulose (Wang et al., 2016). In 1891, cellulose xanthate solution was made from natural cellulose. Because of the high viscosity of cellulose xanthate solution, the cellulose was precipitated again treated with acid. Among regenerated cellulose, Tencel fiber is a cellulose fiber regenerated from coniferous wood pulp and produced by solvent spinning (NMMO). Lyocell fiber takes natural

Surface Science and Advanced Functional Applications of Cellulose Composites
https://doi.org/10.1016/B978-0-443-13675-7.00001-4

cellulose as raw material, uses the aqueous solution of NMMO as the solvent to directly prepare cellulose solution, and produced by dry jet wet spinning. There are active hydroxyl groups on cellulose molecules, which enable regenerated cellulose to be grafted and copolymerized with many other molecules for combined modification, providing promising applications for the development of various cellulose composites.

Nanocellulose refers to cellulose with nanoscale structures (Dufresne, 2013), which can be further divided into three types according to their sources and methods, including cellulose nanocrystals (CNCs), cellulose nanofibers (CNFs), and bacterial nanocellulose (BNC). CNCs with shorter length and higher crystallinity were prepared by treated the biomass cellulose with strong acids such as concentrated sulfuric acid to hydrolyze the materials in the amorphous area, and retain the structural integrity of the crystalline area. TEMPO can selectively oxidize the hydroxyl group on the cellulose surface C6 to carboxyl group, and then use high-strength ultrasound/homogenization treatment to obtain CNFs. Cellulose is an impure multiphase solid, often accompanied by lignin, hemicellulose, and other organic and inorganic small molecular substances. How to efficiently separate cellulose and remove impurities is a research focus of cellulose science. BNC was first discovered by Brown in 1886 by cultured *Acetobacter xylinum* under static conditions. There are many kinds of bacteria that can produce cellulose, among which *A. xylinum* is the most powerful microbial strain for cellulose synthesis. BNC has many advantages of high crystallinity, relatively consistent molecular orientation, high purity, high tensile strength, high Young's modulus, and good biocompatibility. More importantly, different from CNCs and CNFs, BNC has no lignin, hemicellulose, and other impurities. Nanocellulose has excellent characteristics, biological property, and mechanical property. Nanocellulose has rapid development in various applications, especially biomedical applications (Jorfi and Foster, 2015). For example, typical biomedical applications include drug delivery, cell culture, tissue engineering, antibacterial materials, wound dressing and healing, and biomedical implants.

CNF products have gone from the laboratory stage to market. For example, Mitsubishi Pencil has sold pens containing CNF ink in North America, Japanese Paper Company has developed adult diapers containing CNF. Kyoto University has designed a CNF concept vehicle (NCV), which will be displayed during the 2019 Tokyo Auto Show. The body, hood, and translucent roof are all made of CNF-reinforced composite materials, and the weight is only half of that of traditional cars.

There are free hydroxyl groups distributed on the surface of cellulose, which can be modified by a series of physical, chemical, and biological methods to produce various functional materials for special purposes. The development of cellulose composites meets the requirements of sustainable development (Cao et al., 2018; Khalil et al., 2012). However, there are few published books on the cellulose composites for surface science and advanced functional applications.

1.2 Overview of advanced functional cellulose composites

Composite is a new material composed of two or more components. Composite materials not only retain the advantages of the properties of each component, but also can obtain comprehensive properties that cannot be achieved by a single component material through the complementation and correlation of the properties of each component. Composites can be mainly divided into structural composites and functional composites. Structural composites are used as load-bearing structures (Kessler et al., 2003), which are basically composed of reinforcement components that can bear loads and matrix components that can connect the reinforcement into a whole material and play the role of force transmission. Functional composites have many functions, such as conductive, superconducting, semiconducting, magnetic, piezoelectric, damping, wave absorbing, wave transmission, friction, shielding, flame retardant, heat proof, sound absorption, heat insulation, and other functions (Bai et al., 2011). At the same time, functional composites may also have new functions due to the composite effect. Multifunctional composites are the development direction of functional composites.

There are many reports on the advanced functional cellulose composites. A variety of synthesis methods and various applications were employed for the advanced functional cellulose composites. In the chapter entitled "***Microwave-assisted synthesis of cellulose-based (nano) composites and derivatives***" (Ma, 2015a), we reviewed the conversion of cellulose to cellulose-based (nano) composites via the microwave-assisted method, and introduced the concept, process, and mechanism of microwave-assisted method and cellulose-based (nano)composites or derivatives. In the chapter entitled "***Cellulose-based composites: Synthesis, characterization and potential applications***" (Ma, 2015b), a variety of synthesis methods for the preparation of cellulose-based composites are summarized, such as the hydrothermal/solvothermal method, biomimetic mineralization method, microwave-assisted method, (co)precipitation method, and sonochemistry method. It found that carbon-based nanomaterials could be obtained from biomass (Ma et al., 2017b), which have application in the field of biofuel cell. Biomass antibacterials can be acted as extracellular structural layer of microorganism and have an effect on the movement, transmembrane transport, and biochemical reactions of microorganism (Dong et al., 2017.). The synthesis, properties, and biomedical applications of biomass antibacterials were reviewed via some typical examples. We also presented a comprehensive status on the synthesis and unique properties of various magnetic nanocomposites adsorbents for the removal of inorganic, organic, and biological contaminants from water (Liu et al., 2018.). The adsorbents produced by the modification and functionalization of carboxymethyl cellulose are introduced and the cellulose-based (nano)composites as adsorbents are reviewed in detail (Shi et al., 2022). Various cellulose-based nanocarriers such as bacterial cellulose, cellulose acetate, microcrystalline cellulose, carboxymethyl cellulose, cellulose nanocrystals, and cellulose nanofibrils are reviewed in terms of being used in drug delivery systems for cancer treatment (Meng et al., 2017). The preparation, properties, and applications of these cellulose-based functional composites were depicted, including biological applications, water treatment, sensor, reinforcing agent, energy storage materials, Pickering emulsion stabilizer, and other versatile applications (Li et al., 2018). In the review article, the state-of-the-art progress of cellulose-based hydrogels was covered from their preparation methods (including chemical methods and physical methods) and physicochemical properties (such as stimuli-responsive properties, mechanical properties, and self-healing properties) to their biomedical applications, including drug delivery, tissue engineering, wound dressing, bioimaging, wearable sensors, and so on. Fu et al. (2019) and Deng et al. (2022) reviewed the applications of three types of nanocellulose (cellulose nanocrystals, bacterial cellulose, and cellulose nanofibers) and their derivatives in bone tissue composites, such as bone disease treatments, including bone defect repair, cartilage regeneration, osteomyelitis, and osteoporosis.

Lan et al. (2022) reviewed the current development of synthetic routes, major properties, and important applications of nanocellulose-based composites, aerogels, hydrogels, and films for potential environmental application with some examples, e.g., gases purification, contaminants removal, heavy-metal ions removal, etc. Metal−organic frameworks (MOFs), as the star nanomaterial in functional polymers, have a large surface area, high porosity, and adjustable structure. The collaboration of nanocellulose and MOFs is a desirable strategy to make composites especially interesting for multifunctional and multifield applications. Mai et al. (2023) highlighted an up-to-date overview of nanocellulose-based MOFs composites for the sewage treatment, gas separation, energy storage, and biomedical applications. Yuan et al. (2021) summarized the recent advancements of nanocellulose, including cellulose nanofibers (CNFs), cellulose nanocrystals (CNCs), and bacterial cellulose (BC), which are promising for biomedical applications such as biosensing, drug delivery, wound dressing, and tissue engineering. Ma et al. (2018) reviewed the main applications of nanocellulose composites in the areas of biomedical field, environmental field, electrode, and sensor applications. In the chapter entitled "***Nanocellulose and nanocellulose composites: Synthesis, characterization, and potential applications***," nanocellulose composites such as hydrogels, aerogels, electrode materials, photocatalytic materials, antibacterial materials, sustained release, and sensors were reviewed (Ma et al., 2017a).

In this book, we design Introduction as Chapter 1, listing three sections including process and production of cellulose and regenerated cellulose, advanced functional cellulose composites, and surface chemistry and polymeric physics of the cellulosic composite. Before the introduction of advanced functional cellulose composites, it is necessary to summary the process and production of cellulose and regenerated cellulose. The main content of this book is about advanced functional cellulose composites, not cellulose. Therefore, we applied the applications of advanced functional cellulose composites as main chapter. The chapter two-seven included applications of advanced functional cellulose composites in water treatment, electronic devices, energy storage, biomedical, antibacterial, and optical applications. Based on the references, it found that the hydrogel and aerogel of cellulose composite received special attention. Therefore, we listed the hydrogel and aerogel of cellulose composite as chapter eight and nine. Obviously, the advanced functional cellulose composites have widely applications, such as biofuel cell, carbon-based nanomaterials, Pickering emulsion stabilizer, etc. We designed these chapters based on our knowledge. We firm believer that do the things with familiar and do the things you be good at. Of course, some chapter is designed based on the suggestions of editor and reviewers. For example, the chapter about the optical application of advanced functional cellulose composites is added by the reviewers.

1.3 Surface chemistry of the cellulose composites

Interface refers to the boundary area between material phases, which exists between two phases and is about several to dozens of molecular layers thick. Surface chemistry or interfacial chemistry is a science that studies the surface characteristics of substances in multiphase systems and the physical and chemical processes and laws of surface occurrence. This means that the research content of interface chemistry is not limited to chemical processes and laws, but also to the characteristics of interface systems and physical processes and laws. The interface of composites refers to the small area with significant changes in chemical composition between the matrix and reinforcement, which can combine with each other and play a role in load transfer. Interface is the key factor to determine the properties of composite materials, and is the focus in the research field of composites. For a long time, people have made great efforts to obtain composites with predetermined properties by designing and fabricating interfaces with appropriate structures and properties.

The interface structure of composites has a significant influence on the properties. The relationship between the interface structure and the properties of composites is found by characterizing the chemical and physical structures, thickness and morphology, adhesive strength and residual stress of interface phases. The chemical structure of the interface phase of composites includes the constituent elements, valence states, and their distribution. It can be characterized by Auger electron spectroscopy, electron microprobe, X-ray photoelectron spectroscopy, scanning secondary ion mass spectrometer, electron energy loss spectrometer, Fourier infrared spectroscopy, micro-Raman spectroscopy, extended X-ray absorption fine structure spectroscopy, etc. The morphology and thickness of interface phase can be characterized by transmission electron microscopy and scanning electron microscopy. Because of the small area and complex composition of the interfacial phase, it is difficult to obtain all the information of the interfacial phase in composites. The interface effect of composites is to transfer the stress on the matrix to the reinforcement under load, which requires that the interface phase has sufficient bonding strength, and the two phase surfaces that can penetrate each other are prerequisites.

The interface problem is always a challenge with two substances are mixed together. The hydrophilic CNF cannot avoid agglomeration when mixed with most hydrophobic polymers. Therefore, many chemical methods have been proposed to modify CNF, such as oxidation, silylation, esterification, etherification, and polymer grafting. The only purpose of these methods is to enhance the interfacial compatibility between CNF and polymers.

The publications on cellulose composites are very rich, but there are few books devoted to interface issues. The research results and latest progress on interface problems are widely distributed in numerous publications in various disciplines. It is not easy for us to explore the surface chemistry of the cellulose composites. This book still focused on the synthesis, properties, and applications of advanced functional cellulose composites. The surface chemistry of the cellulose composites will receive special attention.

1.4 Chapter summary

In summary, this book has nine chapters, including introduction, hydrogel and aerogel of cellulose composite, providing a comprehensive summary looking at cellulose composites in surface science, and applications of biomedical, antibacterial, optical, water treatment, flexible electronic devices, and energy storage. Surface chemistry of the cellulose composites receives special attention to explore the interaction mechanism among advanced functional cellulose composites. Case studies about cellulose composites are another feature and highlight of this book. Various promising commercial applications of cellulose composites are also summarized. The primary audience will be chemists, chemical engineers, and materials scientists engaged in biomass, cellulose, nanocellulose, composites, surface science, and related application fields of biomedical, antibacterial, water treatment, flexible electronic device, and energy storage.

Acknowledgments

The financial support from the National Natural Science Foundation of China (22478036) and Beijing Nova Program (20230484431) is gratefully acknowledged.

References

Bai, H., Li, C., Shi, G.Q., 2011. Functional composite materials based on chemically converted graphene. Adv. Mater. 23, 1089–1115.

Cao, W.T., Chen, F.F., Zhu, Y.J., Zhang, Y.G., Jiang, Y.Y., Ma, M.G., Chen, F., 2018. Binary strengthening and toughening of MXene/cellulose nanofiber composite paper with nacre-inspired structure and superior electromagnetic interference shielding properties. ACS Nano 12 (5), 4583–4593. https://doi.org/10.1021/acsnano.8b00997.

Deng, J., Song, Q., Liu, S., Pei, W., Wang, P., Zheng, L., Huang, C., Ma, M., Jiang, Q., Zhang, K., 2022. Advanced applications of cellulose-based composites in fighting bone diseases. Compos. B Eng. 245. https://doi.org/10.1016/j.compositesb.2022.110221.

Dong, Y.Y., Ma, M.G., Zhang, X.M., 2017. In: Collins, E. (Ed.), Synthesis and Properties of Biomass-Based Nanomaterials for Antibacterials Applications. Antibacterials: Synthesis, Properties and Biological Activities. Nova Science Publishers, ISBN 978-1-63485-801-4.

Dufresne, A., 2013. Nanocellulose: a new ageless bionanomaterial. Mater. Today 16 (6), 220–227. https://doi.org/10.1016/j.mattod.2013.06.004.

Fu, L.H., Qi, C., Ma, M.G., Wan, P.B., 2019. Multifunctional cellulose-based hydrogels for biomedical applications. J. Mater. Chem. B 7, 1541–1562.

Jorfi, M., Foster, E.J., 2015. Recent advances in nanocellulose for biomedical applications. J. Appl. Polym. Sci. 132 (14). https://doi.org/10.1002/app.41719.

Kessler, M.R., Sottos, N.R., White, S.R., 2003. Self-healing structural composite materials. Compos. Part A-Appl. Sci. Manufacturing 34, 743–753.

Khalil, H.P.S.A., Bhat, A.H., Yusra, A.F.I., 2012. Green composites from sustainable cellulose nanofibrils: a review. Carbohydr. Polym. 87, 963–979.

Klemm, D., Heublein, B., Fink, H.P., Bohn, A., 2005. Cellulose: fascinating biopolymer and sustainable raw material. Angew. Chem. Int. Ed. 44 (22), 3358–3393. https://doi.org/10.1002/anie.200460587.

Lan, G.X., Liu, Y., Zhou, N., Guo, D.Q., Ma, M.G., 2022. Multifunctional nanocellulose-based composites for potential environmental applications. Cellulose 30, 39–60.

Li, Y.Y., Wang, B., Ma, M.G., Wang, B., 2018. Review of recent development on preparation, properties, and applications of cellulose-based functional materials. Int. J. Polym. Sci., 8973643

Liu, S., Ma, M.G., Zhang, X.M., Xu, F., 2018. Magnetic nanocomposite adsorbents. In: Kyzas, G., Mitropouios, A. (Eds.), Composite Nanoadsorbents. Elsevier, 5th December 2018. Paperback ISBN: 9780128141328, 385 pages.

Ma, M.G., 2015a. Cellulose and Cellulose Derivatives: Synthesis, Modification, Nanostructure and Applications. Published by Nova Science Publishers.

Ma, M.G., 2015b. Microwave-assisted synthesis of cellulose-based (nano) composites and derivatives. In: Production of Biofuels and Chemicals with Microwave.

Ma, M.G., Liu, Y.J., Dong, Y.Y., 2017a. In: Thakur, V.K. (Ed.), Nanocellulose and Nanocellulose Composites: Synthesis, Characterization, and Potential Applications. Handbook of Composites from Renewable Materials: Volume 8: Nanocomposites: Advanced Applications. John Wiley & Sons Inc. ISBN10 1119223830; ISBN13 9781119223832.

Ma, M.G., Liu, B., Meng, L.Y., 2017b. Applications of carbon-based nanomaterials in biofuel cell. In: Rai, M. (Ed.), Nanotechnology for Bioenergy and Biofuel Production. Springer, ISBN 978-3-319-45458-0.

Ma, C., Ma, M.G., Li, Z.W., Wang, B., 2018. Nanocellulose composites - property and applications. Paper Biomater. 3, 51–63.

Mai, T., Li, D.D., Chen, L., Ma, M.G., 2023. Collaboration of two-star nanomaterials: the applications of nanocellulose-based metal organic frameworks composites. Carbohydr. Polym. 302. https://doi.org/10.1016/j.carbpol.2022.120359.

Meng, L.Y., Wang, B., Ma, M.G., Zhu, J.F., 2017. Cellulose-based nanocarriers as platforms for cancer therapy. Curr. Pharmaceut. Des. 23 (35), 5292–5300.

Shi, R.J., Wang, T., Lang, J.Q., Zhou, N., Ma, M.G., 2022. Multifunctional cellulose and cellulose-based (nano) composite adsorbents. Front. Bioeng. Biotechnol. 10. https://doi.org/10.3389/fbioe.2022.891034.

Wang, S., Lu, A., Zhang, L.N., 2016. Recent advances in regenerated cellulose materials. Prog. Polym. Sci. 53, 169–206.

Yuan, Q., Bian, J., Ma, M.G., 2021. Advances in biomedical application of nanocellulose-based materials: a review. Curr. Med. Chem. 28 (40), 8275–8295.

Chapter 2

Cellulose composites for biomedical applications

Qi Liu and Mingguo Ma*

MOE Engineering Research Center of Forestry Biomass Materials and Bioenergy, Research Center of Biomass Clean Utilization, Beijing Key Laboratory of Lignocellulosic Chemistry, College of Materials Science and Technology, Beijing Forestry University, Beijing, PR China

Corresponding author. E-mail: mg_ma@bjfu.edu.cn

2.1 Introduction

Degradable polymers mainly include collagen, linear aliphatic polyesters, chitin, cellulose, polyamino acids, polyvinyl alcohol, polyhexyl ester, etc. They can undergo structural damage and performance degradation under the action of biological environments, and their degradation products can be absorbed and utilized by the body through normal metabolism or eliminated from the body. The ideal biomedical materials should be nontoxic, nonallergenic, nonirritating, nongenotoxicity, noncarcinogenic, and other adverse reactions. Therefore, it is crucial to understand the biological reactions of biomedical materials to the human body, including tissue reactions, blood reactions, and immune reactions.

In the review article, our group summarized the recent advancements of nanocellulose, including CNFs, CNCs, and BC (Table 2.1), for biomedical applications, such as biosensing, drug delivery, wound dressing, and tissue engineering (Yuan et al., 2021). Especially, in the mini-review by Meng et al. (2017), they focused on the up-to-date development of cellulose-based nanocarriers, such as BC, CA, MCC, CMC, CNCs, and CNFs, in drug delivery systems for cancer therapy. Fu et al. (2019a,b) reviewed state-of-the-art progress in cellulose-based hydrogels, covering from their preparation methods (including chemical methods and physical methods) and physicochemical properties (such as stimuli-responsive properties, mechanical properties, and self-healing properties) to their biomedical applications, including

Surface Science and Advanced Functional Applications of Cellulose Composites
https://doi.org/10.1016/B978-0-443-13675-7.00002-6

TABLE 2.1 List of abbreviations.

BC	Bacterial cellulose	HX	Hypoxanthine
CNCs	Cellulose nanocrystals	XOD	Xanthine oxidase
CNFs	Cellulose nanofibers	CLC	Carboxylated cellulose
CMC	Carboxymethyl cellulose	PEDOT	Poly(3,4-ethylene dioxythiophene)
CA	Cellulose acetate		
MCC	Microcrystalline cellulose	CNW	Cellulose nanowhiskers
HA	Hydroxyapatite	SF	Silk fibroin
CHA	Carbonated hydroxyapatite	HC	Hydroxyethyl cellulose
ILs	Ionic liquids	GO	Graphene oxide
ATP	Adenosine 50-triphosphate disodium salt	DOX	Doxorubicin
		c-MWCNTs	Carboxylated multiwalled carbon nanotubes
CP	Creatine phosphate disodium salt tetrahydrate		
FBP	D-fructose 1,6-bisphosphate trisodium salt octahydrate	PEG	Polyethylene glycol
		MB	Methylene blue
CQDs	Carbon quantum dots	MO	Methyl orange
GOD	Glucose oxidase		
MOFs	Metal–organic frameworks		
3D	Three-dimensional		
PVA	Polyvinyl alcohol		
RGO	Reduced graphene oxide		
GCE	Glass carbon electrode		
CO	Carbon monoxide		
VAG	Vertically aligned graphene		
PBS	Phosphate buffer saline		
CDs	Carbon dots		
PEI	Polyethyleneimine		

drug delivery, tissue engineering, wound dressing, bioimaging, wearable sensors, and so on. Qian et al. (2023) summarized state-of-the-art manufacturing, structural traits, and applications of BC, along with in situ and ex situ modification techniques and their biomedical applications.

In this chapter, we introduced the synthesis and property of cellulose composites for biomedical applications. We specially paid more attention to hydrothermal method, microwave-assisted method, and ultrasound agitation method. Then, we reviewed the development of advanced functional cellulose composites for biomedical applications, such as biosensors, bone regeneration, drug delivery, and bioimaging. Surface chemistry of advanced functional cellulose composites for biomedical applications received special attention. Finally, the problem and potential directions were suggested based on our knowledge. We try to contribute great content to the book "***Surface Science and Advanced Functional Applications of Cellulose Composites***."

2.2 Synthesis and property of cellulose composites for biomedical applications

As for the synthesis of cellulose composites for biomedical applications, our group did system work. Various methods, such as hydrothermal method, microwave-assisted method, and ultrasound agitation method, have been developed to fabricate cellulose composites.

2.2.1 Hydrothermal method

Hydrothermal synthesis refers to the method of chemical reaction synthesis of materials in aqueous solution at the temperature of 100–1000°C and pressure of 1 MPa ~ 1 GPa. Under subcritical and supercritical hydrothermal conditions, the reaction activity increases due to the reaction being at the molecular level. In general, hydrothermal method is widely used to synthesize nanomaterials. CHA has a composition and structure analogous to the bone apatite, showing higher bioactivity than HA. As early as 2010, we applied hydrothermal method to synthesize cellulose–CHA nanocomposites in NaOH–urea aqueous solution using microcrystalline cellulose, $CaCl_2$, and NaH_2PO_4 (Jia et al., 2010a). The microcrystalline cellulose was dissolute into NaOH–urea aqueous solution to obtain cellulose solution. CHA nanoparticles were dispersed in the cellulose matrix. The size of CHA nanoparticles in cellulose–CHA nanocomposites decreased with the increasing preparation temperature. As the main inorganic component of shells, calcium carbonate ($CaCO_3$) is known to have biological activities of protein-adhesive properties, cell compatibility, and hard tissue compatibility. In the previous work, we developed hydrothermal route to prepare cellulose/$CaCO_3$ bionanocomposites using the microcrystalline cellulose, $Ca(NO_3)_2 \cdot 4H_2O$, and $Na_2SiO_3 \cdot 9H_2O$ in NaOH–urea aqueous solution (Jia et al., 2012a,b). The urea also acts as the $CO_3{}^{2-}$ source for the synthesis of $CaCO_3$. The hydrothermal conditions had an effect on the morphology of the bionanocomposites, and the cellulose/$CaCO_3$ bionanocomposites had good biocompatibility as ideal candidate for practical biomedical applications (Fig. 2.1). Moreover, the hydrothermal route was also applied for the preparation of wood powder/calcite composites in the NaOH/urea solution using the dewaxed wood powder (Ma et al., 2012a,b). Compared with cellulose-based composites, the wood powder composites can utilize all the main components of lignocelluloses. It achieved $CaCO_3$ with the particles, six-side-like shape, and cube-like shape, the increase size of $CaCO_3$ with increasing heating time, and the decreased number of $CaCO_3$ with increasing heating time. Using NaH_2PO_4, it obtained the wood powder/$CaCO_3$ composites as majority phases and hydroxyapatite as a minor phase, the increased number of $CaCO_3$ cubes and congregates with increasing heating time.

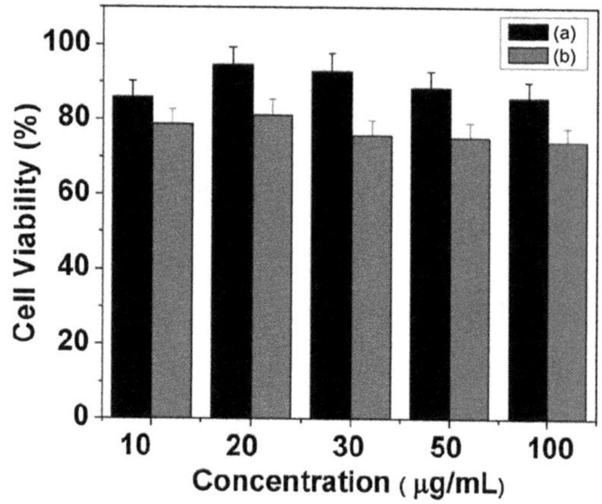

FIGURE 2.1 Viabilities of normal human fibroblasts incubated with the cellulose/$CaCO_3$ bionanocomposites: (a) using microcrystalline cellulose and (b) using cellulose solution. They were determined by survival cells per well relative to that of untreated cells. The error bars stand for standard deviations. *Reproduced from Jia et al. (2012a,b) with permission from Elsevier.*

2.2.2 Microwave-assisted method

Microwave is an electromagnetic wave with a frequency range of 0.3–300 GHz and a wavelength of 0.01–1 m. Microwave synthesis refers to the application of modern synthesis techniques under microwave conditions, utilizing its advantages such as fast heating, homogeneity, and selectivity. In 1986, Gedyse and colleagues discovered that the reaction between 4-cyanophenol salt and benzyl chloride in microwave was 240 times faster than traditional heating reflux. After that, microwave-assisted reaction has become a hot spot in the field of organic chemistry. In the literature, it reported that with the help of microwave technology, the reaction speed is tens or even thousands of times faster than traditional heating methods, and it has the characteristics of simple operation, high yield, easy purification of products, and safety (Zhu and Chen, 2014).

In 2010, the microwave-assisted method was used to synthesize cellulose–HA nanocomposites using microcrystalline cellulose, $CaCl_2$, and NaH_2PO_4 in DMAc solvent at 150°C for 10 min (Ma et al., 2010). HA nanorods were homogenously dispersed in the cellulose matrix. A strong interaction existed between the groups of cellulose and HA nanorods. This method has advantages of being simple, rapid, low-cost, and environmentally friendly. The microwave-assisted method was also applied to synthesize the cellulose–CHA nanocomposites in NaOH–urea aqueous solution by using the microcrystalline cellulose, $CaCl_2$, and NaH_2PO_4 (Jia et al., 2010b). Our group explored the influences of different cellulose types of alkali extraction cellulose and microcrystalline cellulose on the cellulose/$CaCO_3$ composites via the microwave-assisted method (Ma et al., 2012b). The types of cellulose played an important role in the microstructure and morphologies of the cellulose/$CaCO_3$ composites. The sample synthesized using microcrystalline cellulose had better crystallinity than that of the sample using alkali extraction cellulose.

The cellulose fibers and $CaCO_3$ particles were observed using alkali extraction cellulose. However, using microcrystalline cellulose instead of alkali extraction cellulose, the cellulose with irregular shape and $CaCO_3$ microspheres were obtained. Therefore, choosing appropriate cellulose types is very important for the formation of cellulose/$CaCO_3$ composites. Our group investigated the different nanocomposites of cellulose and alkali earth metal fluorides (MF_2, $M = $ Ca, Mg, Sr, Ba) via the microwave-assisted method using cellulose solution pretreated with NaOH/urea (Deng et al., 2015b). Sodium fluoride (NaF) was used as the source of fluoride ion, and the alkali earth metal nitrate ($Ca(NO_3)_2$, $Mg(NO_3)_2$, $Sr(NO_3)_2$, and $Ba(NO_3)_2$) was used as the reaction initiator. The cellulose types played an important role in the recombination of cellulose and alkali earth metal fluorides, and the types of alkali earth metal fluorides also displayed different compound abilities. What's more, the alkali earth metal fluorides on the cellulose substrate had nice biocompatibility and could be candidates for the potential applications of biomedical, drug loading, and releasing fields. The microwave-assisted method was used for the rapid synthesis of lignocellulose/HA nanocomposites using lignocellulose, $CaCl_2$, and NaH_2PO_4 in the NaOH−urea aqueous solution (Fu et al., 2015). Our group investigated the effect of the cellulose content on the synthesis of cellulose/SrF_2 nanocomposites by the microwave-assisted method at 100°C for 20 min (Deng et al., 2016). This synthetic route was also extended to fabricate the other cellulose/alkaline earth metal fluorides (MF_2, $M = $ Ca, Mg, Ba) nanocomposites. What's more, the influence of cellulose/alkaline earth metal fluorides (MF_2, $M = $ Ca, Mg, Sr, Ba) nanocomposites on the tensile properties of paper was also explored. The cellulose content is found to have a certain influence on the morphology, crystal lattice, crystallinity, and thermostability of alkaline earth metal fluorides. The different types of alkaline earth metal fluorides have an effect on the tensile strength of paper.

ILs, as desirable "green" solvents, have aroused increasing interest due to their unique properties, such as high fluidity, low melting temperature, low toxicity, nonflammability, and high ionic conductivity. More importantly, ILs are very good media for absorbing microwaves, thus leading to a very high heating rate. Therefore, microwave-assisted IL method is developed to synthesize functional cellulose composites for biomedical application. Preparation of cellulose-based nanocomposites by microwave-assisted IL method allows the high value-added applications of cellulose by combining three major green chemistry principles: using environmentally preferable solvents, environmentally friendly method, and renewable biomaterials. For example, microwave-assisted IL method was applied to prepare cellulose/calcium silicate nanocomposites in ethylene glycol (Jia et al., 2011a). It found that the additive of IL favored the composite of cellulose and calcium silicate and the weight loss of nanocomposites was decreased with the increasing IL concentrations. Moreover, microwave-assisted IL method was applied to prepare cellulose/CHA nanocomposites at 130°C for 30 min (Ma et al., 2011). The cellulose in nanocomposites displayed cellulose type II crystalline structure. IL acts simultaneously as a solvent and a microwave absorber for the synthesis of cellulose/CHA nanocomposites. Cellulose/CHA nanocomposites with rough surface were formed in IL by microwave-assisted method, in contrast to the cellulose/HA nanocomposites with flake-like cellulose and rod-like HA prepared in *N,N*-dimethylacetamide by the microwave heating. Furthermore, the rapid and green microwave-assisted method for the synthesis of the cellulose/calcium silicate nanocomposites in ILs and recycled ILs (Jia et al., 2011b). These calcium silicate nanoparticles or nanosheets were homogenously dispersed in the cellulose matrix. The experimental results confirm that the ILs can be used repeatedly. It observed the slight differences using ILs and recycled ILs. The microwave-assisted ILs method was developed to fabricate cellulose/F-substituted HA nanocomposites (Jia et al., 2012b). F-substituted HA particles were embedded in the cellulose matrix or dispersed on the surface of cellulose matrix. The number of F-substituted HA in nanocomposites increased with increasing heating time. The cellulose/CaF nanocomposites were obtained at high NaF concentration. The ILs concentration played an important role in the dispersion of particles. Microwave-assisted IL method is applied to the preparation of cellulose/$CaCO_3$ hybrid nanocomposites in the alkali extraction cellulose using $CaCl_2$ and Na_2CO_3 as starting reactants (Ma et al., 2013). The IL acts as the excellent solvent for absorbing microwave and the dissolution of cellulose, and the synthesis of cellulose/$CaCO_3$ nanocomposites. The morphologies of $CaCO_3$ changed from polyhedral to cube to particle with increasing cellulose concentration. Cytotoxicity experiments demonstrated that the cellulose/$CaCO_3$ nanocomposites had good biocompatibility and could be a candidate for the biomedical applications. Microwave-assisted IL method is used for the synthesis of $CaCO_3$-filled wood powder nanocomposites in the NaOH/urea solution (Ma et al., 2014). The presence of $(NH_4)_2SO_4$ played an important role in the morphology and dispersion of $CaCO_3$ in the nanocomposites. Microwave-assisted IL method was used to compound cellulose/alkali earth metal fluorides (MF_2, $M = $ Ca, Mg, Sr, Ba) nanocomposites (Deng et al., 2015a). Interestingly, cellulose/CaF_2 and cellulose/MgF_2 were synthesized through this method, while cellulose/SrF_2 and cellulose/BaF_2 could not be synthesized. IL ([Bmim][BF_4]) was used for dissolving microcrystalline cellulose and providing the source of fluoride ionic, and the alkali earth metal nitrate ($Ca(NO_3)_2$, $Mg(NO_3)_2$, $Sr(NO_3)_2$, and $Ba(NO_3)_2$) was used as the reaction initiator. The different heating modes have influence on the morphology and property. The different temperature and heating time also have a certain influence on the morphology and crystallinity of calcium fluoride. The microwave-assisted hydrothermal method was reported for the synthesis of cellulose/HA nanocomposites using an inorganic phosphorus source (sodium dihydrogen phosphate dihydrate ($NaH_2PO_4 \cdot 2H_2O$)), or organic phosphorus sources (ATP, CP, or FBP) (Fu et al., 2016). The phosphate sources played a critical role on the phase, size, and morphology of the minerals in the nanocomposites. The HA was obtained by using $NaH_2PO_4 \cdot 2H_2O$ as phosphorus source, while all the ATP, CP, and FBP led to the by-product, calcite. It obtained the HA nanostructures with various morphologies, including nanorods, pseudo-cubic, pseudo-spherical, and spherical nanoparticles, by varying the phosphorus sources or adjusting the reaction parameters.

2.2.3 Ultrasound agitation method

Ultrasonic wave is a kind of mechanical wave with extremely short wavelength, which is generally shorter than 2 cm in air. The ultrasonic synthesis is to use the high-frequency vibration of ultrasound to promote chemical reactions, which are completed through liquid acoustic cavitation. Acoustic cavitation refers to the formation, oscillation, growth, contraction to disintegration of small bubble nuclei in liquids, and their triggering of physical and chemical changes. There are four types of ultrasonic effects: mechanical effects, cavitation effects, thermal effects, and chemical effects. Ultrasonic cleaning has the advantages of high efficiency, low cost, and simplicity.

As early as 2013, our group investigated the influences of synthesis strategies of microwave-assisted method and ultrasound agitation method on the $CaCO_3$ crystals on the cellulose substrate (Fu et al., 2013a). The phase of vaterite spheres with the diameter of about 320–600 nm was obtained by ultrasound agitation method (Fig. 2.2); meanwhile, the mixed phases of calcite and vaterite with the diameter of about 0.82–1.24 μm were observed by microwave-assisted method. The ultrasound agitation method does more favor to the synthesis of $CaCO_3$ crystals with uniform morphology and size, compared with microwave-assisted method. The $CaCO_3$ crystals on the cellulose substrate synthesized by ultrasound agitation method had good biocompatibility as a candidate for the biomedical applications. Then, we explored the growth mechanism of vaterite on the cellulose matrix via a sonochemistry process (Fu et al., 2013b). The influences of synthetic parameters including the ultrasound agitation method, MCC, solvent, and reactant on the phase and shape of vaterite were investigated in detail. All the experimental parameters work together to induce the synthesis of vaterite phase spheres. After that, a green sonochemical route was used to synthesize cellulose/vaterite nanocomposites in the absence of chemical additives as application in protein adsorption (Fu et al., 2019a). The phase of $CaCO_3$ crystals can be easily controlled by adjusting the concentration of cellulose, and the pure vaterite nanospheres can be obtained at the cellulose concentration of 4.05 and 8.10 mg/mL. The vaterite nanospheres were further formed by self-assembly of rice-shaped vaterite nanoparticles with 75 nm in length and 42 nm in diameter. The as-prepared cellulose/vaterite nanocomposites had excellent cytocompatibility and a relatively high protein adsorption ability (247.5 mg/g) using hemoglobin as a model protein. The cellulose/vaterite nanocomposites

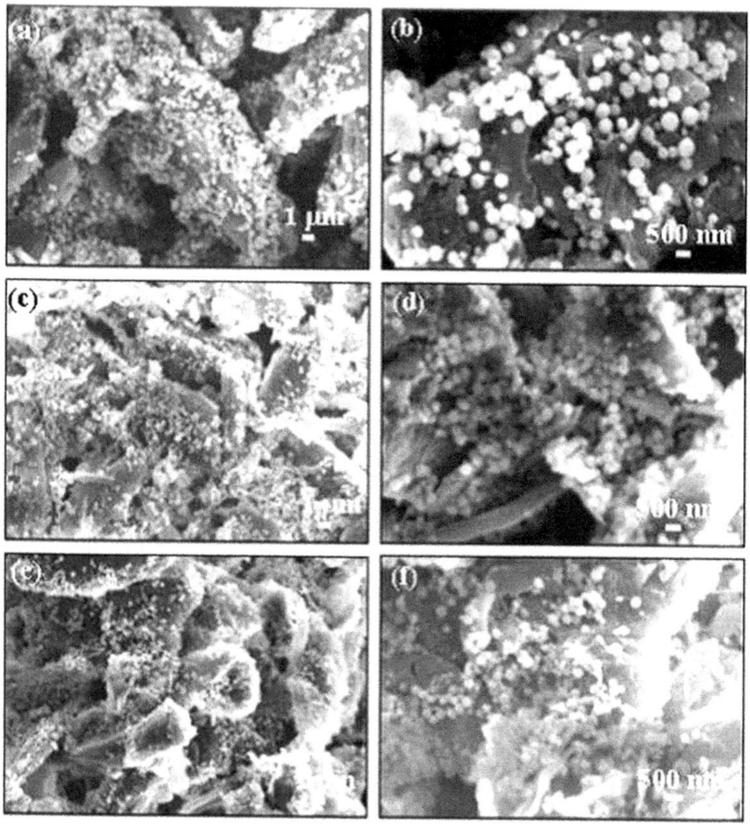

FIGURE 2.2 SEM images of the $CaCO_3$ crystals prepared by ultrasound agitation method for different times: (a,b) 10 min; (c,d) 20 min; (e,f) 30 min. *Reproduced from Fu et al. (2013a) with permission from Elsevier.*

are promising for biomedical applications such as protein drug delivery. Moreover, the sonochemical route was also developed for the rapid synthesis of cellulose/HA nanocomposites in NaOH/urea aqueous solution for applications in biomedical fields, such as tissue engineering and protein/drug delivery (Fu et al., 2018). The in vitro behavior of the cellulose/HA nanocomposites was studied to evaluate the biological response of the nanocomposites following immersion in simulated body fluid for various periods (maximum of 28 days). The HA crystals formed on the surface of the nanocomposites were carbonate-containing apatite. The HA nanosheets assembly of nanorods was mineralized on the surface of the nanocomposites, and maximum mass of the nanocomposites was reached 1.82 times of initial mass after 28 days of soaking. The cellulose/HA nanocomposites had good cytocompatibility and showed a relatively high protein adsorption ability of 321.5 mg/g using hemoglobin as a model protein.

2.3 Advanced functional cellulose composites for biomedical applications

In the review article by Du et al. (2019), they summarized CNCs and CNFs-based hydrogels and their biomedical applications hydrogels, including drug delivery, wound dressings, and tissue engineering scaffolds. Lizundia et al. (2020) reviewed cellulose nanocrystal–based multifunctional nanohybrids for biomedical applications including tissue engineering, wound dressing, drug delivery, and cancer therapy. Pandey (2021) reviewed pharmaceutical and biomedical applications of CNFs in controlled drug delivery, excipient, wound healing dressing material, anticancer, antimicrobial, and transdermal drug delivery. Especially, in the review article by Abhishek Pandey, he reported some interesting results. For example, it reported the release of indomethacin in 15–30 days and itraconazole up to 3 months, CNFs wound dressing with high biocompatibility and rapid epithelialization of burn wounds in 11–21 days, and 14 miscellaneous biomedical applications of CNFs for blood vessel, nucleus pulposus replacement, enzyme immobilization, cardiac, ophthalmic, and neural tissue engineering.

2.3.1 Cellulose composites for biosensors

Biosensor is an instrument that is sensitive to biogenic substance and converts their concentration into electrical signals for detection. It is an analytical tool or system composed of biosensitive materials as recognition elements (including enzyme, antibody, antigen, microorganism, cell, tissue, nucleic acid, and other bioactive substances), physical and chemical transducers (such as oxygen electrode, photosensitive tube, field-effect transistor, piezoelectric crystal, etc.), and signal amplification devices. Cellulose composites can meet the requirements of breathability and wearability. When applied in wearable biosensors, they can monitor people's physical conditions in real-time, including vital signs such as pulse, respiration, limb movements, analysis of components such as sweat and saliva, and detection of exhaled matter. Compared with traditional biosensors, wearable biosensors based on cellulose composite intelligent fibers can be used for on-site real-time monitoring, playing an important role in disease prevention, improving clinical outcomes and quality of life, improving productivity, reducing medical burden, and reducing medical costs.

Yu et al. (2019) fabricated an optical fiber glucose biosensor based on fluorescent CQDs-GOD/CA complex sensitive film by the dip-coating method. The fiber-optic biosensor exhibited high sensitivity and repeatability for continuous online detection of low concentration glucose, allowing visualization of real-time glucose fluctuations over a period of time. The relationship between relative fluorescence intensity ratio and glucose concentration complied well with the modified Stern–Volmer equation in the range of

10–200 μmol/L with the detection limit of 6.43 μM, and in the range of 10–100 nmol/L with the detection limit of 25.79 nM, respectively. Traceyet al. (2020) presented (non)sulfated CNC/magnetite thin films to act as dermal and oral glucose biosensors. The biocompatible hybrid systems exhibited peroxidase-like activity, indicated by an almost instant color change when in the presence of glucose and ABTS. Both types of biosensors detected glucose concentrations as low as 5 mM, corresponding to the level of glucose in biological fluids. The hybrids can be used to determine glucose levels by using sweat and saliva—nonblood bodily secretions, which tend to be slightly to moderately acidic and have relatively low glucose levels. Li et al. (2021) synthesized a highly flexible electrode for effective enzyme immobilization by encapsulating enzyme into the MOFs and robustly anchored to the CA nanofiber membrane as self-powered glucose biosensor. The biosensor demonstrated excellent stability over 15 h of continuous long-term monitoring. The remarkable stability of self-powered glucose biosensor could inspire the application of enzymatic biosensors in biometrics, chronic disease management, and clinical diagnosis. Gunatilake et al. (2021) proposed a 3D titanium dioxide nanotubes/alginate hydrogel scaffold for the detection of sweat biomarkers, lactate, and glucose in artificial sweat. It carried out rapid colorimetric detection (blue color optical signal) for both lactate and glucose biomarkers in artificial sweat at 4 and 6 min, respectively. The superhydrophilicity and the capillarity of the titanium dioxide nanotubes facilitated the rapid transfer of the artificial sweat components throughout the sensor scaffold, decreasing the detection times. The scaffold was integrated on a cellulose paper to demonstrate the adaptability of the material to other matrixes, obtaining fast and homogenous colorimetric detection of lactate and glucose in the paper substrate when image analysis was performed. The titanium dioxide nanotubes/alginate hydrogel scaffolds had biocompatibility, the efficient immobilization of biological enzymes/colorimetric assays, and the quick optical signal readout behavior. Campo et al. (2022) studied the impact of nine different polymer modifications on the storage and release of enzymes and nanoparticles in paper-based biosensors using CMC, PVA, carrageenan, and chitosan. Polymer reservoirs resulted in colorimetric glucose biosensors with an ultralow limit of detection of 0.005 mM. Reservoirs concentrated antibody-decorated nanoparticles and decreased 10 times the limit of detection of a model immunosensor. It detected the pathogen *Klebsiella pneumoniae* in urine samples from hospital patients at the infectious dose threshold rapidly and with high specificity. Dong et al. (2023) reported the synthesis of Pt/RGO/CNC nanohybrids with a porous network structure for nonenzymatic electrochemical glucose sensing (Fig. 2.3). CNCs acted as a biocompatible framework for wrapped RGO and monodispersed Pt nanoparticles, effectively preventing the restacking of graphene during reduction. It achieved the superior glucose sensing performance of Pt/RGO/CNC-modified GCE with a linear concentration range from 0.005 to 8.5 mM and a low detection limit of

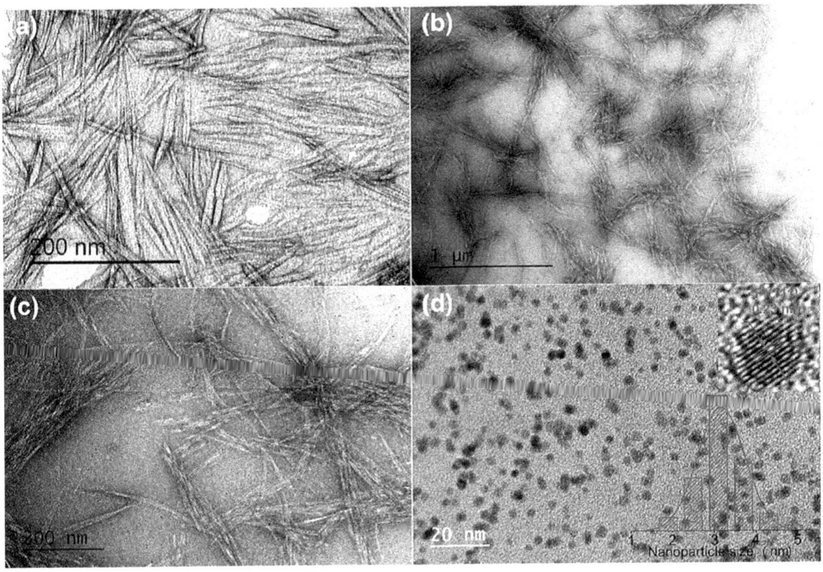

FIGURE 2.3 TEM images of CNCs (a), Pt/RGO/CNCs (b,c), and monodisperse Pt NPs in Pt/RGO/CNCs (d). The size distribution and the HRTEM image of Pt NPs (inset of d). *Reproduced from Dong et al. (2023) with permission from Elsevier.*

2.1 μM. The Pt/RGO/CNC/GCE showed remarkable sensitivity, selectivity, durability, and reproducibility. The CNCs-based bioinorganic nanohybrids could be a promising electrode material in electrochemical biosensors.

Gan et al. (2019) achieved a hierarchical spacing strategy of DNA fluorescence probes for ion detection via locating them separately on rod-like CNCs and further isolating CNCs by pregrafting long molecular chains. It brought a vast linear detection range from 10 nmol/L to 5 μmol/L of Hg^{2+} concentration. It adjusted the length of PtBA molecular chains via controlled radical polymerization, increasing the linear sensitivity coefficient of Hg^{2+} detection by ca. 2.5 times. It developed a hierarchical spacing strategy to separate DNA fluorescent probes on CNCs and detect ion concentration linearly. He et al. (2022) designed an antifouling electrochemical biosensor based on the interpenetrating network of Y-DNA scaffold and polyaniline hydrogel for ATP detection. It observed a synergistic antifouling effect between superhydrophilic CNCs and zwitterion polypeptide sequence (Pep). The hydrogel sensor had a wide linear range of 0.1 pM−1 μM for ATP detection and a low detection limit of 0.025 pM (S/N = 3). For ATP detection in actual serum samples, it achieved the recovery of 99.5%−106% for the sensor, and the relative standard deviation of 0.4%−2.88%. Lin et al. (2019) proposed a paper-based biosensor (DA-PBs) for equipment-free and naked-eye analysis of NF-κB p50 through using dopamine as the signal-responsive material based on the Exo III-mediated cycling amplification reaction. In the presence of target NF-κB p50, it experienced dopamine consumption and aminochrome production correlated with the red signal increase due to the different color of them by virtue of hemin/G-quadruplex-catalyzed oxidation of dopamine. The DA-PBs readily served as a powerful solid biosensor for NF-κB p50 biosensing, with the advantages of low cost, visualization, portable and disposable characteristics, applied in the analysis of NF-κB p50 in biological matrix. The DA-PBs can also be readily applied in the analysis of other biomolecules through simply changing the target recognition DNA structure.

Colozza et al. (2019) developed a wearable electrochemical biosensor for the rapid and on-site detection of mustard agents. It carried out mustard agent detection by monitoring their inhibitory effects toward the choline oxidase enzyme, through the amperometric measurement of the enzymatic by-product hydrogen peroxide. A carbon black/Prussian blue nanocomposite was used as a bulk modifier of the conductive graphite ink constituting the working electrode, allowing for the electrocatalysis of the hydrogen peroxide reduction. After having verified the detecting capability toward a mustard agent simulant, it achieved the applicability of the origami-like biosensor for the rapid and real-time detection of real sulfur mustard, obtaining limits of detection equal to 1 mM and 0.019 g min/m^3 for liquid and aerosol phase, respectively. Ivanova et al. (2020) presented a versatile template-assisted synthesis of nanostructured CNC-templated tin dioxide thin films for gas sensing. It obtained highly porous thin films with pore dimensions in the range of 10–20 nm and high surface area of 46–64 m^2/g. A resistive gas sensor based on CNC-templated tin dioxide showed high sensitivity to CO in ppm concentrations and low cross-sensitivity to humidity. It observed the response to the analyte gas and the signal decay after gas exposure within a few seconds, faster than in standard SnO_2-based CO sensors.

Wang et al. (2020) reported the use of photonic annealing on a flexographically printed graphene-ethyl cellulose composite to produce VAG biosensors for ultrasensitive detection of algal toxins in drinking water (Fig. 2.4). The biosensor exhibited a limit of detection of 1.2 ng/L for microcystin-LR in local tap water. An ultrasensitive VAG biosensor is suitable for low-cost mass production using an integrated roll-to-roll flexographic printing with rapid photonic annealing technique. Ghasemi et al. (2020) used BC/polypyrrole/TiO_2–Ag nanocomposite film to detect and measure the growth of five pathogenic bacteria as diagnostic tools. It increased the sensitivity of the sensors with the increasing concentration of polypyrrole and TiO_2–Ag. The biosensors are portable, and the response time of these sensors is very short for target analysis. Farooq et al. (2020) reported the development of an electrochemical biosensor through the immobilization of bacteriophage in surface-modified BC matrix. It observed 11.7 ± 1.2 phage particles·μm^{-2} immobilized in the BC matrix and antistaphylococcal activity by producing clear lytic zone and reduced bacterial growth. It detected 3 CFU/mL and

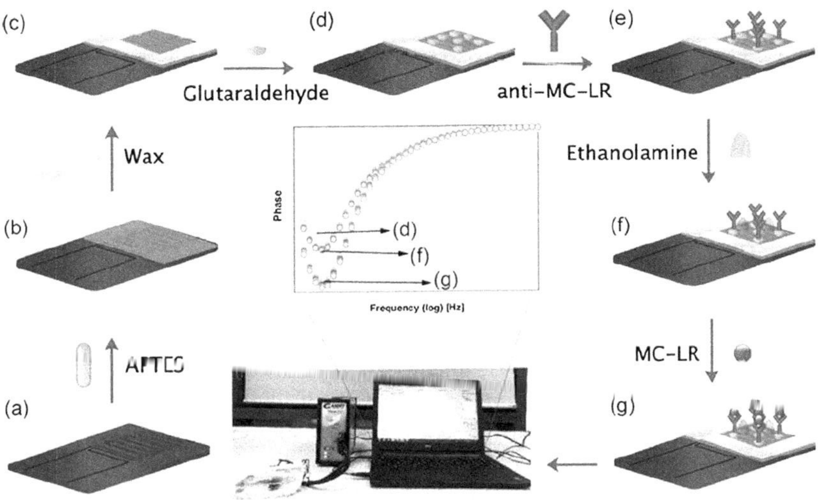

FIGURE 2.4 Surface functionalization procedure of the VAG biosensor from (a) to (g) and nonfaradaic EIS measurement at the step (d), (f), and (g), respectively. *Reproduced from Wang et al. (2020) with permission from American Chemical Society.*

5 CFU/mL of *S. aureus* in PBS and milk, respectively, within 30 min at neutral pH and showed stability over 6-week at 4°C. The biosensor showed high specificity for *S. aureus*, both in pure and mixed cultures of nonhost bacteria, and effectively discriminated live *S. aureus* in a mixture of live/dead cells. Mustafa and Andreescu (2020) reported the development of an enzyme-based paper biosensor, which can monitor fish freshness and predict spoilage. The biosensor measured the release of HX, an indicator of meat and fish degradation, using the enzymatic conversion of HX by XOD. It obtained linearity in the micromolar concentration range with a detection limit of 3.7 μM for HX. The biosensor has high selectivity toward HX. It achieved the applicability of the biosensor by following fish degradation over time and measuring HX concentrations ranging from 117 ± 9 to 198 ± 5 μM within 24 h of degradation, at levels that are comparable with those measured by a commercial enzymatic kit for HX detection. The paper-based HX biosensor has potential as a robust reagentless device for real-time monitoring of food freshness.

Wang et al. (2019) explored probe-free and liquid cellulose biosensor via a one-pot process for Cu^{2+} ions detection by naked eyes. The as-prepared biosensor displayed specific recognition towards Cu^{2+} ions in aqueous solution, even in complex urine and serum samples within 10 s, showing fast response characteristic. The detection limit of Cu^{2+} ions, urine, and serum is as low as 1.185, 1.9309, and 1.9154 ppm, respectively. The cellulose biosensor could act as an excellent candidate to sense Cu^{2+} ions in biological samples including urine and serum from SD rats with satisfying accuracies and biocompatibility. Wang et al. (2021a,b,c) developed a dual-function CNF-based bionic biosensor with good biocompatibility for detecting Ag^+ and acetylcholinesterase (AChE) by grafting deoxyribonucleic acid (DNA) onto CNF. The Ag^+ ions captured by the biosensor acted as recognition sites for the detection of AChE. The CNF-based bionic biosensor could detect Ag^+ concentrations as low as 10^{-6} nM in the presence of interference metal ions (Hg^{2+}, Ba^{2+}, Cd^{2+}, Mg^{2+}, Mn^{2+}, Pb^{2+}, and Zn^{2+}). The CNF-DNA-AgNCs exhibited excellent sensitivity and selectivity for trace levels (0.053 mU/mL) of AChE in the presence of interference reagents. Bi et al. (2020) reported cellulose/inorganic hybrid rattle-type microspheres using microalgae as natural chemical reservoirs, microreactors, and matrix for various applications in immunoassay, biosensors, enrichment and separation of target substances, drug delivery, and environmental remediation. It obtained rattle-type microspheres with simultaneously controllable mesoporous outer shells (19.4–46.3 nm) and multicomponent nanocores (i.e., $Ca_5(PO_4)_3(OH)$ and $Fe_3O_4/MgFe_2O_4$) using microalgae as single-source precursors by adjusting only pH and temperature. The rattle-type microspheres-mediated immunosensor showed ultrahigh sensitivity for the detection of trace microcystin-LR in complex real water samples with a limit of quantitation of 0.05 ng/mL, which is a 10-fold improvement compared with conventional enzyme linked immunosorbent assays.

Ling et al. (2019) explored two forms of nanocellulose as transducer surfaces to enable colorimetric detection of HNE and a wide range of inflammatory diseases. It utilized a DES to mediate formation of cotton cellulose nanocrystals (DCNCs) to prepare a peptide–cellulose conjugate as a protease sensor of HNE. DCNCs showed greater degree of substitution of HNE tetrapeptide and sensitivity to the elastase than WCNFs, despite the smaller surface area and pore sizes. It observed the higher crystallinity and larger crystallite sizes of DCNCs, indicating the well-arranged cellulose chains for immobilization of the tetrapeptide on (110) lattice reflections of cellulose crystals. The sensitivity of DCNCs-based colorimetric sensor was less than 0.005 U/mL. Liu et al. (2021) synthesized CLC dissolved in [Bmim]Cl as a soft template combined with the functional group ($-NH_2$, $-SH$)-grafted PEDOT derivatives by an in situ polymerization method. CLC increased the degree of conjugation of the polymers in the composites and led to the formation of the fibrous and sheet structure of the composite. The electrode modified by PEDOT-MeNH$_2$/CLC and PEDOT-MeSH/CLC had detection limits ($S/N = 3$) of 0.38 (0.1–380 μM) and 0.042 (0.02–520 μM) for the detection of guanine (G), while for the detection of UA, the corresponding detection limits ($S/N = 3$) were 0.255 (5–400 μM) and 0.085 (0.4–650 μM). The PEDOT-MeSH/CLC-based electrochemical sensor possessed a wide linear response with the detection limits of 0.077 (0.6–620 μM) and 0.144 (0.2–460 μM) for the simultaneous detection of UA and G, respectively.

Khattab et al. (2020) fabricated colorimetric nanocomposite film sensor by incorporating tricyanofuran-hydrazone spectroscopic probe into CNW/urease enzyme matrix. It found solvatochromic behavior and a reversible color change of the tricyanofuran-hydrazone probe solution in acetone under acid/base conditions. It observed through visible color shift from light yellow to pink for the film biosensor when exposed to urea in aqueous media. It explored the photophysical properties of the tricyanofuran-hydrazone probe, including solvatochromism and pH sensory. Wang et al. (2021a,b,c) reported on robust SF gels fabricated by incorporating CNCs (SF/CNC) as a "tough" unit and photopolymerization of acrylamide as an "elastic" segment. The gel shows robust mechanical stability (88.8% of initial stress after 1000 compression cycles) and excellent adhesion to various materials. The connected gel can recover its ionic conductivity within 20 s and be stretched to a maximum strain of 498% after healing for 10 h with an efficiency of 95.2%. This multifunctional gel sensor can sensitively detect different toxic gases and small-scale and large-scale human motions in real-time. It obtained the sensitivity of GF = 3.84 at 0%–200% strain. The gel with 5 wt% thermochromic pigments as a visual temperature indicator can quickly reflect abnormal human body temperature according to the color change.

2.3.2 Bone regeneration of cellulose composites

Bone regeneration refers to the implantation of scaffolds and/or bioactive substances into the bone defect site, coordinating osteogenic and osteoclast activity, and promoting bone tissue regeneration. The main factors affecting bone regeneration include the differentiation and proliferation ability of primitive cells, bioactive substances that regulate bone formation, microenvironments suitable for cell growth, and the nutritional components of the body itself. The mechanisms that promote bone regeneration mainly include osteogenesis, induced osteogenesis, and conduction osteogenesis.

Miles et al. (2019) reported in vitro assessment of differentiation, matrix deposition, and compressive load bearing of scalable mesenchymal stem cells–derived bone tissue modules. It encapsulated bone marrow mesenchymal stem cells and HA microgranules in polyelectrolyte complex membranes composed of chondroitin 4-sulfate, CMC, and chitosan. Microcapsules facilitated alkaline phosphatase secretion and deposition of bone specific proteins (osteocalcin and osteopontin) by encapsulated mesenchymal stem cells over 28 days of osteogenic culture. It observed cell-deposited mineral covering the surfaces of the HA microgranules and interior of the microcapsule membrane. Microcapsule mineralization greatly enhanced the yield stresses of Osteo and Osteo + HA fused constructs (10.4 ± 4.4 and 6.4 ± 2.8 MPa). Cai et al. (2023) reported injectable poly(lactide-co-glycolic acid) nanofiber into calcium phosphate bone cement with CMC with high strength, antiwashout property, and controlled biodegradability for minimally invasive bone regeneration. It afforded a sustainable provision of Ca^{2+} bioactive components of cement with porous structure to improve synergistically new bone formation and osteo-integration in vivo by using a rat model of femur condyle defect. The good minimally invasive bone regeneration may come from the synergistic enhanced osteogenic effect of calcium ion enrichment and the improved revascularization capacity contributed from the porosity as well as the lactic acid released from poly(lactide-co-glycolic acid) nanofiber.

Ribeiro et al. (2022) applied platelet lysate and biomineralized CNCs as highly elastic and bioactive bone biomimetic scaffolds. The nanocomposite scaffolds showed high microporosity (94%–95%), high elasticity (recover from 75% strain cycles), injectability, and modulated platelet-derived growth factors sequestration and release. It increased alkaline phosphatase activity (up to 10-fold) and upregulated the expression of bone-related markers (up to twofold), without osteogenic supplementation, demonstrating the osteoinductive properties. The scaffolds promoted the chemotaxis of endothelial cells and enhanced the expression of endothelial markers, showing proangiogenic potential. Patel et al. (2022) reported 3D-printable biodegradable chitosan/silk fibroin/cellulose nanoparticle scaffolds with improved rheological and recovery strength for bone regeneration via M2 macrophage polarization. A significant shift from M1 → M2 macrophages polarization occurred in the scaffolds treated groups than the control after 3 days of incubation, showing its immune-modulatory potential. It observed enhanced bone regeneration in the calvaria defect rat model.

Kima and Tiwari (2020) used the in situ–modified gas-foamed technique to synthesize functional 3D multilayered fibrous polycaprolactone/cellulose electrospun scaffold containing calcium-based particles for bone regeneration (Fig. 2.5). 3D fibers exhibited superior mechanical properties (up to twofold) and thermal stability, compared with the 2D/3D counterparts. Ca-3D@PCL-CL24 exhibited 170% and 215% cell growth, compared with lower 140% and 170% for 3D@PCL-CL at 7 and 14 days, respectively. The Ca-3D@PCL-CL24 exhibited the capability of cellular infiltration up to 20 μm. The synergistic approach of 3D porous layered structures together with functionalized $Ca(OH)_2$ particles played major roles in showing significant improvements. Zhang et al. (2023) reported protocatechualdehyde modified CA membrane for bone regeneration through immunomodulation. The membrane showed a significant radical scavenging activity, suppressing the inflammatory response and creating a favorable immune microenvironment for osteogenesis and mineralization. Addition of metal ions could further stimulate the osteogenic differentiation of stem cells and accelerate bone regeneration both in vitro and in vivo. Abdelaziz et al. (2021) reported biodegradable nanoparticles-in-nanofibers based membranes for guided periodontal tissue and bone regeneration with enhanced antibacterial activity. HA improved the cell viability by around 50% for both types of nanofibrous scaffolds. Ag

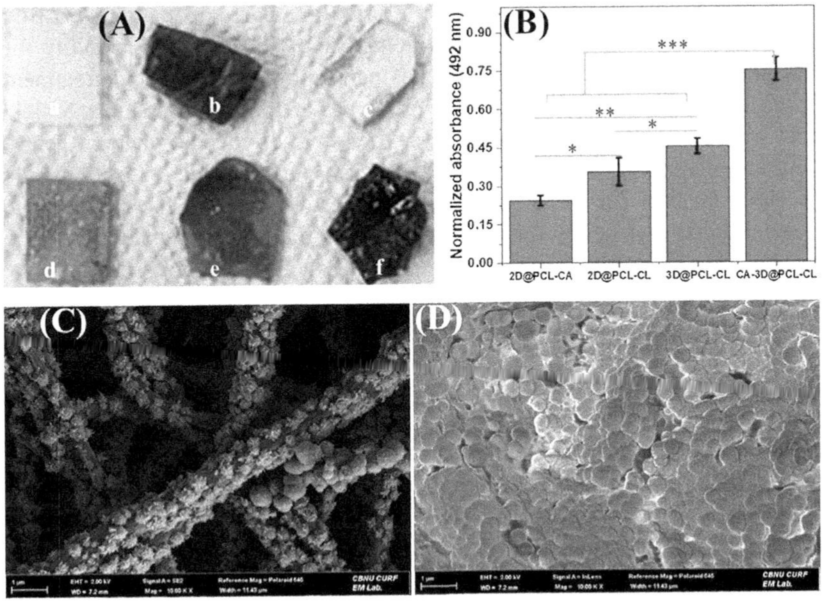

FIGURE 2.5 ARS stained non-SBF-treated 2D@PCL-CL (Aa) and Ca-3D@PCL-CL24 (Ab) while "Ac," "Ad," "Ae," and "Af" represent the ARS stained 2D@PCL-CA, 2D@PCL-CL, 3D@PCL-CL, and Ca-3D@PCL-CL24 of 5-day SBF treatment. Absorbance showed by ARS extracted from the scaffolds treated with SBF solution for 5 days (B), the FE- SEM images of the SBF-treated Ca-3D@PCL-CL24 of 5 (C) and 10 days (D) at 10 K X magnification. *Reproduced from Kima and Tiwari (2020) with permission from Elsevier.*

nanoparticles improved the antibacterial activity with 40 mm inhibition zone after 32 days. The nanofibrous scaffolds showed a desirable degradation profile with losing around 40%–70% of its mass in 8 weeks. The nanofibrous membranes are promising scaffolds for both GTR and GBR applications. Wu et al. (2021) in vitro and in vivo studied biodegradable nanoclay mineral-reinforced macroporous nanocomposite scaffolds for in situ bone regeneration by chemical cross-linking of hydroxyethyl cellulose/soy protein isolate and montmorillonite modification. The nanocomposites showed enhanced mechanical and physical properties within the ideal range for bone tissue engineering. Sustained sequential release of Ca and Mg ions during the degradation of the nanocomposites provided essential signaling cues for guiding bone regeneration. The nanocomposites possessed satisfactory cytocompatibility, offering a highly desirable microenvironment for cellular behaviors, including adhesion, proliferation, and spreading, as well as promoting the mineralization and osteoblastic differentiation of rat bone marrow stem cells. When employed to repair critical-sized cranial defects in rat models, the nanocomposites demonstrated reliable biosafety and significant osteointegration and bone-forming ability in vivo.

Wang et al. (2021a,b,c) prepared TEMPO-mediated oxidation (TO-BC) and maleic acid treatment BC-based gelatin gel with good cell viability for preosteoblast to promote in vivo bone regeneration. Maleic acid treatment improved the mechanical and wetting properties and Young's modulus. Maleic acid treatment BC outperformed in expressing osteogenic marker genes, improving alkaline phosphatase activity, and forming mineralized nodule in vitro. Gelatin scaffolds showed better performance in increasing bone mineral density of newly formed bone and the trabecular thickness for bone regeneration of rat calvaria, without damaging the viscera and serum in vivo.

2.3.3 Drug delivery of cellulose composites

Drug delivery system refers to a technical system that comprehensively regulates the distribution of drugs in organisms. Its goal is to deliver an appropriate amount of drugs to the correct location at the appropriate time, thereby increasing drug utilization efficiency, improving efficacy, reducing costs, and reducing toxic side effects. The materials involved in drug delivery systems mainly include inorganic materials, polymer materials, stabilizers, blockers, and accelerators. Polymer materials can be divided into natural polymer materials (such as gelatin, gum arabic, alginic acid, albumin, chitosan, starch), semisynthesis polymer materials (such as cellulose derivatives), and synthetic polymer materials. Khine and Stenzel (2020) reviewed surface-modified cellulose nanomaterials for drug delivery. They summarized the use of individual cellulose nanomaterials modified with therapeutic agents or targeting ligands as drug delivery systems via surface modification techniques ranging from the covalent attachment of small molecules to the grafting of polymers. Moreover, in the review article by Lugoloobi et al. (2021), they discussed the extraction, properties, modification, in vivo/in vitro medical applications, biocompatibility, challenges and future perspectives of CNCs in cancer diagnostics and treatment.

Zhang et al. (2020) reported a pinecone-inspired smart CNFs microcage with open network shells for nano/microparticle drug delivery. Open network shells can intelligently open/close triggered by lesion stimuli, making the therapy "always on-demand." The microcages exhibited integrated properties of superior structural stability, superhydrophilicity, and pH-triggered, smart across-shell transport of silver nanoparticles and silicate nanoplatelets, enabling extraordinary antiinfection and bone regeneration. Chen et al. (2022) reported a bionic 3D CNF-based nanocage wound dressing for NIR-triggered multiple synergistic therapy of tumors and infected wounds. The wound dressing had a high drug-loading capacity of doxorubicin (400 mg/g) and indocyanine green (25 mg/g). The NIR- and pH-responsive bionic "on/off" switches of the dressing enabled a controllable drug release onto the wound area. The dual NIR-triggered wound dressing with photothermal conversion performance possessed good antibacterial properties against *E. coli*, *S. aureus*, and drug-resistant *S. aureus*. The bionic wound dressing could adapt and treat irregular postoperative skin tumor wounds and drug-resistant bacterial infection via the synergistic therapy of photothermal, photodynamic, and chemotherapy.

Liu et al. (2020a,b) reported mesoporous polydopamine@GO/CNF composite hydrogel with an encapsulation structure for controllable drug release and toxicity shielding (Fig. 2.6). Polydopamine nanoparticles exhibited a high drug loading ratio up to 35 wt% for tetracycline hydrochloride. GO encapsulated polydopamine nanoparticles for extending the drug release time and reinforcing the physical strength of the hydrogel. The composite hydrogel had five times greater of mechanical strength, compared with that of the CNF hydrogel. The drug release time of the composite hydrogel was 3 times and 7.2 times longer than that of the polydopamine/CNF hydrogel and CNF hydrogel, respectively. The rate of drug release could be accelerated by near-infrared (NIR) light irradiation and pH value change. Jeong et al. (2022) developed the donepezil hydrochloride–reinforced CNC hydrogel with pH control for sustained drug delivery through subcutaneous injection. The cross-linked gel exhibited immediate gelation, injection capability through a single syringe, improved viscoelasticity, and shear-thinning properties. It obtained a higher half-life and mean residence time and low maximum drug concentration values in the group after subcutaneous injection. Zong et al. (2022) reported intelligent TEMPO-oxidized CNF hydrogel with both redox and thermoresponse for controlled drug delivery. It obtained the maximum cumulative release rate of DOX of 39.56% and the berberine of 99.50% after 60 h using tumor therapeutic drugs as model drugs for in vitro release studies. The swelling and transparency of hydrogels showed dramatic changes in the range of 30–40°C. Zhu et al. (2022) used TEMPO-oxidized CNF (TOCNF) to form nanocellulose microspheres with high dye adsorption and drug encapsulation by the emulsion method. The hydrangea-like nanocellulose microspheres presented a stable 3D network porous structure and exhibited excellent adsorption properties. The microspheres had a high adsorption capacity for

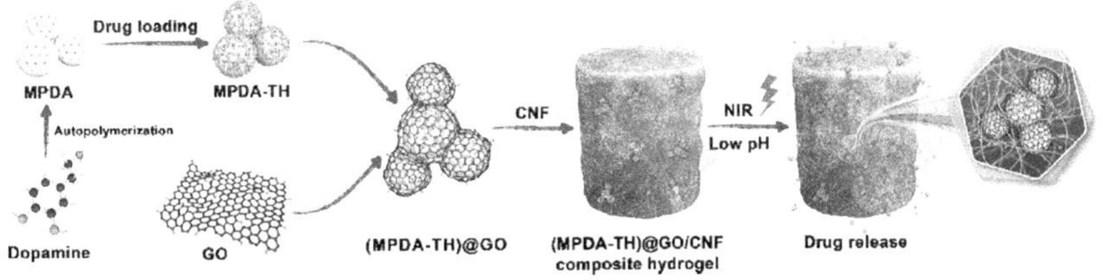

FIGURE 2.6 Schematic illustration of the preparation of the MPDA@GO/CNF composite hydrogel and its mechanism of drug release by a combination of multiple stimuli (pH change and NIR light irradiation). *Reproduced from Liu et al. (2020a,b) with permission from American Chemical Society.*

MB and MO with the optimal adsorption capacity of 412.1 and 286.5 mg/g under neutral conditions, respectively. The microspheres displayed excellent adsorption selectivity on MB/MO mixed dyes, which could be used to selectively adsorb MB. It obtained the encapsulation rate of the positively charged drug doxorubicin hydrochloride (DOX) of 93% and the drug loading capacity of 34.5%.

Zhang et al. (2019) developed a target drug delivery system of BC-based composites to enhance cancer targeting under a static magnetic field and laser irradiation to inhibit tumor growth, which was used for the synchronization of in situ transdermal administration, a magnetic field, and laser and biological targeting. Laser-activatable and magnetic-targeting laser-sensitized magnetic nanoparticles exerted a synergistic effect on breast cancer inhibition. Beekmann et al. (2020) presented BC production modified using water-soluble PEG on pilot-scale for tailor-made antiinflammatory drug delivery systems. It obtained up to 41.5% ± 3.0% higher transparency of the dressing, 40.6% ± 3.8% increased loading capacity, and 9% increased total release of the anti-inflammatory model drug diclofenac sodium. It found spray loading as material efficient alternative to absorption loading, allowing a significant reduction in loading time. Ojagh et al. (2021) reported BC-based composites for drug delivery. It obtained potential application in drug delivery system using MB as a model drug where the loading capacity and swelling ratio for the samples were as follows: BC/CMC > BC/carboxymethylated diethylaminoethyl cellulose > BC. The MB loading data were fitted using Langmuir and Freundlich equations, and kinetic behavior of the release was described by Higuchi and Korsmeyer–Peppas models. Park et al. (2021) proposed a BC nanofibers-reinforced hybrid hydrogel system for mechanical compression-responsive on-demand drug release through a polyacrylamide mesh. The swelling kinetics depended on the mechanical properties determined by the BC nanofibers aspect ratio. The tailored composition of BC nanofibers with polyacrylamide retarded drug release by a factor of two compared with polyacrylamide only while enabling on-demand drug release in response to the applied compressive stress. Singh et al. (2021) proposed a mucoadhesive drug delivery platform of BC adhesive composites with extended soft tissue adhesion capability of up to 7 days for on-site management of oral wound (Fig. 2.7). The bioadhesive composite had an adhesion strength of 7–17 kPa and duration exceeding of 48 h in wet conditions under sustained shear forces, while other mucoadhesives based on hydrophilic macromolecules exhibited adhesion strength of 0.5–5 kPa and last only a few hours.

Yan et al. (2020) reported stable and biocompatible cellulose-based $CaCO_3$ porous microspheres with a highly specific surface area of 363 m^2/g for tunable pH-responsive drug delivery. $CaCO_3$ microspheres retained high porosity for over 4 weeks under ethanol storage conditions, showing outstanding stability. The porous microspheres as potential drug carriers

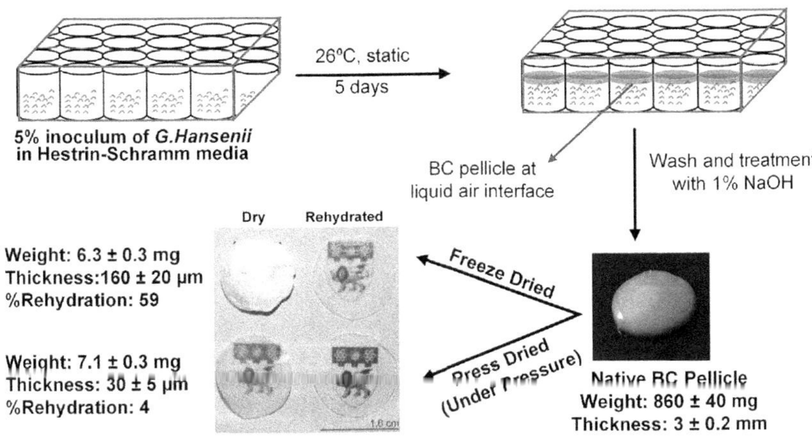

FIGURE 2.7 Schematic representation of BC production and drying processes. BC pellicles are grown for 5 days and form at the air–liquid interface. Washing with distilled water and treatment with 0.25 N NaOH at 60°C for 30 min remove the bacterial cells from BC, confirmed by residual protein assays. Freeze drying and press drying result in similar water content removal. %Rehydration represents water absorbed by the dry films upon swelling as a percentage of initial weight of native BC. Data presented as mean ± SD, $n = 3$. *Reproduced from Singh et al. (2021) with permission from Elsevier.*

showed good encapsulation efficiency, excellent pH responsiveness (pH = 2.0 $\gg$ 7.4), and good biocompatibility, which could control the drug release at a specific location.

2.3.4 Bioimaging of functional cellulose composites

Bioimaging is an important research tool for understanding the tissue structure of organisms and elucidating their various physiological functions. It uses optical or electron microscopes to directly obtain microstructure images of biological cells and tissues and analyzes the obtained images to understand various physiological processes of biological cells. In recent years, with the development of optical imaging technology, especially the introduction of digital imaging technology and computer image analysis technology, biological imaging technology has become an indispensable method in cell biology research.

Liu et al. (2020a,b) reported oxidized nanocellulose photoluminescent nitrogen-doped fluorescent CDs for Fe^{3+} ions detection and fluorescent probe for bioimaging of Vero cells (Fig. 2.8). It obtained the overall yield of 16.1% and quantum yield of 30.3% for nitrogen-doped CDs from dissolving pulp with an oxidation treatment, attributing to the enhanced chemical accessibility of cellulose by the oxidation process. The N-CDs showed a graphite-like structure rich in carboxyl and amino functional groups, which were well

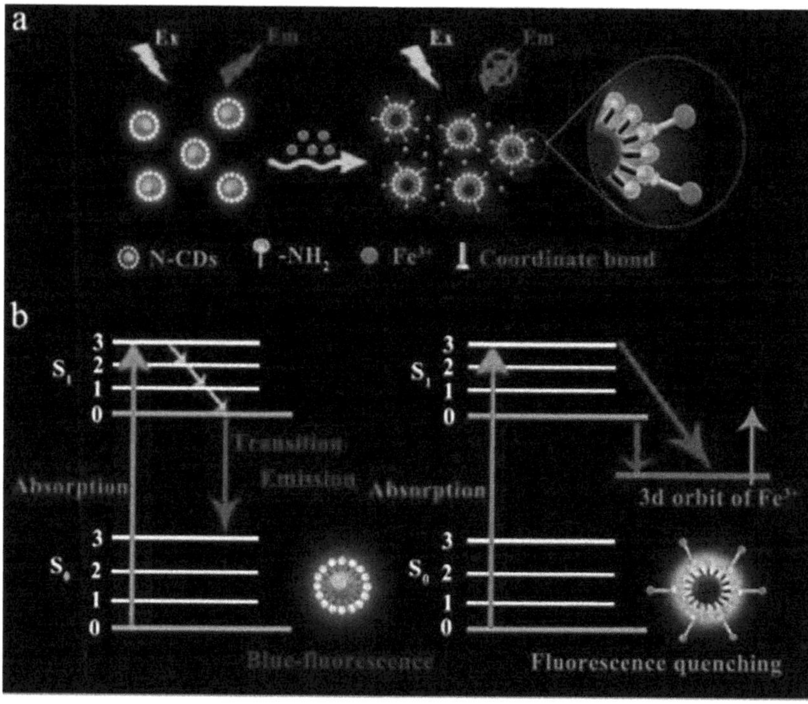

FIGURE 2.8 (a) Schematic diagram of the N-CDs coordinated with Fe^{3+} ions and (b) electron transfer process. *Reproduced from Liu et al. (2020a,b) with permission from Elsevier.*

monodispersed spherical nanoparticles with a mean diameter of 1.86 nm. Liu et al. (2022) prepared nitrogen and phosphorus codoped CQDs from cellulose pulp using a hydrothermal method as a probe for pH detection, metal ions detection, and cell imaging. N,P-CQDs-500 exhibited unique selectivity toward Fe(II), meanwhile the N,P-CQDs-1000 was sensitive to pH as a pH probe (Fig. 2.9). The N,P-CQDs-3500 showed excellent bioimaging ability in five cancer cells, good quenching effect on Fe(III) ions, good fluorescence performance in suspension cells (Raji and RPMI8226), and superior selectivity for adherent cells (H1650).

The development of future biological imaging technology not only needs to further improve the resolution of images but also needs to enhance the real-time and continuous imaging, to achieve continuous tracking of individual biological functional molecules in vivo, record their physiological processes in detail, and fully reveal their biological functions. In addition, the application of biological imaging technology in clinical medical diagnosis is also receiving increasing attention, and the development of noninvasive in vivo imaging technology is an important prerequisite for its widespread application in disease diagnosis.

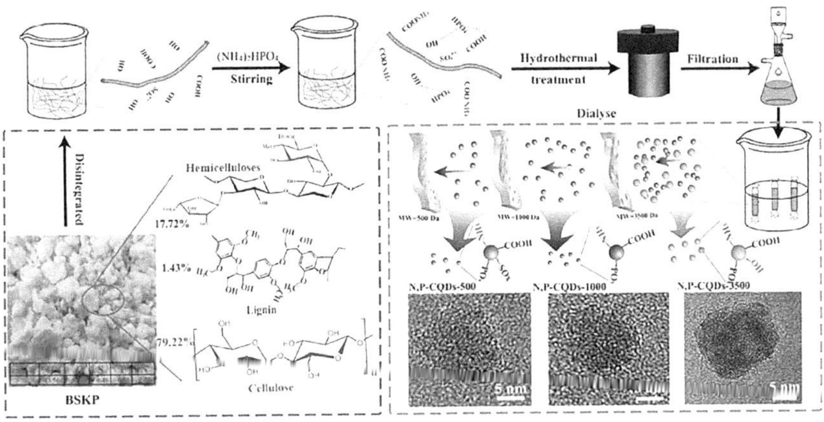

FIGURE 2.9 Schematic illustration of fabrication and fractionation of N,P-CQDs. *Reproduced from Liu et al. (2022) with permission from Elsevier.*

2.4 Chapter summary

In summary, we reviewed hydrothermal, microwave, and ultrasound agitation synthesis of cellulose composites for biomedical applications. Moreover, we summarized various biomedical applications of advanced functional cellulose composites, such as biosensors, bone regeneration, drug delivery, and bioimaging. Significant progress has been made in this field.

Based on our knowledge, the main problems and directions of advanced functional cellulose composites for biomedical applications included high production cost, insufficient strength, poor stability, lack of elasticity, and interface compatibility. The production process of nanocellulose requires high technology and equipment investment, which leads to relatively high production costs and limits their applications in biomedical fields. Although nanocelluloses have a certain mechanical strength, their strength may not meet the requirements in certain high-strength applications, and further research and improvement are needed. Nanocelluloses are easily affected by moisture and heat and may lose their original properties and functions. Therefore, additional protective measures are needed. In the applications that require large-scale movement, biomedical applications of cellulose composites may be limited due to a lack of sufficient elasticity. When cellulose composites are applied in the biomedical fields, there are still interface compatibility issues with the skin and wounds.

There are currently some studies and technologies attempting to address the aforementioned issues, for example, by improving production processes, using new cellulose composites, and other methods to reduce costs, increase strength and stability, and improve the surface properties. In addition, there are research exploring how to endow cellulose composites with better elasticity to adapt to a wider range of biomedical applications. Although there are some problems with cellulose composites, with the continuous progress of technology and in-depth research, these problems are expected to be gradually solved, allowing cellulose composites to play their unique advantages in more biomedical fields.

Acknowledgments

The financial support from Beijing Nova Program (20230484431) is gratefully acknowledged.

References

Abdelaziz, D., Hefnawy, A., Al-Wakeel, E., El-Fallal, A., El-Sherbiny, I.M., 2021. New biodegradable nanoparticles-in-nanofibers based membranes for guided periodontal tissue and bone regeneration with enhanced antibacterial activity. J. Adv. Res. 28, 51–62. https://doi.org/10.1016/j.jare.2020.06.014.

Beekmann, U., Schmölz, L., Lorkowski, S., Werz, O., Thamm, J., Fischer, D., Kralisch, D., 2020. Process control and scale-up of modified bacterial cellulose production for tailor-made anti-inflammatory drug delivery systems. Carbohydr. Polym. 236. https://doi.org/10.1016/j.carbpol.2020.116062.

Bi, L., Chen, Y.-P., Wang, C., Su, J., Pan, G., 2020. Microalgae-derived cellulose/inorganic nanocomposite rattle-type microspheres as an advanced sensor for pollutant detection. Chem. Eng. J. 395. https://doi.org/10.1016/j.cej.2020.125073.

Cai, P., Lu, S., Yu, J., Xiao, L., Wang, J., Liang, H., Huang, L., Han, G., Bian, M., Zhang, S., Zhang, J., Liu, C., Jiang, L., Li, Y., 2023. Injectable nanofiber-reinforced bone cement with controlled biodegradability for minimally-invasive bone regeneration. Bioact. Mater. 21, 267–283. https://doi.org/10.1016/j.bioactmat.2022.08.009.

Campo, M.M.G., Alba-Patiño, A., Palomino, C., Bauza, M., Rojo-Molinero, E., Oliver, A., Turnes, G., Rica, 2022. Boosting the sensitivity of paper-based biosensors with polymeric water-soluble reservoirs. Sensor. Actuator. B Chem. 354.

Chen, R., Zhao, C., Chen, Z., Shi, X., Zhu, H., Bu, Q., Wang, L., Wang, C., He, H., 2022. A bionic cellulose nanofiber-based nanocage wound dressing for NIR-triggered multiple synergistic therapy of tumors and infected wounds. Biomaterials 281. https://doi.org/10.1016/j.biomaterials.2021.121330.

Colozza, N., Kehe, K., Dionisi, G., Popp, T., Tsoutsoulopoulos, A., Steinritz, D., Moscone, D., Arduini, F., 2019. A wearable origami-like paper-based electrochemical biosensor for sulfur mustard detection. Biosens. Bioelectron. 129, 15–23. https://doi.org/10.1016/j.bios.2019.01.002.

Deng, F., Fu, L.H., Ma, M.G., 2015a. Microwave-assisted rapid synthesis and characterization of CaF_2 particles-filled cellulose nanocomposites in ionic liquid. Carbohydr. Polym. 121, 163–168.

Deng, F., Fu, L.H., Zhang, X.M., Ma, M.G., 2015b. Comparative study on the nanocomposites of cellulose and alkali earth metal fluorides (MF_2, M = Ca, Mg, Sr, Ba) via microwave-assisted method. Sci. Adv. Mater. 7, 509–517.

Deng, F., Dong, Y.Y., Liu, S., Wang, B., Ma, M.G., Du, X., 2016. Revealing the influences of cellulose on cellulose/SrF2 nanocomposites synthesized by microwave-assisted method. Ind. Crop. Prod. 85, 258–265. https://doi.org/10.1016/j.indcrop.2016.03.018.

Dong, L.L., Ren, S., Zhang, X., Yang, Y., Wu, Q., Lei, T., 2023. In-situ synthesis of Pt nanoparticles/reduced graphene oxide/cellulose nanohybrid for nonenzymatic glucose sensing. Carbohydr. Polym. 303, 120463.

Du, H., Liu, W., Zhang, M., Si, C., Zhang, X., Li, B., 2019. Cellulose nanocrystals and cellulose nanofibrils based hydrogels for biomedical applications. Carbohydr. Polym. 209, 130–144. https://doi.org/10.1016/j.carbpol.2019.01.020.

Farooq, U., Ullah, M.W., Yang, Q., Aziz, A., Xu, J., Zhou, L., Wang, S., 2020. High-density phage particles immobilization in surface-modified bacterial cellulose for ultra-sensitive and selective electrochemical detection of Staphylococcus aureus. Biosens. Bioelectron. 157. https://doi.org/10.1016/j.bios.2020.112163.

Fu, L.H., Dong, Y.Y., Ma, M.G., Li, S.M., Sun, R.C., 2013a. Compare study $CaCO_3$ crystals on the cellulose substrate by microwave-assisted method and ultrasound agitation method. Ultrason. Sonochem. 20, 839–845.

Fu, L.H., Dong, Y.Y., Ma, M.G., Yue, W., Sun, S.L., Sun, R.C., 2013b. Why to synthesize vaterite polymorph of calcium carbonate on the cellulose matrix via sonochemistry process? Ultrason. Sonochem. 20, 1188–1193.

Fu, L.H., Xie, Y.M., Bian, J., Ma, M.G., Tian, C.H., Jin, X.J., 2015. Microwave-assisted rapid synthesis of lignocellulose/hydroxyapatite nanocomposites. Mater. Lett. 159, 51–53.

Fu, L.H., Liu, Y.J., Ma, M.G., Zhang, X.M., Xue, Z.M., Zhu, J.F., 2016. Microwave-assisted hydrothermal synthesis of cellulose/hydroxyapatite nanocomposites. Polymers 8, 316.

Fu, L.H., Qi, C., Liu, Y.J., Cao, W.T., Ma, M.G., 2018. Sonochemical synthesis of cellulose/hydroxyapatite nanocomposites and their application in protein adsorption. Sci. Rep. 8, 8292.

Fu, L.H., Qi, C., Hu, Y.R., Mei, C.G., Ma, M.G., 2019a. Cellulose/vaterite nanocomposites: sonochemical synthesis, characterization, and their application in protein adsorption. Mater. Sci. Eng. C 96, 426–435.

Fu, L.H., Qi, C., Ma, M.G., Wan, P.B., 2019b. Multifunctional cellulose-based hydrogels for biomedical applications. J. Mater. Chem. B 7, 1541–1562.

Gan, L., Wang, Y., Zhang, M., Xia, X., Huang, J., 2019. Hierarchically spacing DNA probes on bio-based nanocrystal for spatial detection requirements. Sci. Bull. 64, 934–940.

Ghasemi, S., Bari, M.R., Pirsa, S., Amiri, S., 2020. Use of bacterial cellulose film modified by polypyrrole/TiO_2-Ag nanocomposite for detecting and measuring the growth of pathogenic bacteria. Carbohydr. Polym. 232. https://doi.org/10.1016/j.carbpol.2019.115801.

Gunatilake, U.B., Garcia-Rey, S., Ojeda, E., Basabe-Desmonts, L., Benito-Lopez, F., 2021. TiO_2Nanotubes alginate hydrogel scaffold for rapid sensing of sweat biomarkers: lactate and glucose. ACS Appl. Mater. Interf. 13 (31), 37734–37745. https://doi.org/10.1021/acsami.1c11446.

He, X.Y., Sun, N., Jia, H., Hou, M., Tan, Z., Lu, X., 2022. Antifouling electrochemical biosensor based on conductive hydrogel of DNA scaffold for ultrasensitive detection of ATP. ACS Appl. Mater. Interf. 14, 40624–40632.

Ivanova, A., Frka-Petesic, B., Paul, A., Wagner, T., Jumabekov, A.N., Vilk, Y., Weber, J., Schmedt Auf Der Günne, J., Vignolini, S., Tiemann, M., Fattakhova-Rohlfing, D., Bein, T., 2020. Cellulose nanocrystal-templated tin dioxide thin films for gas sensing. ACS Appl. Mater. Interf. 12 (11), 12639–12647. https://doi.org/10.1021/acsami.9b11891.

Jeong, D.I., Kim, S., Kim, M.H., Yoon, I.S., Lee, S.H., Kim, D.D., Cho, H.J., 2022. Donepezil hydrochloride-reinforced cellulose nanocrystal-aggregated gel structure for long-acting drug delivery. Carbohydr. Polym. 296. https://doi.org/10.1016/j.carbpol.2022.119887.

Jia, N., Li, S.M., Ma, M.G., Sun, R.C., Zhu, J.F., 2010a. Hydrothermal synthesis and characterization of cellulose-carbonated hydroxyapatite nanocomposites in NaOH-urea aqueous solution. Sci. Adv. Mater. 2, 210–214.

Jia, N., Li, S.M., Zhu, J.F., Ma, M.G., Xu, F., Wang, B., Sun, R.C., 2010b. Microwave-assisted synthesis and characterization of cellulose-carbonated hydroxyapatite nanocomposites in NaOH-urea aqueous solution. Mater. Lett. 64, 2223–2225.

Jia, N., Li, S.M., Ma, M.G., Sun, R.C., 2011a. Microwave-assisted ionic liquid preparation and characterization of cellulose/calcium silicate nanocomposites in ethylene glycol. Mater. Lett. 65, 918–921.

Jia, N., Li, S.M., Ma, M.G., Sun, R.C., Zhu, L., 2011b. Green microwave-assisted synthesis of cellulose/calcium silicate nanocomposites in ionic liquids and recycled ionic liquids. Carbohydr. Res. 346, 2970–2974.

Jia, N., Li, S.M., Ma, M.G., Sun, R.C., Zhu, J.F., 2012a. Hydrothermal fabrication, characterization, and biological activity of cellulose/$CaCO_3$ bionanocomposites. Carbohydr. Polym. 88, 179–184.

Jia, N., Li, S.M., Ma, M.G., Sun, R.C., 2012b. Rapid microwave-assisted fabrication of cellulose/F-substituted hydroxyapatite nanocomposites using green ionic liquids as additive. Mater. Lett. 68, 44–46.

Khattab, T.A., Fouda, M.M.G., Rehan, M., Okla, M.K., Alamri, S.A., Alaraidh, I.A., AL-ghamdi, A.A., Soufan, W.H., Abdelsalam, E.M., Allam, A.A., 2020. Novel halochromic cellulose nanowhiskers from rice straw: visual detection of urea. Carbohydr. Polym. 231. https://doi.org/10.1016/j.carbpol.2019.115740.

Khine, Y.Y., Stenzel, M.H., 2020. Surface modified cellulose nanomaterials: a source of non-spherical nanoparticles for drug delivery. Mater. Horiz. 7 (7), 1727–1758. https://doi.org/10.1039/c9mh01727e.

Kim, S.E., Tiwari, A.P., 2020. Three dimensional polycaprolactone/cellulose scaffold containing calcium-based particles: a new platform for bone regeneration. Carbohydr. Polym. 250, 116880.

Li, X., Feng, Q., Lu, K., Huang, J., Zhang, Y., Hou, Y., Qiao, H., Li, D., Wei, Q., 2021. Encapsulating enzyme into metal-organic framework during in-situ growth on cellulose acetate nanofibers as self-powered glucose biosensor. Biosens. Bioelectron. 171, 112690.

Lin, H.Y., Wang, X., Wu, J.H., Li, H., Li, F., 2019. Equipment-free and visualized biosensor for transcription factor rapid assay based on dopamine-functionalized cellulose paper. J. Mater. Chem. B 7, 5461–5464.

Ling, Z., Xu, F., Edwards, J.V., Prevost, N.T., Nam, S., Condon, B.D., French, A.D., 2019. Nanocellulose as a colorimetric biosensor for effective and facile detection of human neutrophil elastase. Carbohydr. Polym. 216, 360–368. https://doi.org/10.1016/j.carbpol.2019.04.027.

Liu, Y.Y., Fan, Q., Huo, Y., Liu, C., Li, B., Li, Y., 2020a. Construction of a Mesoporous polydopamine@GO/cellulose nanofibril composite hydrogel with an encapsulation structure for controllable drug release and toxicity shielding. ACS Appl. Mater. Interf. 12, 57410–57420.

Liu, Z.Z., Chen, M., Guo, Y., Zhou, J., Shi, Q., Sun, R., 2020b. Oxidized nanocellulose facilitates preparing photoluminescent nitrogendoped fluorescent carbon dots for Fe^{3+} ions detection and bioimaging. Chem. Eng. J. 384, 123260.

Liu, H., Jamal, R., Abdiryim, T., Simayi, R., Liu, L., Liu, Y., 2021. Carboxylated cellulose as a soft template combined with PEDOT derivatives in [BMIM]Cl: a competent biosensor for detection of guanine and uric acid in the blood. ACS Sustain. Chem. Eng. 9 (17), 5860–5871. https://doi.org/10.1021/acssuschemeng.0c09259.

Liu, W., Jiang, C., Zhang, L., Li, X., Hou, Q., Ni, Y., 2022. Valorization of cellulose pulp derived carbon quantum dots by controllable fractionation. Ind. Crop. Prod. 188, 115560.

Lizundia, E., Puglia, D., Nguyen, T.-D., Armentano, I., 2020. Cellulose nanocrystal based multifunctional nanohybrids. Prog. Mater. Sci. 112. https://doi.org/10.1016/j.pmatsci.2020.100668.

Lugoloobi, I., Maniriho, H., Jia, L., Namulinda, T., Shi, X., Zhao, Y., 2021. Cellulose nanocrystals in cancer diagnostics and treatment. J. Contr. Release 336, 207–232. https://doi.org/10.1016/j.jconrel.2021.06.004.

Ma, M.G., Zhu, J.F., Jia, N., Li, S.M., Sun, R.C., Cao, S.W., Chen, F., 2010. Rapid microwave-assisted synthesis and characterization of cellulose-hydroxyapatite nanocomposites in N,N-dimethylacetamide solvent. Carbohydr. Res. 345 (8), 1046–1050. https://doi.org/10.1016/j.carres.2010.03.004.

Ma, M.G., Jia, N., Li, S.M., Sun, R.C., 2011. Nanocomposites of cellulose/carbonated hydroxyapatite by microwave-assisted fabrication in ionic liquid: characterization and thermal stability. Iran. Polym. J. (Engl. Ed.) 20, 413–421.

Ma, M.G., Fu, L.H., Li, S.M., Zhang, X.M., Sun, R.C., Dai, Y.D., 2012a. Hydrothermal synthesis and characterization of wood powder/$CaCO_3$ composites. Carbohydr. Polym. 88, 1470–1475.

Ma, M.G., Fu, L.H., Sun, R.C., Jia, N., 2012b. Compare study on the cellulose/$CaCO_3$ composites via microwave-assisted method using different cellulose types. Carbohydr. Polym. 90, 309–315.

Ma, M.G., Dong, Y.Y., Fu, L.H., Li, S.M., Sun, R.C., 2013. Cellulose/$CaCO_3$ nanocomposites: microwave ionic liquid synthesis, characterization, and biological activity. Carbohydr. Polym. 92, 1669–1676.

Ma, M.G., Deng, F., Yao, K., Tian, C.H., 2014. Microwave-assisted synthesis and characterization of $CaCO_3$ particles-filled wood powder nanocomposites. Bioresources 9, 3909–3918.

Meng, L.Y., Wang, B., Ma, M.G., Zhu, J.F., 2017. Cellulose-based nanocarriers as platforms for cancer therapy. Curr. Pharmaceut. Des. 23 (35), 5292–5300. https://doi.org/10.2174/1381612823666171031111950.

Miles, K.B., Maerz, T., Matthew, H.W.T., 2019. Scalable MSC-derived bone tissue modules: in vitro assessment of differentiation, matrix deposition, and compressive load bearing. Acta Biomater. 95, 395–407.

Mustafa, F., Andreescu, S., 2020. Paper-based enzyme biosensor for one-step detection of hypoxanthine in fresh and degraded fish. ACS Sens. 5 (12), 4092–4100. https://doi.org/10.1021/acssensors.0c02350.

Ojagh, S.M.A., Vahabzadeh, F., Karimi, A., 2021. Synthesis and characterization of bacterial cellulose-based composites for drug delivery. Carbohydr. Polym. 273, 118587.

Pandey, A., 2021. Pharmaceutical and biomedical applications of cellulose nanofibers: a review. Environ. Chem. Lett. 19 (3), 2043–2055. https://doi.org/10.1007/s10311-021-01182-2.

Park, D., Kim, J.W., Shin, K., Kim, J.W., 2021. Bacterial cellulose nanofibrils-reinforced composite hydrogels for mechanical compression-responsive on-demand drug release. Carbohydr. Polym. 272. https://doi.org/10.1016/j.carbpol.2021.118459.

Patel, D.K., Dutta, S.D., Hexiu, J., Ganguly, K., Lim, K.T., 2022. 3D-printable chitosan/silk fibroin/cellulose nanoparticle scaffolds for bone regeneration via M2 macrophage polarization. Carbohydr. Polym. 281. https://doi.org/10.1016/j.carbpol.2021.119077.

Qian, H., Liu, J., Wang, X., Pei, W., Fu, C., Ma, M., Huang, C., 2023. The state-of-the-art application of functional bacterial cellulose-based materials in biomedical fields. Carbohydr. Polym. 300. https://doi.org/10.1016/j.carbpol.2022.120252.

Ribeiro, J.P., Domingues, R.M.A., Babo, P.S., Nogueira, L.P., Reseland, J.E., Reis, R.L., Gomez-Florit, M., Gomes, M.E., 2022. Highly elastic and bioactive bone biomimetic scaffolds based on platelet lysate and biomineralized cellulose nanocrystals. Carbohydr. Polym. 292. https://doi.org/10.1016/j.carbpol.2022.119638.

Singh, J., Tan, N.C.S., Mahadevaswamy, U.R., Chanchareonsook, N., Steele, T.W.J., Lim, S., 2021. Bacterial cellulose adhesive composites for oral cavity applications. Carbohydr. Polym. 274. https://doi.org/10.1016/j.carbpol.2021.118403.

Tracey, C.T., Torlopov, M.A., Martakov, I.S., Vdovichenko, E.A., Zhukov, M., Krivoshapkin, P.V., Mikhaylov, V.I., Krivoshapkina, E.F., 2020. Hybrid cellulose nanocrystal/magnetite glucose biosensors. Carbohydr. Polym. 247, 116704.

Wang, R.J., Zhang, L.W., Liu, R., Liu, L., Yao, J.M., 2019. Ultra-fast and probe-free cellulose biosensor for visual detection of Cu^{2+} ions in biological samples. Carbohydr. Polym. 223. https://doi.org/10.1016/j.carbpol.2019.115117.

Wang, L., Zhang, W., Samavat, S., Deganello, D., Teng, K.S., 2020. Vertically aligned graphene prepared by photonic annealing for ultrasensitive biosensors. ACS Appl. Mater. Interf. 12 (31), 35328–35336. https://doi.org/10.1021/acsami.0c08036.

Wang, X.C., Tang, S., Chai, S., Wang, P., Qin, J., Pei, W., Bian, H., Jiang, Q., Huang, C., 2021a. Preparing printable bacterial cellulose based gelatin gel to promote in vivo bone regeneration. Carbohydr. Polym. 270, 118342.

Wang, L., Guo, W., Zhu, H., He, H., Wang, S., 2021b. Preparation and properties of a dual-function cellulose nanofiber-based bionic biosensor for detecting silver ions and acetylcholinesterase. J. Hazard Mater. 403, 123921.

Wang, C., Zhu, M., Yu, H.Y., Abdalkarim, S.Y.H., Ouyang, Z., Zhu, J., Yao, J., 2021c. Multi-functional biosensors made with self-healable silk fibroin imitating skin. ACS Appl. Mater. Interf. 13, 33371–33382.

Wu, M., Chen, F., Wu, P., Yang, Z., Zhang, S., Xiao, L., Deng, Z., Zhang, C., Chen, Y., Cai, L., 2021. Nanoclay mineral-reinforced macroporous nanocomposite scaffolds for in situ bone regeneration: in vitro and in vivo studies. Mater. Des. 205. https://doi.org/10.1016/j.matdes.2021.109734.

Yan, G.H., Feng, Y., Gao, Z., Zeng, X., Hong, W., Liu, W., Sun, Y., Tang, X., Lei, T., Lin, L., 2020. Stable and biocompatible cellulose-based $CaCO_3$ microspheres for tunable pH-responsive drug delivery. ACS Sustain. Chem. Eng. 7.

Yu, S., Ding, L., Lin, H., Wu, W., Huang, J., 2019. A novel optical fiber glucose biosensor based on carbon quantum dots-glucose oxidase/cellulose acetate complex sensitive film. Biosens. Bioelectron. 146. https://doi.org/10.1016/j.bios.2019.111760.

Yuan, Q., Bian, J., Ma, M.G., 2021. Advances in biomedical application of nanocellulose-based materials: a review. Curr. Med. Chem. 28, 8275–8295.

Zhang, L.K., Du, S., Wang, X., Jiao, Y., Yin, L., Zhang, Y., Guan, Y.Q., 2019. Bacterial cellulose based composites enhanced transdermal drug targeting for breast cancer treatment. Chem. Eng. J. 370, 749–759.

Zhang, S.C., Zhou, S., Liu, H., Xing, M., Ding, B., Li, B., 2020. Pinecone-inspired nano-architectured smart microcages enable nano/microparticle drug delivery. Adv. Funct. Mater. 30, 2002434.

Zhang, Q.Y., Tan, J., Huang, K., Nie, R., Feng, Z.Y., Zou, C.Y., Li, Q.J., Chen, J., Sheng, N., Qin, B.Q., Gu, Z.P., Liu, L.M., Xie, H.Q., 2023. Polyphenolic-modified cellulose acetate membrane for bone regeneration through immunomodulation. Carbohydr. Polym. 305, 120546.

Zhu, Y.J., Chen, F., 2014. Microwave-assisted preparation of inorganic nanostructures in liquid phase. Chem. Rev. 114 (12), 6462–6555. https://doi.org/10.1021/cr400366s.

Zhu, H.F., Zhu, E., Xie, Y., Liu, D., Hu, Y., Shi, Z., Xiong, C., Yang, Q., 2022. Hydrangea-like nanocellulose microspheres with high dye adsorption and drug encapsulation prepared by emulsion method. Carbohydr. Polym. 296, 119947.

Zong, S., Wen, H., Lv, H., Li, T., Tang, R., Liu, L., Jiang, J., Wang, S., Duan, J., 2022. Intelligent hydrogel with both redox and thermo-response based on cellulose nanofiber for controlled drug delivery. Carbohydr. Polym. 278. https://doi.org/10.1016/j.carbpol.2021.118943.

Chapter 3

Cellulose nanocomposites for antibacterial applications

Lianhua Fu,[1] Yanyan Dong[2] and Mingguo Ma[3,*]

[1]*School of Biomedical Engineering, Shenzhen University Medical School, Shenzhen University, Shenzhen, Guangdong, PR China;* [2]*Institute of Environment and Sustainable Development in Agriculture, Chinese Academy of Agricultural Sciences, Beijing, PR China;* [3]*MOE Engineering Research Center of Forestry Biomass Materials and Bioenergy, Research Center of Biomass Clean Utilization, Beijing Key Laboratory of Lignocellulosic Chemistry, College of Materials Science and Technology, Beijing Forestry University, Beijing, PR China*

[*]*Corresponding author: E-mail: mg_ma@bjfu.edu.cn*

3.1 Introduction

In recent years, antibacterial materials have received increasing attention and become a research hotspot. It is reported that in the past decade, there have been approximately 71,430 papers using "***antibacterial***" as the "***Title***" on Web of Science. Biomass is the most abundant renewable natural polysaccharide on the earth, with unique properties, such as chemical stability, mechanical strength, biocompatibility, and biodegradability. Antibacterial biomass materials may come from various types of biomass, including polysaccharides, peptides, and glycopeptide. Antibacterial biomass materials combine the advantages of biomass and enhancers to use biomass as a substrate and generate new properties through synergistic effects.

In general, inorganic antibacterial raw materials mainly include silver, copper, zinc oxide, and titanium dioxide. Among the aforementioned antibacterial materials, silver is the best antibacterial material among inorganic antibacterial materials. As a precious metal, silver is relatively expensive. Copper has its certain biological toxicity, and zinc oxide has relatively weak antibacterial properties. Titanium dioxide has good applications in the antibacterial field due to its good antibacterial properties, high stability, and low price. In addition, one of the main problems faced by inorganic antibacterial

Surface Science and Advanced Functional Applications of Cellulose Composites
https://doi.org/10.1016/B978-0-443-13675-7.00003-8

materials is that their inorganic particles are not resistant to water washing. Cellulose is one of the most abundant glycans and renewable resources on the earth, which has biodegradability, biocompatibility, and nontoxicity (Lam et al., 2012). Cellulose has a multihydroxyl structure, which can attach and fix silver to the surface of cellulose through weak interaction. With the addition of cellulose, it reduces the amount of silver and solve easy washing of silver ions, improving the utilization of silver.

As for antibacterial cellulose-based nanocomposites, our group did system work and synthesized a series of nanocomposites, including cellulose/Ag, cellulose/AgCl, cellulose/AgBr, cellulose/Ag/AgCl, cellulose/TiO_2, cellulose/CuO, cellulose/ZnO, etc. In this chapter, we introduced the synthesis and property of advanced functional cellulose composites for antibacterial applications. Five typical examples were selected to present antibacterial cellulose-based nanocomposites, such as cotton stalk cellulose/Ag, cellulose/Ag/AgCl, cellulose/AgCl, Ag@Fe_3O_4@nanocellulose, and cellulose/TiO_2 nanocomposites. We try to contribute great content to this book "***Surface Science and Advanced Functional Applications of Cellulose Composites***."

3.2 Synthesis and property of advanced functional antibacterial cellulose-based nanocomposites

3.2.1 Cotton stalk cellulose/silver nanocomposites

China is one of the countries with the highest cotton production in the world. According to the announcement of the National Bureau of Statistics, the total cotton production in China was 5.65 million tons in 2015 and 5.343 million tons in 2016. Under the huge cotton production, a large number of cotton stalk have become agricultural waste. In vast rural areas, cotton stalks are often used by farmers as fuel or directly burned into fertile fields, with low utilization value. Additionally, a small amount is used for papermaking or composting. Therefore, the effective conversion of cotton stalks to prepare cellulose-based nanocomposites can greatly improve the utilization value of cotton stalks. It is of great significance for achieving the resource utilization of cotton stalks. In addition, the microwave-assisted heating method belongs to bulk heating without temperature gradient. Compared with the traditional heating method, the microwave-assisted hydrothermal (MWH) method has the characteristics of high reaction conversion rate, short reaction time, energy saving, and green environmental protection (Zhu and Chen, 2014; Nadagouda et al., 2011). The reaction can be completed in a few min to dozens of min, while the traditional hydrothermal method generally requires heating for a few hours or even days. Moreover, nanostructures with narrower size distribution and better crystallinity can be prepared by MWH method (Nadagouda et al., 2011).

After being exposed to sunlight, the cotton stalk is peeled and cut into small strips of about 5 cm. It is then placed in an oven and dried at 60°C for

24 h, which was grinded into powder using a miniature plant crusher (model FZ102) and used a sieve to select powder ranging from 40 to 60 mesh. The selected powder is degreased by toluene ethanol treatment and delignified by sodium chlorous acid treatment to obtain holocellulose. Holocellulose was dispersed into NaOH aqueous solution and extract for 4 h at 60°C. After the extraction is completed, it is cooled and filtered. The filter residue is washed with a large amount of water until the filtered water is neutral. Then, it is rinsed with ethanol and dried in an oven for 24 h to obtain cellulose. The filtrate was adjusted pH to 5.5–6.0 using acetic acid. At room temperature, cotton stalk cellulose was dispersed in water, stirred and added $AgNO_3$ solution and of hemicellulose solution. The obtained mixture was heated to 160°C for 10 min using an MWH method to obtain cotton stalk cellulose/Ag nanocomposite.

According to the thermal stability characteristics of cellulose and hemicellulose, the microwave heating temperature range was selected ranging from 120 to 180 °C. The XRD patterns of samples prepared with hemicellulose concentration of 5 mg mL^{-1} under microwave temperature conditions of 120–180°C for 10 min observed that the color of cotton stalk cellulose without MWH treatment is white, and the color changes after MWH treatment. As the reaction temperature increases from 120 to 160°C, the color of the sample changes from light yellow to dark brown. However, further increasing the temperature to 180°C results in a dark gray color of the sample. In the XRD patterns, all samples at $2\theta = 15.6$ degrees and 22.5 degrees show characteristic peaks of cellulose I (JCPDS 03–0289), while other diffraction peaks correspond to the diffraction pattern of cubic crystalline silver (JCPDS 03–0289). When the microwave reaction temperature is 120°C, a weak diffraction peak of silver appeared at $2\theta = 38.0°$, and the Ag content in the sample detected by ICP was 2.21 wt%. When the reaction temperature is 140°C, the diffraction peak of silver increases, and the Ag content in the sample increases to 4.72 wt%. When the temperature is raised to 160°C, the diffraction peaks corresponding to the (200) and (220) crystal planes of silver ($2\theta = 44.2$ degrees and 64.3 degrees) appear, indicating that AgNPs with high crystallinity can be obtained in a short reaction time (10 min) at 160°C. In previous reports, under the same other conditions but without the addition of hemicellulose, samples prepared by microwave heating at 160°C for 60 min could only observe weak peak at $2\theta = 37.9°$ (Fu et al., 2016). Therefore, under microwave heating conditions, the addition of hemicellulose can effectively reduce Ag^+ ions and prepare AgNPs with higher crystallinity. When the temperature continues to rise to 180°C, the diffraction peak of silver becomes higher and sharper, while the diffraction peak intensity of cellulose further weakens, indicating that AgNPs in the composite material have higher crystallinity. The experimental results show that the microwave heating temperature has an impact on the content and crystallinity of Ag in the sample. Increasing the reaction temperature increased the content of Ag in the sample

(Ag content is 2.21 wt%, 4.72 wt%, 7.92 wt%, and 11.43 wt%, respectively), and increased the crystallinity. Therefore, increasing the temperature is beneficial for accelerating the reaction between hemicellulose and Ag^+ ions, obtaining composites with high Ag content and crystallization.

The samples prepared by microwave heating at 120–180°C for 10 min exhibit similar FT-IR patterns. Among them, the absorption peaks located at 3400 cm^{-1} and 1371 cm^{-1} belong to the stretching and bending vibrations of the −OH group in the sample, respectively. The absorption peak located at 1633 cm^{-1} comes from the adsorbed water in the sample, and the absorption peak located at 2899 cm^{-1} belongs to the stretching vibrations of the CH_2 and C−H groups. The bending vibrations of CH_2 and C−OH skeleton vibrations occur at 1432 cm^{-1} and 1112 cm^{-1}, respectively. The characteristic absorption peaks of β-glycosidic bond appeared at 1165 cm^{-1} and 895 cm^{-1}, while the characteristic peak of C−O−C skeleton vibration of pyran ring is at 1052 cm^{-1}. Compared with cotton straw cellulose without microwave treatment, a weak absorption peak can be observed at 1735 cm^{-1} in the sample, which is the characteristic absorption peak of C=O in hemicellulose acetyl group (Sun et al., 2011). Cellulose and hemicellulose often form hydrogen bond connection, so this peak may come from the residue of hemicellulose (Bian et al., 2014). From the FT-IR results, it found that the samples prepared at different microwave heating temperatures show characteristic peaks of cellulose, indicating that the cellulose structure of cotton straws can be well maintained after MWH treatment.

To investigate the effect of hemicellulose concentration, we prepared cotton straw cellulose/Ag nanocomposites using hemicellulose concentrations of 0.5, 1, and 10 mg mL^{-1} under microwave heating at 160°C for 10 min. Ag content in the sample was calculated through ICP analysis to be 3.15 wt% (sample 5), 3.73 wt% (sample 6), and 10.49 wt% (sample 7), respectively. The XRD pattern of the sample shows that the characteristic diffraction peak of cellulose is located at $2\theta = 15.4$ degrees and 22.1 degree, and the other diffraction peaks correspond to the (111), (200), and (220) crystal planes of cubic silver (JCPDS 03-0289), respectively. As the concentration of hemicellulose in the reaction system increases, the diffraction peak intensity of silver in the sample gradually increases. Therefore, AgNPs with higher crystallinity can be obtained at higher hemicellulose concentrations. It found that there was no significant difference in the morphology of the composite material. Only a very small amount of AgNPs can be observed in sample 5. When the hemicellulose concentration is increased to 1 mg mL^{-1}, the number of AgNPs increases. When the hemicellulose concentration continued to increase to 10 mg mL^{-1}, the number of AgNPs further increased, but some AgNPs agglomerated. From the aforementioned results, it concluded that the concentration of hemicellulose has an impact on the content of Ag in cotton stalk cellulose/Ag nanocomposites and the dispersibility of AgNPs on cotton stalk cellulose matrix.

Based on the previous experimental results, we further investigated the effect of microwave heating time on cotton stalk cellulose/Ag nanocomposites. Samples were prepared by reacting hemicellulose at a concentration of 5 mg mL^{-1} under microwave heating at 160°C for 30 and 60 min, respectively. It observed that compared with the sample 3 (microwave heating for 10 min), the diffraction peak intensity of cellulose in sample 8 and sample 9 is weakened, the diffraction peak located at $2\theta = 15.6$ degrees almost disappears, while the diffraction peak intensity of silver significantly increases. The results showed that under the same other conditions, prolonging the reaction time resulted in better crystallinity of AgNPs in the composite. It observed that compared with sample 3, the distribution of AgNPs in sample 8 and sample 9 is denser, and some AgNPs also agglomerate. The Ag content in samples 8 and 9 was calculated by ICP detection to be 15.42 wt% and 23.10 wt%, respectively. The formation of metal nanoparticles includes nucleation, growth, and aggregation processes. When the nuclear speed is fast, the growth of the product is inhibited, and smaller particles can be obtained. When the growth rate is fast, larger particles are obtained (Fu et al., 2013).

To further investigate the effect of heating time on the material, we calculated the particle size of AgNPs in the sample using the Scheler formula. According to the Scheler formula, the particle sizes of AgNPs in the samples prepared by microwave heating at 160°C for 10, 30, and 60 min with a hemicellulose concentration of 5 mg mL^{-1} were 13.5, 15.3, and 16.0 nm, respectively. The results indicate that extending the reaction time obtains AgNPs with larger particle sizes. Therefore, prolonging the reaction time is beneficial for the growth and aggregation of AgNPs. In previous reports, similar results were observed in the preparation of cotton stalk cellulose/Ag nanocomposites using glucose as a reducing agent during hydrothermal reaction for 3–12 h (Dong et al., 2014).

The thermal stability of cotton stalk cellulose/Ag nanocomposites was analyzed using TG-DSC. It find that the samples prepared using hemicellulose concentration of 5 mg mL^{-1} under microwave heating at 160°C for 10–60 min exhibit similar weight loss trends. The quality loss mainly occurs at 300–400°C, caused by the thermal degradation of cotton stalk cellulose. The mass loss at around 100°C is due to the loss of adsorbed moisture in the material, while the mass loss in the range of 200–300°C may be caused by the degradation of residual hemicellulose in the sample (Sun et al., 2011). In the corresponding DSC curves, four endothermic peaks appeared at 98/200/267/338, 122/201/264/336, and 90/203/264/336°C, which are consistent with the mass loss situation in the TG curve. From room temperature to 600°C, the total mass loss of the samples was 75.46 wt% (sample 3, 10 min), 66.30 wt% (sample 8, 30 min), and 64.58 wt% (sample 9, 60 min), respectively. The Ag content in the sample is 7.92 wt% (sample 3, 10 min), 15.42 wt% (sample 8, 30 min), and 23.10 wt% (sample 9, 60 min), respectively. It can be inferred that the microwave reaction time affects the thermal stability of the material, and the longer the reaction time within a certain range, the higher the Ag content in the sample, and the better the thermal stability of the material.

The antibacterial properties of cotton stalk cellulose/Ag nanocomposites play an important role in their application in biomedicine (Fu et al., 2018). Therefore, we used *Staphylococcus aureus* ATCC 6538 as the gram-positive model bacteria and *Escherichia coli* HB101 as the gram-negative model bacteria to study the antibacterial properties of the materials. Fig. 3.1 shows the antibacterial ring test results of samples prepared using hemicellulose

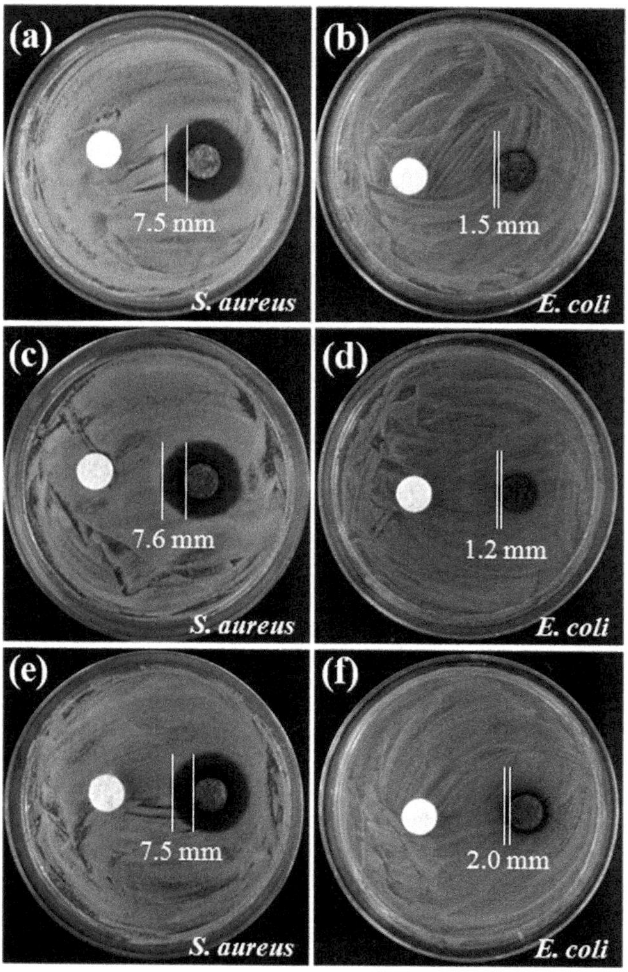

FIGURE 3.1 The antibacterial activities of the cellulose/Ag nanocomposites: (a,b) sample 3; (c,d) sample 8; (e,f) sample 9. The samples in the left of the cultures are pure cellulose and the right are the cellulose/Ag nanocomposites. *Reproduced from Fu et al. (2018) with permission from MDPI.*

concentration of 5 mg mL^{-1} under microwave heating at 160°C for 10 min (Fig. 3.1a and b, sample 3), 30 min (Fig. 3.1c and d, sample 8), and 60 min (Fig. 3.1e and f, sample 9). It observed that no antibacterial rings were found around the pure cotton stalk cellulose in the blank control group, indicating that the cotton stalk cellulose itself has no antibacterial properties. All cotton stalk cellulose/Ag nanocomposite samples showed obvious antibacterial rings around them, indicating that the prepared cotton stalk cellulose/Ag nanocomposite material has good antibacterial properties. After measurement, the size of the antibacterial ring of the sample against *S. aureus* was 7.5 mm (sample 3, Fig. 3.1a), 7.6 mm (sample 8, Fig. 3.1c), and 7.5 mm (sample 9, Fig. 3.1e), respectively. The antibacterial performance was significantly improved, compared with the cellulose/Ag nanocomposites prepared in previous work (under the same test conditions, the antibacterial ring of the material against *S. aureus* was only 1.0–2.0 mm) (Dong et al., 2015). From Fig. 3.1b,d and f, it can be seen that the size of the antibacterial ring of the material against *E. coli* is 1.5 mm (sample 3), 1.2 mm (sample 8), and 2.0 mm (sample 9), respectively. It has been found that the cotton stalk cellulose/Ag nanocomposite prepared in this section has better antibacterial effect on *S. aureus* than *E. coli*, which may be due to the different sensitivity of gram-positive bacteria and gram-negative bacteria to materials. The cell wall of gram-negative bacteria has an outer membrane, and the content of peptidoglycan in the cell wall is low, so *E. coli* is not vulnerable to damage by cotton stalk cellulose/Ag nanocomposites (Yuan et al., 2014; Feng et al., 2000), thus showing a small inhibition ring phenomenon. Similar results were also observed in previous research work (Fu et al., 2016). Although the antibacterial ring of the material against *E. coli* is only 1.2–2.0 mm, its antibacterial performance is also equivalent to that of cellulose/Ag/AgCl nanocomposites prepared by hydrothermal reaction at 160°C for 12 h (Dong et al., 2015). The aforementioned experimental results show that the cotton stalk cellulose/Ag nanocomposite prepared in this section has good antibacterial effect against *S. aureus* (gram-positive bacteria) and *E. coli* (gram-negative bacteria).

It can also be observed that the size of the bacteriostatic ring of the three samples against *S. aureus* is close (7.5, 7.6, and 7.5 mm, respectively), while the size of the bacteriostatic ring against *E. coli* is quite different (1.5, 1.2, and 2.0 mm, respectively). It proposed that the antibacterial performance of silver antibacterial materials is influenced by the particle size, morphology, specific surface area, and Ag content of AgNPs in the material (Rai et al., 2009; Sureshkumar et al., 2010), and under the same Ag content, the smaller the particle size of AgNPs, the higher their antibacterial performance (Zhu et al., 2009). From the previous experimental results, it can be seen that the content of Ag and the particle size of AgNPs in the cotton stalk cellulose/Ag nanocomposites prepared in this section increase with the increase of microwave heating time. The particle sizes of AgNPs in the samples are 13.5 nm (sample 3, 10 min), 15.3 nm (sample 8, 30 min), and 16.0 nm (sample 9, 60 min), respectively. The Ag content is 7.92 wt% (sample 3, 10 min), 15.42 wt% (sample 8, 30 min), and 23.10 wt% (sample 9, 60 min), respectively. Therefore, the similarity in the antibacterial rings of the three samples against *S. aureus* may be due to the combined effects of AgNPs particle size and Ag content in the samples. The bacteriostatic ring of the three samples to *E. coli* is quite different, which may be affected by other factors besides AgNPs particle size and Ag content, such as the difference between *E. coli* and *S. aureus* in cell wall structure, and the release of Ag^+ ions in the samples (Dong et al., 2015). In addition, the aggregation of AgNPs can reduce or even disappear their antibacterial properties (Sureshkumar et al., 2010). Therefore, the dispersion of AgNPs on cotton stalk cellulose matrix may also affect the antibacterial properties of the sample. Based on the FE-SEM results, it was observed that AgNPs were uniformly dispersed on the cotton stalk cellulose matrix after 10 min of microwave heating, while extending the microwave heating time to 30 or 60 min resulted in varying degrees of aggregation of AgNPs.

The minimum inhibitory concentrations (MICs) of cotton stalk cellulose/Ag nanocomposites (sample 3, sample 8, and sample 9) against *S. aureus* ATCC 6538 were determined by agar dilution method. The fine screening results of MICs are shown in Fig. 3.2. It has been observed that the MICs of the samples prepared by microwave heating for 10, 30, and 60 min are 1, 2, and 2 mg mL^{-1}, respectively. Fig. 3.3 is an enlarged view of the area within the purple box in Fig. 3.2. In previous studies, the MICs of holocellulose/Ag nanocomposites prepared by hydrothermal reaction at 160°C for 12 h against *S. aureus* were 2.5–20 mg mL^{-1} (Fu et al., 2016). The three samples tested in

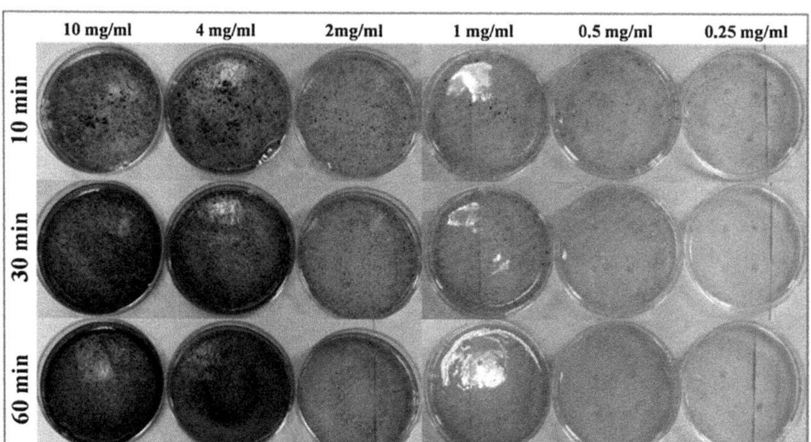

FIGURE 3.2 The second screening of MICs for the products prepared by microwave–hydrothermal method with hemicellulose concentration of 5 mg mL^{-1} at for different times against ***Staphylococcus aureus***. *Reproduced from Fu et al. (2018) with permission from MDPI.*

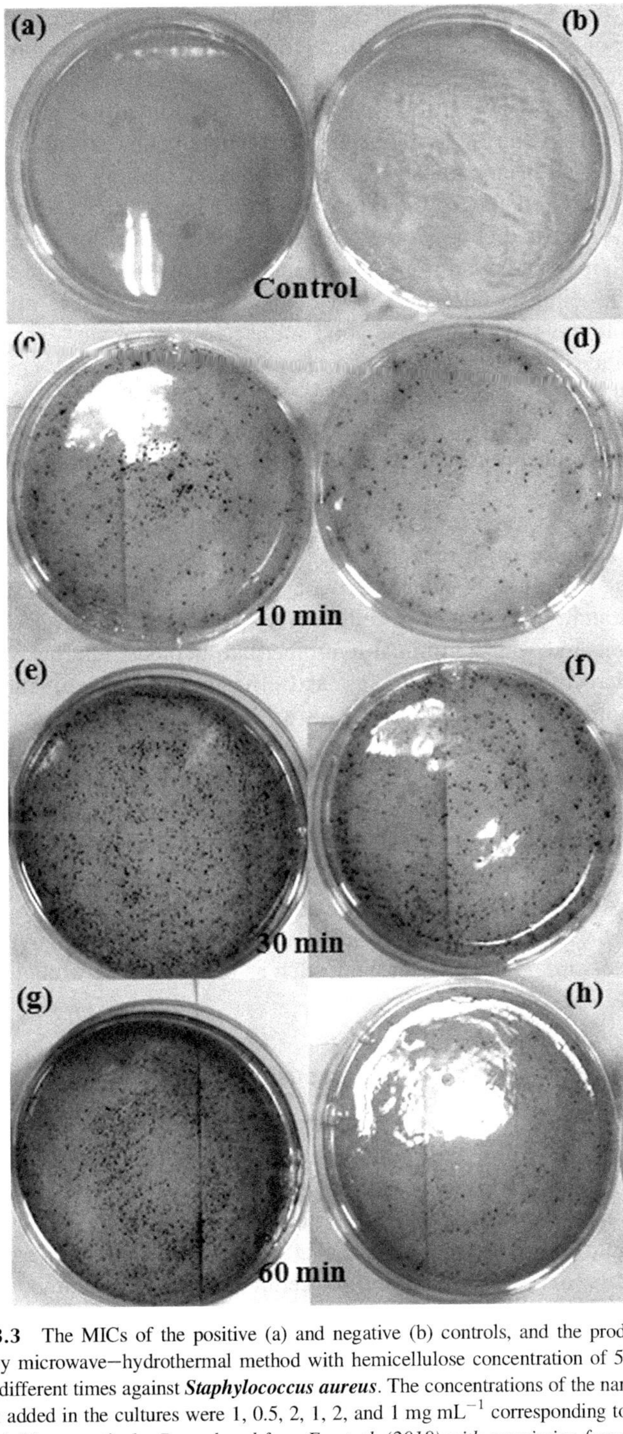

FIGURE 3.3 The MICs of the positive (a) and negative (b) controls, and the products (c ~ h) prepared by microwave–hydrothermal method with hemicellulose concentration of 5 mg mL^{-1} at 160°C for different times against ***Staphylococcus aureus***. The concentrations of the nanocomposites suspension added in the cultures were 1, 0.5, 2, 1, 2, and 1 mg mL^{-1} corresponding to (c), (d), (e), (f), (g), and (h), respectively. *Reproduced from Fu et al. (2018) with permission from MDPI.*

this study showed better antibacterial performance, which may be related to the release of Ag^+ ions in the material. The cellulose chain contains abundant −OH, which makes the cellulose negatively charged in the water surface. A large number of −OH can form intramolecular and intermolecular hydrogen bond network structures, giving cellulose excellent colloidal stability, and the growth of metal nanoparticles can be controlled through the electrostatic interaction between the −OH group and metal nanoparticles (Sureshkumar et al., 2010; Stenstad et al., 2008; Shin et al., 2008). Therefore, we speculate that AgNPs may be bound to the hydrogen bond network structure formed by cotton stalk cellulose OH, thereby achieving controlled release of Ag^+ ions and enabling the material to have good antibacterial properties. The results of bacteriostatic ring test and MICs test show that the cotton stalk cellulose/Ag nanocomposite has good antibacterial activity against *S. aureus* and *E. coli*. Therefore, the cotton stalk cellulose/Ag nanocomposite prepared in this study has potential application value in the biomedical field, such as wound dressing or bandage.

In this section, cotton stalk cellulose/Ag nanocomposites were prepared using hemicellulose as a reducing agent with microwave rapid heating technology. The effects of hemicellulose concentration, microwave heating time, and temperature on cotton stalk cellulose/Ag nanocomposites were studied in detail. Compared with traditional hydrothermal methods, microwave heating has the characteristics of being fast and efficient. The experimental results indicate that hemicellulose can serve as an effective reducing agent for Ag^+ ions, and the particle size and content of AgNPs in the material, as well as the dispersion of AgNPs on cotton stalk cellulose matrix, are influenced by hemicellulose concentration, microwave heating time, and temperature. The results of inhibition ring test and MICs test showed that the cotton stalk cellulose/Ag nanocomposite showed good antibacterial performance against *S. aureus* (gram-positive bacteria) and *E. coli* (gram-negative bacteria). Therefore, the cotton stalk cellulose/Ag nanocomposites have good application prospects in the biomedical field, such as being used as wound dressings or bandages.

3.2.2 Cellulose/Ag/AgCl nanocomposites

Cellulose is currently one of the most abundant biomass in nature (Klemm et al., 2005). There are various antibacterial composites, including cellulose/silver nanoparticles (Ferraria et al., 2009; He et al., 2003; Zhu et al., 2009), porous or nonporous cellulose/silver nanostructures (He et al., 2005), and cellulose/gold−silver alloy nanoparticles (Shin et al., 2008). There are also some reports on the preparation of cellulose/silver nanocomposites. For example, Maneerung et al. implanted silver nanoparticles into bacterial cellulose to obtain a medical wound dressing with excellent antibacterial properties (Maneerung et al., 2008). Pinto et al. prepared cellulose/silver nanocomposites using two completely different methods and two different sources of cellulose substrates (plant cellulose and bacterial cellulose), which had good antibacterial properties against rod-shaped bacteria, *S. aureus*, and *Klebsiella pneumoniae* (Pinto et al., 2009). Maria et al. studied the effects of different reducing agents (hydrazine, hydroxylamine, and ascorbic acid) and polyvinylpyrrolidone (PVP) on the synthesis of bacterial cellulose/colloidal silver nanocomposites (Maria et al., 2010). Sureshkumar et al. endowed bacterial cellulose/silver composites with magnetic antibacterial properties using a high-speed mixer as a reactor and polyethylene as a reducing agent (Sureshkumar et al., 2010).

As is well known, the microwave heating is achieved through the interaction of dipole moments between molecules and the synergistic effect of high-frequency electromagnetic radiation. DMAc molecules have a large dipole moment, making them a good solvent for microwave heating. The microwave heating method has a promoting effect on the crystallinity of silver and the dissolution of cellulose. Moreover, ascorbic acid is widely used as a reducing agent in the preparation of metal nanostructures (Biçer and Şişman, 2010; Miranda et al., 2006). The synthesis of silver nanostructures using ascorbic acid as a reducing agent has been reported (Wu et al., 2005; Sondi et al., 2003). Ascorbic acid was selected as a reducing agent due to its mild reactivity. However, how to tightly combine inorganic silver nanomaterials with cellulose and evenly disperse inorganic silver nanoparticles on the surface of cellulose still needs to be solved.

Lithium chloride was added to a DMAc solution, forming a lithium chloride/DMAc solution system under magnetic stirring. Afterward, microcrystalline cellulose was dissolved in the previously formed lithium chloride/DMAc solution system, preparing the cellulose solution under magnetic stirring conditions. The dissolved cellulose solution was added to a DMAc solution, then silver nitrate and ascorbic acid were added to the colloidal solution to obtain cellulose/silver nanocomposites. The colloidal solution was heated to a specific temperature by microwave and kept at this temperature for a certain reaction time, and then free cooling to room temperature.

Samples were prepared using cellulose solution, silver nitrate, and ascorbic acid in DMAc solution under microwave heating at 150°C for 40 min. The product sample appeared is diffraction peaks at $2\theta = 23°$ and 14.5°-16.5°, which correspond to the cellulose crystal I. The other diffraction peaks are all diffraction crystal planes of cubic silver, belonging to (111), (200), and (220) crystal planes (JCPDS 04-0783). No diffraction peaks of any impurity mixture can be observed. The results indicated the synthesis of cellulose/silver nanocomposite. When the amounts of ascorbic acid were 0.352 g, 1.000 g, and 2.000 g, respectively, the X-ray diffraction spectra of cellulose-based nanocomposites prepared by microwave-assisted method showed that in addition to the diffraction peak of cellulose crystal I, cubic silver was also observed on the diffraction crystal planes of (111), (200), and (220), as well as silver chloride on the characteristic diffraction crystal planes of (111), (200), (220), (311), and (222). The inadequate ascorbic acid is not sufficient to completely reduce silver nitrate to elemental silver. Due to the presence of chloride ions in the solution, the characteristic diffraction peak of silver chloride can be observed on the X-ray diffraction pattern. As the amount of ascorbic acid increased, the diffraction peak intensity of cubic crystalline silver increased, while the diffraction peak intensity of silver chloride weakened. When the amount of ascorbic acid is 3.000 g, no characteristic diffraction peaks of silver chloride can be observed on the X-ray diffraction pattern. To investigate the effect of ascorbic acid as a reducing agent on the product, the sample was heated in an oil bath to 150°C for 40 min. When the total amount of ascorbic acid increased to 1.000 g, the diffraction peak intensity of silver chloride crystal plane (220) was greater than that of silver crystal plane (200), indicating that the microwave heating treatment method had a significant promoting effect on the phase conversion of silver chloride to silver. Moreover, in the oil bath heating reaction, when the amount of ascorbic acid increased to 3.000 g, the (111), (200), and (220) crystal planes of silver chloride still existed, but the diffraction peak intensity decreased. On the contrary, pure cellulose/silver nanocomposites were obtained by microwave heating method. This result further confirmed that the heating method had a certain impact on the phase change of nanocomposites.

There are some typical characteristic absorption peaks of the infrared spectra of nanocomposites. The peak at 1045 cm^{-1} is the stretching vibration of the C—O—C of the glucose unit. A typical peak at 2898 cm^{-1} can be attributed to the C—H group. A large absorption peak at 3300—3450 cm^{-1} is the stretching vibration of hydroxyl and hydrogen bonds. The spectral peak at 1434 cm^{-1} corresponds to the symmetric bending vibration of CH_2. It can be seen that the infrared spectrum of the nanocomposite exhibited a characteristic absorption peak of microcrystalline cellulose. Due to the presence of carbon dioxide in the atmosphere, a series of peaks at 2130—2400 cm^{-1} are generated by carbon dioxide in the atmosphere. It can be observed from the infrared spectrum that the characteristic absorption peak of cellulose appears in the infrared spectrum of the nanocomposite.

The TG/DSC results show the effect of temperature on the thermal stability behavior of cellulose/silver nanocomposites. Thermogravimetric analysis (TGA) is conducted under air conditions to study the thermal stability of the product. Due to the presence of free and bound water in the environment, the thermogravimetric curve of cellulose/silver nanocomposites shows a small mass loss at around 110°C. From the thermogravimetric curve, it can be seen that mass loss mainly occurs in two stages, corresponding to the thermal degradation of cellulose and complete thermal decomposition. The total mass loss of different samples is 94.5%, 89.2%, 42.7%, and 65.3%, respectively. The differential scanning calorimetry curves of all samples show two distinct thermal peaks at (331 and 480), (348 and 511), (330 and 456), and (340 and 431) °C, respectively. The results confirm the thermogravimetric and differential scanning calorimetry well curves of nanocomposites. The TG/DSC spectra of nanocomposites prepared by oil bath heating method indicated that the total mass loss of different samples was 95.0%, 92.4%, 84.4%, and 76.2%, respectively. The total mass loss is quite high, indicating that the thermal stability of the products prepared under oil bath heating conditions is worse, confirming that the microwave heating treatment method is more conducive to the synthesis of nanocomposites. The differential scanning calorimetry curves show two distinct heat peaks at (332 and 484), (345 and 520), (338 and 480), and (344 and 492) °C, respectively. The nanocomposites synthesized by the oil bath heating method exhibit similar thermal peaks between 330 and 350°C. However, compared with the composite materials prepared under microwave-assisted heating, the thermogravimetric curve of the composite materials prepared under oil bath heating increased from 456/431 to 480/492°C, indicating a decrease in the thermal stability of cellulose in the products prepared by microwave assisted heating. The TG/DSC results further confirm that different heating methods not only have a certain impact on the phase of the nanocomposites, but also have a certain impact on the thermal stability of the materials.

Scanning electron microscopy (SEM) images of cellulose/silver nanocomposites prepared under microwave heating to 150°C for 40 min indicated that the diameter of silver nanoparticles was about 250 nm (Li et al., 2011), and they were widely and uniformly dispersed in the cellulose matrix (Fig. 3.4). In higher magnification SEM image, it was clearly shown that silver nanoparticles had a polyhedral microstructure. The cellulose-based nanocomposites prepared using different ascorbic acid concentrations under microwave heating to 150°C for 40 min indicated that the product had a larger nanoparticle size and a wider particle size distribution. For example, the average particle size of silver chloride in cellulose/silver chloride silver nanocomposites prepared using 0.352 g of ascorbic acid was 30 nm. The average particle size of the nanoparticles in the sample prepared using 1.000 g

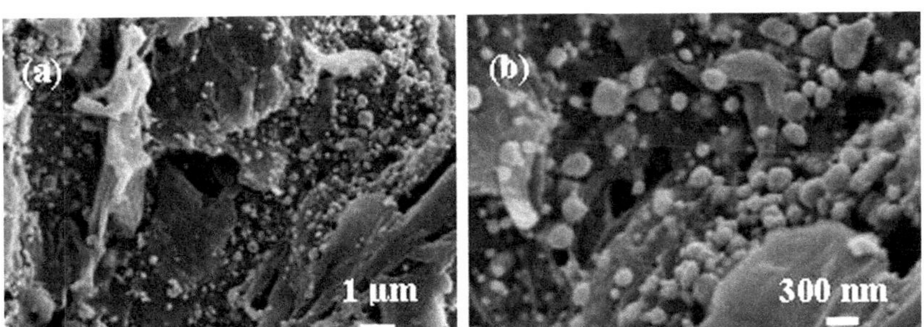

FIGURE 3.4 SEM images of the typical cellulose-silver nanocomposites prepared by microwave heating at 150°C for 40 min. *Reproduced from Li et al. (2011) with permission from Elsevier.*

ascorbic acid was approximately 56 nm. When the concentration of ascorbic acid increased to 2.000 g, nonuniform sized nanoparticles were observed. When the total amount of ascorbic acid increased to 3.000 g, under the same microwave heating time and temperature, there were uniform silver nanoparticles distributed on the cellulose matrix. The aforementioned SEM results showed that the concentration of ascorbic acid not only had a certain impact on the phase of the prepared nanocomposites but also had a certain impact on the size and morphology of the nanocomposites. When the heating time was 10 min, polyhedral silver nanoparticles dispersed in the cellulose matrix were observed. As the heating time increased to 20 min, the density of polyhedral silver particles also increased. As the heating time increased to 40 min, it observed from the scanning electron microscope that more and more silver nanoparticles were distributed in the cellulose matrix. In summary, sufficient microwave reaction heating time was conducive to the nucleation and growth of silver particles.

The CP/MAS ^{13}C solid-state nuclear magnetic resonance spectrum of the prepared cellulose/silver nanocomposite, microcrystalline cellulose, and cellulose/silver chloride nanocomposites indicated that microcrystalline cellulose exhibited characteristic skeleton peaks of cellulose at (62.5 + 65.0), (72.4 + 74.8), (84.2 + 88.8), and 105.0 ppm, corresponding to the C6, C2-C3-C5, C4, and C1 positions of carbon elements, respectively (Zawadzki and Wisniewski 2002). Among them, the small peaks at 62.5 and 84.2 ppm were irregular regions of C6 and C4 in the carbon element. However, at the nuclear magnetic peaks of 65.0 and 88.8 ppm, there were regular regions of carbon elements C6 and C4. The CP/MAS ^{13}C solid-state nuclear magnetic resonance spectra of cellulose/silver chloride nanocomposites and cellulose/silver nanocomposites had similar characteristic nuclear magnetic resonance peaks, compared with cellulose. However, the relative peak intensities of the regular regions of carbon elements C6 and C4 in the nanocomposites decreased, indicating that the formation of nanoparticles composite with cellulose led to a decrease in the crystallinity of cellulose. The interaction between silver colloidal particles and cellulose matrix prevented the formation of an ordered, high crystallinity arrangement of cellulose chains, thereby causing a decrease in cellulose strength. Similar phenomena have been reported in literature, such as a decrease in the crystallinity of cellulose in the synthesis of cellulose/cadmium sulfide nanocomposites (Ruan et al., 2005), as well as changes in the crystallinity of cellulose in the preparation of cellulose/iron oxide nanocomposites (Liu et al., 2008), and cellulose/calcium silicate nanocomposites (Li et al., 2010).

In DMAc solution, the cellulose matrix used ascorbic acid as a reducing agent to reduce silver nitrate to silver nanoparticles. Under the action of microwave heating, silver rapidly undergoes nuclear reaction. Microwave heating provided energy for the reaction and ensured the formation and growth of crystal nuclei. Because microwave heating was an efficient and fast heating method, cellulose/silver nanocomposites can be rapidly synthesized. It can be observed from the scanning electron microscope that silver nanoparticles were embedded in the cellulose matrix, and no single silver nanoparticles were observed, meaning that there was indeed a strong mutual awesome in the cellulose/silver nanocomposites, which made silver and cellulose combine. When the concentration of ascorbic acid was 0.352 g, only a small portion of silver ions were reduced to elemental silver, and most silver ions reacted with chloride ions in the system to form silver chloride. Finally, cellulose/silver chloride silver nanocomposites were obtained. It can be seen that as the concentration of ascorbic acid increased, the amount of silver obtained also increased. When the total amount of ascorbic acid increased to 2.000 g, the main phase was silver, while silver chloride appeared as a secondary phase. When the concentration of ascorbic acid further increased to 3.000 g, silver ions were completely reduced to silver nanoparticles, thus preparing cellulose/silver nanocomposites. The aforementioned characterization showed that microwave heating treatment method had great advantages in shortening reaction time. In addition, the experimental results showed that different phases were obtained by traditional methods (oil bath heating) and microwave heating. Some literature reported that microwave heating had a certain impact on the morphology, size, and particle size distribution of prepared composite materials due to its unique thermal and nonthermal microwave effects (Dallinger and Kappe, 2007; Natile et al., 2005). It is generally believed that microwave effects (thermal and nonthermal effects) existed in microwave heating reactions (Dallinger and Kappe, 2007), but the exact reason why microwave radiation heating can improve the chemical reaction process is still unknown, and efforts need to be made to overcome this challenge.

The antibacterial properties of cellulose-based nanocomposites against *E. coli* and *S. aureus* are shown in Fig. 3.5. Sample 1, sample 2, and blank reference sample (microcrystalline cellulose) each take 0.15 g of compressed

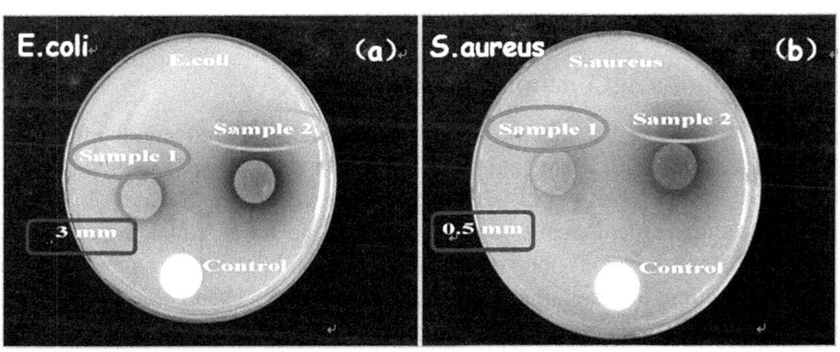

FIGURE 3.5 Antimicrobial activities of the cellulose-based nanocomposites: (a) ***Escherichia coli*** and (b) ***Staphylococcus aureus.***

tablets. Among them, sample 1 was a cellulose/silver chloride nanocomposite prepared with 1.000 g of ascorbic acid microwave-assisted heating, and sample 2 was a cellulose/silver nanocomposite prepared with 3.000 g of ascorbic acid microwave assisted heating. The blank reference sample (microcrystalline cellulose) was shown in the control in Fig. 3.5. The samples were sterilized for 2 h at 120°C in a high-pressure sterilization pot before undergoing antibacterial testing. Antibacterial test results showed that there was no bacteriostatic ring around microcrystalline cellulose. The size of bacteriostatic ring of sample one against *E. coli* and *S. aureus* was 3 and 0.5 mm, respectively, while there was no obvious bacteriostatic ring around sample 2, just pigment diffusion. The antibacterial test results showed that microcrystalline cellulose had no antimicrobial activity against *E. coli* and *S. aureus*. Compared with cellulose/silver nanocomposites, cellulose/silver chloride silver nanocomposites prepared by microwave-assisted heating at 150°C and reaction time of 40 min had better antibacterial performance. The reason why silver nanoparticles had good antibacterial effects on bacterial microorganisms was because they can enter the cell wall or membrane of microorganisms, causing the aerobic metabolic enzymes of microorganisms to lose activity, ultimately causing bacteria to be unable to metabolize and die. At present, there were four widely recognized antibacterial mechanisms of silver: through electrostatic adsorption, inorganic antibacterial ions can react with the DNA of microorganisms (Feng et al., 2000; Feng et al., 1998; Liau et al., 1997), silver ions catalyze microbial cell walls or membranes to kill bacteria under the catalysis of light (Inoue et al., 2002; Davies and Etris, 1997), silver ions simultaneously have their common composite effect to cause microorganisms to lose activity.

In summary, cellulose/silver chloride silver and cellulose/silver nanocomposites were prepared using microwave method by adjusting the amount of ascorbic acid in LiCl/DMAc. Microwave-assisted heating is more conducive to the phase conversion of silver chloride to silver, compared with oil bath heating. It found that cellulose/silver/AgCl nanocomposites had better antibacterial properties than that of cellulose/silver nanocomposites.

3.2.3 Cellulose/AgCl nanocomposites

At present, there is increasing attention to cellulose-based nanocomposites (Eichhorn et al., 2010; Zhang et al., 2007; Habibi et al., 2010; Liu et al., 2008; Luo et al., 2009). Zhang et al. (2010) prepared adjustable-thickness bacterial cellulose/gold nanocomposites using polyethylene as a reducing agent and coupling agent using a one-step biological template synthesis method in aqueous solution. Sureshkumar, Siswanto, and Lee also used polyethylene as a reducing agent to synthesize magnetic cellulose/silver nanocomposites with excellent antibacterial properties (Sureshkumar et al., 2010). As is well known, LiCl/DMAc solution is a good solvent system for dissolving cellulose

(Terbojevich et al., 1985). This section mainly focuses on the research of cellulose/silver chloride nanocomposites. In this section, cellulose was pretreated with LiCl/DMAc solution, and cellulose/silver chloride nanocomposites were prepared by microwave method using silver nitrate as the silver source and DMAc as the solvent. DMAc solution not only served as a reaction solvent system but also served as a good microwave radiation absorber, and lithium chloride was used for reaction to generate silver chloride nanoparticles.

Microcrystalline cellulose and lithium chloride were slowly added them to an *N,N*-dimethylacetamide (DMAc) solution, and heated in an oil bath under magnetic stirring conditions to 90°C for 3 h to obtain a white cellulose solution. Continuing under magnetic stirring, the cellulose solution and silver nitrate were added to a DMAc solution. The mixed solution was heated to 150°C for a certain reaction time in a microwave reactor.

Fig. 3.6 shows the X-ray diffraction spectra of the samples prepared by microwave heating at 110, 130, and 150°C for 20 min, respectively (Li et al., 2012). It can be seen that the three samples obtained at different reaction temperatures all have similar characteristic diffraction peaks. The diffraction peaks at $2\theta = 14.8°$, 22.5°, and 34.5° belong to the typical characteristic absorption peaks of microcrystalline cellulose crystal I. The other characteristic peaks on the X-ray diffraction pattern belong to the typical characteristic peaks of silver chloride (JCPDS 31-1238), which belong to the (111), (200), (220), (311), and (222) planes of silver chloride, respectively. No characteristic peaks related to silver were found. The XRD results indicated that all samples

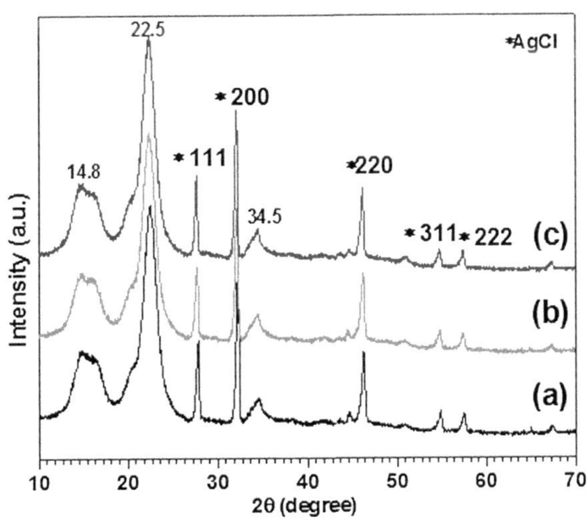

FIGURE 3.6 XRD patterns of the cellulose/AgCl nanocomposites prepared by microwave heating at different temperatures for 20 min: (a) 110°C; (b) 130°C; (c) 150°C. *Reproduced from Li et al. (2012) with permission from Elsevier.*

existed in the form of cellulose and silver chloride crystals, indicating the synthesis of cellulose/silver chloride nanocomposites. Fourier transform infrared spectroscopy analysis was performed on the cellulose/silver chloride nanocomposites. The infrared spectra of cellulose/silver chloride nanocomposites obtained by microwave heating at 110, 130, and 150°C for 20 min, respectively, indicated that the infrared spectra of all three samples exhibited characteristic absorption peaks of cellulose. The absorption peak at 3449 cm^{-1} is caused by the stretching vibration of the cellulose hydroxyl group, while the uneven stretching vibration of the C—H bond on the pyran ring resulted in a characteristic absorption peak at 2905 cm^{-1} in the infrared spectrum. The absorption peak at 1623 cm^{-1} is caused by the bending vibration of adsorbed water, while the absorption peak at 1433 cm^{-1} is due to the presence of methylene. The bending vibration of hydroxyl groups causes an absorption peak at 1377 cm^{-1} in the infrared spectrum (Kokot et al., 1997), while the infrared absorption peak at 1318 cm^{-1} is caused by the skeleton vibration of the C—C and C—O bonds (Pappas et al., 2002). The asymmetric stretching of the C—O bond exhibited an infrared characteristic absorption peak at 1169 cm^{-1}, and a strong absorption peak at 1065 cm^{-1} is attributed to the skeletal vibration of the C—O—C bond on the pyran ring. The connection of β-glycosidic bond between glucose units caused a small characteristic peak at 896 cm^{-1} in the infrared spectrum (Sun et al., 2002). The characteristic peak of cellulose at 3449 cm^{-1} shifts to a lower wavelength range with increasing heating time, showing a trend of broadening. This phenomenon is caused by the interaction between cellulose and silver chloride nanoparticles.

The thermal stability of cellulose/silver chloride nanocomposites showed that the composite material exhibited approximately 93.21% mass loss in the range of 217—460°C, corresponding to the two exothermic absorption peaks of the DSC curve at 333 and 448°C. This is due to the complete thermal degradation of cellulose in the composite material. Microcrystalline cellulose exhibited weight loss in the range of 250—380°C, and corresponded to an exothermic absorption peak at 353°C in the DSC curve (Li et al., 2010). From the aforementioned data, it can be inferred that the thermal degradation temperature of the composite material was higher than that of microcrystalline cellulose, and there were two exothermic absorption peaks. This also indicated that the preparation of the composite partially destroyed the crystal structure of cellulose and improved the thermal stability of the cellulose/silver chloride nanocomposite. This phenomenon of two exothermic absorption peaks observed in the DSC curve had also been reported in the literature on the synthesis of cellulose-based nanocomposites, such as cellulose/hydroxyapatite nanocomposites (Ma et al., 2010), cellulose/silver nanocomposites, and cellulose/carbonate-containing hydroxyapatite nanocomposites (Jia et al., 2010). Cellulose-based nanocomposites exhibited two exothermic absorption peaks in the DSC curve, while pure microcrystalline cellulose only had one exothermic peak. There were literature reports that silver chloride crystals exhibited good thermal stability from room temperature to 500°C (Feng et al., 2006). From the aforementioned TG/DSC research results, it can be concluded that the interaction between cellulose and silver chloride endowed the composite material with good thermal stability.

The microstructure of the composite material was observed by scanning electron microscopy. The SEM images of cellulose/silver chloride nanocomposites prepared under microwave heating at 150°C for 20 min indicated that silver chloride nanoparticles were uniformly dispersed on the surface of cellulose in the composite material, and cellulose with a fibrous structure could be well observed, and silver chloride nanoparticles adhered to the surface of cellulose. Silver chloride crystals were dispersed on the surface of cellulose, and some individual silver chloride nanoparticles could also be clearly observed. The formation of silver chloride nanocrystals occurred almost simultaneously with the precipitation of cellulose, resulting in a uniform distribution of silver chloride nanoparticles on the surface of cellulose. Microwave radiation is a simple and fast heating method, and it can easily disrupt the thermal equilibrium in a solution. In a relatively short time range, cellulose precipitates from the solution and almost simultaneously forms silver chloride with silver cations and chloride anions in the reaction system, which is a rapid nucleation and growth process. The rapid nucleation process mentioned before enables silver chloride nanoparticles to have a good distribution on the surface of cellulose. Compared with the sample heated by microwave reaction for 20 min, there was not much difference in morphology between the nanocomposites prepared by microwave heating for 10 min. However, there were some scattered silver chloride nanoparticles on the surface of cellulose. When the reaction temperature was reduced to 110°C, similar phenomena could also be observed. After microwave heating at 150°C for 20 min, silver chloride nanoparticles were formed and uniformly dispersed in the cellulose matrix. Therefore, higher microwave heating reaction temperature and longer microwave heating time are beneficial for the nucleation and growth of silver chloride nanoparticles in composite materials.

The antibacterial properties of cellulose/silver chloride nanocomposites play a crucial role in their biomedical applications. Fig. 3.7 shows the antimicrobial activities of the composites against *E. coli* and *S. aureus*, respectively. *E. coli* and *S. aureus* are two kinds of common bacteria, which widely exist in infected wounds. In the detection of antimicrobial activity, pure microcrystalline cellulose is used as a blank reference sample. It can be clearly observed that no antibacterial circles were found around the blank reference sample microcrystalline cellulose, indicating that microcrystalline cellulose does not possess any antibacterial properties. Compared with the blank sample, there is a significant phenomenon of antibacterial ring around the cellulose/silver chloride nanocomposite. According to the observation and measurement, the size of the bacteriostatic zone of the composite against *E. coli* (Gram stain negative) and *S. aureus* (Gram stain positive) is 4.5 mm.

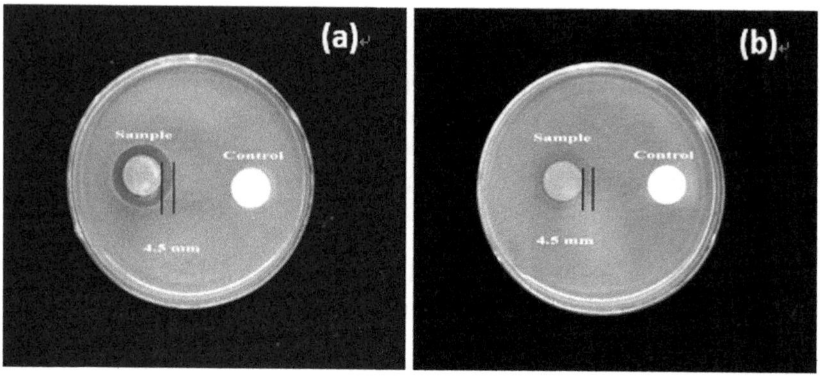

FIGURE 3.7 Antimicrobial activities of the cellulose/AgCl nanocomposites: (a) ***Escherichia coli*** and (b) ***Staphylococcus aureus.*** *Reproduced from Li et al. (2012) with permission from Elsevier.*

The aforementioned results also indicate that the cellulose/silver chloride nanocomposite has relatively good antimicrobial activity. Due to the absence of antibacterial circles around the blank reference sample, the aforementioned observation results can also confirm that the antibacterial effect is due to the presence of composite species of silver chloride particles, rather than microcrystalline cellulose. Due to the advantages of high reaction rate, fast reaction time, greatly improved reaction selectivity, and energy savings, cellulose/silver chloride nanocomposites can be prepared under 10 min of microwave heating time. In this reaction system, the addition of lithium chloride not only provides Cl^- for the reaction, but also is one of the components that make up the dissolved cellulose in the LiCl/DMAc system. Ag^+ and Cl^- generate AgCl, which also leads to a rapid nucleation and growth process, allowing silver chloride nanoparticles to be well distributed in the cellulose matrix.

In summary, cellulose/silver chloride nanocomposites were prepared using microwave-assisted method in LiCl/DMAc pretreatment. Based on the principle that the nucleation and growth of silver chloride nanoparticles occur synchronously with cellulose precipitation, LiCl/DMAc simultaneously serves as a solvent for cellulose, providing raw materials for the preparation of silver chloride. These results indicated that silver chloride has good dispersibility on cellulose matrix, and cellulose/silver chloride nanocomposites have good thermal stability and antibacterial properties.

3.2.4 Multifunctional silver@Fe_3O_4@nanocellulose composites

Combining the advantages of multiple components, nanocomposites synthesized from two or more nanomaterials typically exhibit unique functions in various fields, such as catalysis, electronics, photography, optics, and antibacterial. As we all know, Fe_3O_4 nanoparticles have excellent magnetic separation performance and have good application prospects in supercapacitor electrode materials, renewable catalysts for catalytic oxidation of alcohols, and magnetic carriers for protein separation and wastewater treatment (Wang et al., 2013). Silver has strong antibacterial activity against 650 bacteria and relatively low toxicity to the human body (Das et al., 2012; Sambhy et al., 2006; Sharma et al., 2009). It can be used as an antibacterial material in various fields, including water treatment (Song et al., 2016), textile manufacturing, food packaging and storage (Peter et al., 2016), and biomedical fields (Luo et al., 2014). Ag and Fe_3O_4 have received increasing attention in the past few decades. For example, microwave-assisted method was used to synthesize Ag@Fe_3O_4@carboxymethyl cellulose (CMC) composites as a nanocatalyst for water hydrogenation and aldehyde oxidation. Ag@TiO_2@Fe_3O_4 has enhanced photocatalytic and cycling performance, which can be used for the degradation of ampicillin and marine environmental pollution (Zhao et al., 2016).

It reported that multiple methods can be used to synthesize Ag@Fe_3O_4 nanocomposites. By depositing Au and Ag on the surface of Fe_3O_4 nanoparticles in aqueous solution at room temperature, Au@Fe_3O_4 and Ag@Au@Fe_3O_4 nanocomposites with core−shell structure were synthesized. Using silver nanowires as nucleation sites for Fe_3O_4 growth in aqueous solution, one-dimensional core−shell Ag@Fe_3O_4 heterojunction nanowires were achieved using a precipitation method. In view of the requirements of green chemistry and sustainable development strategy, it is increasingly urgent to develop green synthesis methods of nanomaterials. Nanocellulose can be used as a carrier and reducing agent to synthesize Ag@Fe_3O_4 nanocomposites (Xiong et al., 2013). Ag−Fe_3O_4 core−shell nanocomposites can be obtained by using grape stem extract waste as a green solvent, reducing agent, and sealing agent (Venkateswarlu et al., 2015). MWH method, as a green and convenient method, has been widely used to synthesize inorganic nanoparticles. For example, the MWH ionic liquid method was used to synthesize $ZnFe_2O_4$ nanoparticles with better photocatalytic performance than phenol (Cao et al., 2009). MWH method synthesized cubic $CuFe_2O_4$ nanoparticles with magnetic properties (Phuruangrat et al., 2016). MWH method synthesized CuO crystals using cellulose as the substrate (Yao et al., 2016). This section introduced the MWH method to synthesize Ag@Fe_3O_4@CNC nanocomposites and discussed the phase, morphology, magnetic properties, and thermal stability of Ag@Fe_3O_4@CNC nanocomposites in detail. This study used neutral deionized water as a solvent throughout the entire system, and the green low-cost synthesis route did not require any chemical reducing agents.

Fe_3O_4 particles are prepared by hydrothermal method using hexahydrate of iron chloride, sodium acetate, ethylene glycol, and ethylenediamine. CNC was added to deionized water to form a CNC solution. Ferric oxide was added to deionized water to form a solution of ferric oxide. Silver nitrate was added to deionized water to form a silver nitrate solution. All CNC solution, ferric oxide solution, and silver nitrate solution were mixed under magnetic stirring conditions for 10 min, which were placed in a polytetrafluoroethylene lined reactor into a microwave–hydrothermal synthesis instrument at 180°C for different times.

The schematic diagram of the possible formation process of Ag@Fe_3O_4@NCC nanocomposites is shown in Fig. 3.8. Cellulose is composed of β-(1,4) glycosidic bond and contains rich hydroxyl groups. In the literature, it reported that cellulose has a certain reduction ability, which can reduce silver ions to simple silver (Padalkar et al., 2010). Cellulose also exhibits reducibility under microwave conditions. Under microwave conditions, heavy metals, such as gold, platinum, copper, silver, and palladium, were prepared using an aqueous solution of carboxymethyl cellulose. In this section, silver was prepared by MWH method using CNC as reducing agent. Silver and ferric oxide were combined with CNC through hydrogen bonding and electrostatic bonding. The X-ray diffraction spectrum of the sample synthesized by hydrothermal method observed the (220), (311), (400), (422), (511), and (440) surfaces of the sample, all of which can be attributed to iron tetroxide (JCPDS No. 19-0629). Ferric oxide is used to prepare Ag@Fe_3O_4@CNC nanocomposites. The X-ray diffraction spectra of the synthesized samples at microwave–hydrothermal reaction times of 10 min, 20 min, and 30 min indicated that all samples showed similar diffraction peaks, which were be attributed to iron tetroxide and silver, indicating that silver elemental was prepared from silver ions using CNC.

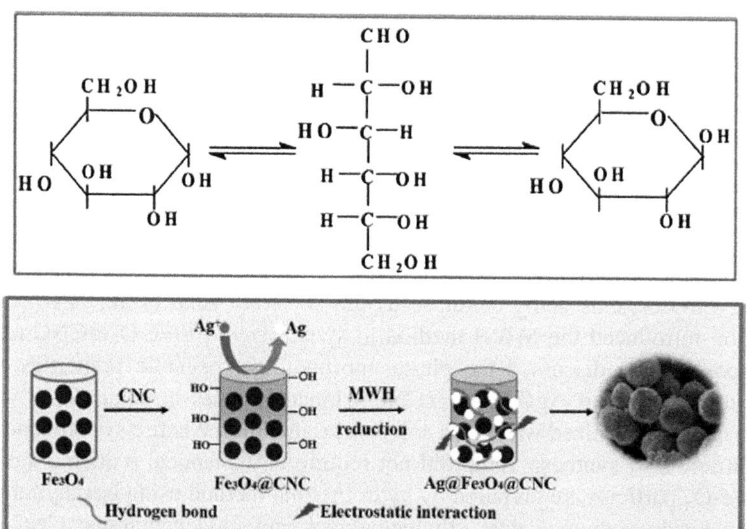

FIGURE 3.8 Schematic illustration of the formation process of Ag@Fe_3O_4@NCC nanocomposites synthesized by microwave-assisted hydrothermal method. *Reproduced from Dong et al. (2017) with permission from Springer.*

The SEM images and particle size distribution of Fe_3O_4 and Ag@Fe_3O_4@CNC nanocomposites are shown in Fig. 3.9. Ferric oxide has a spherical structure and a smooth surface, with an average particle size of 275 ± 15 nm. After adding silver and CNC, the spherical structure of ferric oxide was retained, but its surface became less smooth than before. CNC is surrounded by iron trioxide, and a portion of the CNC is also embedded inside the CNC. The silver nanoparticles obtained by reduction using CNC as a reducing agent are dispersed on the surface of Fe_3O_4 and CNC. When the reaction time is 10 min, the average particle size of silver@Fe_3O_4 is 279 ± 26 nm. When the reaction time is 20 min, the average particle size of silver@Fe_3O_4 is 283 ± 18 nm. When the reaction time is 30 min, a large number of silver nanoparticles appear and are uniformly connected to the surface of iron tetroxide and CNC. It has been observed that there are a large number of silver@Fe_3O_4 spherical structures with sizes of approximately 100–400 nm, which have a higher average particle size of 287 ± 36 nm. These results indicate that the length of heating time has a certain impact on the size of the sample, and the longer the reaction time, the more favorable it is for the generation of large-sized silver@Fe_3O_4. Using CNC as both a reducing agent and a stabilizer, most of the iron oxide particles are spherical, while silver particles have good dispersibility and stability (Song et al., 2016).

Samples were also synthesized using CNC solutions of different volumes. No CNC solution is added, all diffraction peaks can be attributed to Fe_3O_4. When the volumes of the CNC solution are 5 and 10 mL, the diffraction peaks can be attributed to the mixed phase of iron tetroxide (the main phase) and silver (the secondary phase). Based on the aforementioned results, CNC plays an important role in the synthesis process of silver and exhibits excellent reduction ability. These results are consistent with previous reports (Benaissi et al., 2010; Huang and Yang, 2004). Different synthesis methods were used to prepare samples, while other conditions remained consistent. The research results show that the samples prepared by microwave–hydrothermal method have different results from those prepared by traditional hydrothermal method. The sample prepared by microwave–hydrothermal method is composed of the mixed phase of ferric oxide and silver, but the silver diffraction peak is weak. It is difficult to find the silver diffraction peak from the sample prepared by traditional hydrothermal method. The corresponding scanning electron microscope images indicated that there are significant differences in the morphology of the samples prepared by different methods. Only a few CNC fibers are found around Fe_3O_4, and the average particle size of the obtained sample is about 288 ± 15 nm. The SEM image of the sample synthesized by traditional hydrothermal method indicated that only a few CNC fibers were observed, and the surface of Fe_3O_4 was smoother, with a slightly lower average particle size of about 279 ± 10 nm (similar to the particle size of Fe_3O_4 precursor). The corresponding EDX photoelectron spectra demonstrated that all samples showed strong signal peaks for carbon, oxygen, iron,

FIGURE 3.9 SEM images of Fe_3O_4 nanoparticles synthesized by hydrothermal method at 180°C (a); $Ag@Fe_3O_4@CNC$ nanocomposites synthesized by MW hydrothermal method for different times: (b) 10 min; (c) 20 min, (d) 30 min. *Reproduced from Dong et al. (2017) with permission from Springer.*

and silver elements from CNC, iron tetroxide, and silver crystals. The concentrations of silver in the samples synthesized using microwave–hydrothermal, microwave, and hydrothermal methods were 20.2%, 7.6%, and 15.6%, respectively. As a reducing agent in the reaction system, the amount of CNC plays an important role in the synthesis of silver. Reducing the amount of CNC can significantly reduce the yield of silver, leading to a sharp weakening of the diffraction peak of silver, which is consistent with the X-ray diffraction results. Based on the aforementioned experimental results, the MWH method is necessary in the synthesis process of silver elemental in this section. This can be explained by the fact that in the synthesis process of silver, neither microwave-assisted nor traditional hydrothermal methods can provide sufficient energy to aggregate iron tetroxide and CNC. The deposited CNC is easily lost from Fe_3O_4, leading to a decrease in stability and weakened reduction performance. However, combining the advantages of microwave-assisted method and traditional hydrothermal method, MWH method can provide high energy for Fe_3O_4@CNC composites in a short time.

The thermal stability of Ag@Fe_3O_4@CNC nanocomposites was determined by DTA and TGA methods. The DTA and TGA curves of samples synthesized by MWH at different times observed that the sample has two stages of weight loss, accompanied by its exothermic peak: the first exothermic stage (180–250°C) may be caused by the thermal degradation of CNC, and the second exothermic stage (280–350°C) may be caused by the complete degradation of CNC. The total mass loss of the corresponding samples is 64.7% and 64.3%, respectively. The thermal stability of samples synthesized by microwave-assisted method and traditional hydrothermal method was also studied. The total mass loss of the sample synthesized by hydrothermal method is about 77.8%, and the total mass loss of the sample synthesized by microwave-assisted method is about 66.7%. For the samples synthesized by hydrothermal method, the corresponding exothermic peaks are approximately 204.2 and 442.3°C, respectively. For the samples synthesized by microwave-assisted method, the corresponding exothermic peaks are approximately 198.6 and 350°C, respectively. The aforementioned experimental results indicate that compared with the other two synthesis methods, the reaction time of microwave–hydrothermal heating slightly affected the thermal stability of Ag@Fe_3O_4@CNC nanocomposites. In this section, compared with microwave-assisted method and hydrothermal method, the samples synthesized by MWH method have the highest thermal stability.

The hysteresis regression curves of samples synthesized by different methods indicated that the saturation magnetization ranges from −10,000 to 10,000 Oe. Ag@Fe_3O_4@CNC nanocomposites exhibit superparamagnetic and ferromagnetic behavior. The saturated magnetization of the sample prepared by microwave-assisted method is 11.31 emu g^{-1}, the saturated magnetization of the sample prepared by hydrothermal method is 8.95 emu g^{-1}, and the saturated magnetization of the sample prepared by MWH method is 12.34 emu g^{-1}.

These results further confirmed that MWH method can improve the magnetic properties of Ag@Fe_3O_4@CNC nanocomposites. In the literature, the saturated magnetization of Fe_3O_4 synthesized using cellulose-based composites as precursors is 8.3 and 17.2 emu g^{-1}. The obtained sample can be easily separated from the dispersion using an external magnetic field. This visual experiment demonstrates that the sample has ferromagnetic properties and has potential applications in the fields of dye removal and magnetic adsorption of wastewater.

An azo dye methylene blue (MB) was used to study the performance of Ag@Fe_3O_4@CNC nanocomposites for removing pollutants. The initial color of the dye is blue, and once the dye molecules are adsorbed, it will become colorless. The initial methylene blue solution has a clear absorption peak at 664 nm, which corresponds to the n-p * transition. Ag@Fe_3O_4 @CNC nanocomposites and excessive $NaBH_4$ desorption agent were added, the adsorption peak at 664 nm significantly decreased within 30 min, indicating that Ag@Fe_3O_4@CNC nanocomposites have excellent adsorption properties. It reported that Ag nanoparticles show excellent catalytic activity and selectivity in many catalytic reactions (Chi et al., 2012; Patel et al., 2007). From the experimental results, it concluded that the elimination of dyes in water follows the mechanisms of catalytic reduction and physical adsorption.

The recyclability of Ag@Fe_3O_4@CNC nanocomposites is an important factor for more economical processes, and the efficiency of repeated cyclic reaction usage of the prepared samples was studied. Due to its good magnetic properties, Ag@Fe_3O_4@CNC nanocomposites can be easily separated from the solution through an external magnetic field. The nanocomposites were thoroughly washed with distilled water and reused for MB degradation. After repeating this process three times, Ag@Fe_3O_4@CNC nanocomposites remain stable and exhibit high reactivity, indicating that Ag@Fe_3O_4@CNC nanocomposites have excellent recyclability. The color of the MB solution after adding Ag@Fe_3O_4@CNC nanocomposites is significantly lighter.

The antibacterial activity of Ag@Fe_3O_4@CNC nanocomposites against *S. aureus* and *E. coli* is shown in Fig. 3.10. Pure CNC was used as a control, and no antibacterial circles were observed, indicating that pure CNC had no antibacterial activity. The inhibition circles of Ag@Fe_3O_4@CNC nanocomposites against *S. aureus* and *E. coli* are 2.8 and 2.4 mm, respectively. These results confirm that Ag@Fe_3O_4 crystals exhibit good antimicrobial activity. In addition, the size of inhibition circles for *S. aureus* and *E. coli* is different, which may be due to the difference between the cell walls of gram-negative and gram-positive bacteria (Feng et al., 2000). In previous studies, silver@silver chloride@cellulose composite was used as bacteriostatic agent, and the size of bacteriostatic zone against *S. aureus* and *E. coli* was 2.0 and 2.5 mm, respectively (Dong et al., 2014). Fig. 3.10c shows the possible antimicrobial mechanisms of Ag@Fe_3O_4@CNC nanocomposites. Ag ions

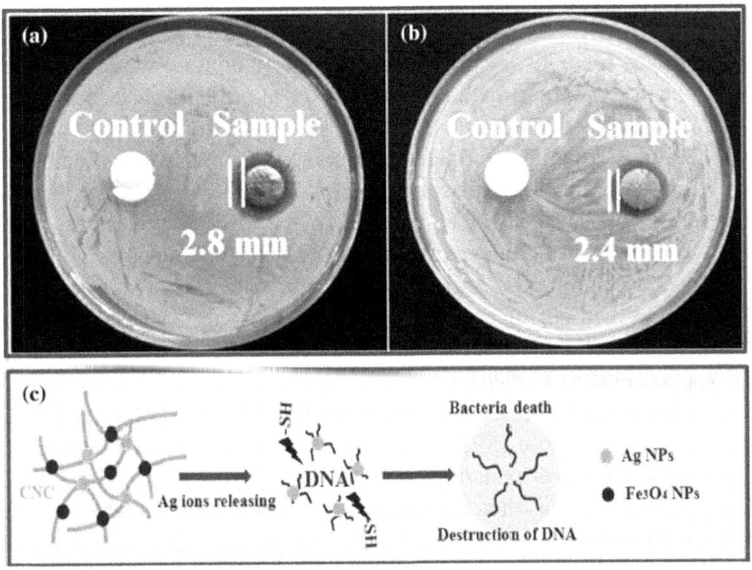

FIGURE 3.10 Antimicrobial activities of Ag@Fe_3O_4@CNC nanocomposites synthesized by MWH method for 30 min: (a) ***Staphylococcus aureus***; (b) ***Escherichia coli***; (c) The schematic illustration of possible antimicrobial mechanism of Ag@Fe_3O_4@CNC nanocomposites. *Reproduced from Dong et al. (2017) with permission from Springer.*

are slowly released from silver nanoparticles, which then pass through the bacterial membrane and react with–SH in the protease, leading to protease inactivation and ultimately bacterial death. In addition, Ag ions can be directly released into the protein to prevent DNA synthesis, leading to bacterial death. Finally, the released silver ions return to cellulose, which will be very important for improving the antibacterial efficiency of Ag ions. Due to its excellent antibacterial activity, the Ag@Fe_3O_4@CNC nanocomposites have potential application prospects in the fields of biomedicine and public health.

In summary, Ag@Fe_3O_4@CNC nanocomposites were prepared by MWH method. CNC exhibits excellent reducibility and plays an important role in the synthesis of silver particles. The longer the reaction time, the more conducive it is to synthesize Ag@Fe_3O_4 particles. Ag@Fe_3O_4@CNC nanocomposites have a narrow average particle size ranging from 275 ± 15 nm to 287 ± 13 nm. Ag@Fe_3O_4@CNC nanocomposites with good superparamagnetic and ferromagnetism enhance their ability to adsorb and degrade dyes, and also increase their application potential in wastewater treatment. It also showed excellent antimicrobial activity against *S. aureus* and *E. coli*. Ag@Fe_3O_4@CNC nanocomposites showed good application prospects in the biomedical field and public health field.

3.2.5 Cellulose/TiO_2 nanocomposites

There are many methods for preparing cellulose/titanium dioxide composites, such as phase separation method (Nagaoka et al., 2005), surface hydrolysis method, and sol–gel conversion method (Uddin et al., 2007; Zeng et al., 2010). Cellulose/titanium dioxide nanocomposites with excellent tensile strength can be prepared by dissolving cellulose in NaOH/urea systems with mass fractions of 7.5% and 11%, respectively. However, there are few reports on the preparation of cellulose/titanium dioxide nanocomposites with excellent antibacterial properties using hydrothermal methods. This section is to explore a new method for preparing cellulose/TiO_2 antibacterial composites with good antibacterial properties, opening up new avenues for high value-added applications of biomass.

Under the condition of continuous magnetic stirring at room temperature, microcrystalline cellulose (MCC) was added to sodium hydroxide/urea solution, where sodium hydroxide/urea was prepared in a mass ratio of 7:12. The resulting suspension was then refrigerated for 12 h in a refrigerator at −12°C. The resulting cellulose solution was used to prepare cellulose/TiO_2 nanocomposites. Tetrabutyl titanate and cellulose solution were added to a high-pressure reactor of polytetrafluoroethylene. The reactor is placed at 180°C for 24 h of reaction. For comparative experiments, tetrabutyl titanate and cellulose solution were taken to prepare cellulose/TiO_2 nanocomposites. The other experimental steps were similar with the aforementioned experiments.

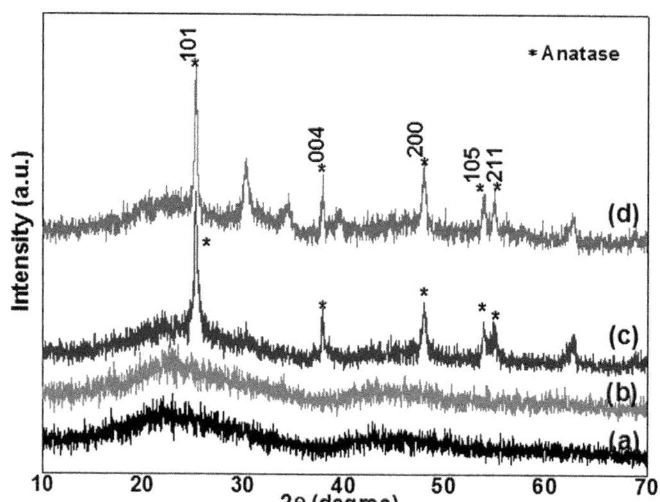

FIGURE 3.11 XRD patterns of the composites prepared using (a and b) LiCl/DMAc solution and (c and d) NaOH/Urea solution by hydrothermal method at 180°C for 24 h: (a and c) tetra-n-butyl titanate (5 mL); (b and d) tetra-n-butyl titanate (1 mL). *Reproduced from Li et al. (2013) with permission from Elsevier.*

Fig. 3.11c shows the X-ray diffraction pattern of a sample prepared by hydrothermal reaction at 180°C for 24 h using cellulose solution dissolved in a sodium hydroxide/urea system and tetrabutyl titanate. It can be observed that the characteristic X-ray diffraction peaks of this sample are all characteristic peaks of anatase (JCPDS 21-1272). Due to the high intensity of the X-ray diffraction peak of anatase, no characteristic peak of cellulose was observed. However, when the concentration of tetra-n-butyl titanate is reduced from 5 to 1 mL, it can be observed from the curve in Fig. 3.11d that the main characteristic peak is still anatase, but it can also be observed that two impurity peaks appeared at $2\theta = 30.2°$ and 34.5°. This result also indicates that the concentration change of tetrabutyl titanate has a significant impact on the synthesis of cellulose/titanium dioxide composite materials. However, as shown in Fig. 3.11a and b, in the product synthesized by dissolving cellulose in a LiCl/DMAc solution system and synthesizing it with tetrabutyl titanate, no characteristic diffraction peaks of titanium dioxide were observed in the X-ray diffraction pattern. This result indicated that the titanium dioxide in the product was amorphous. Based on the analysis of the X-ray diffraction results mentioned before, it has been found that different cellulose solution systems had a significant impact on the phase of the final composite. Using the sodium hydroxide/urea system to dissolve cellulose, the infrared characteristic peaks attributed to microcrystalline cellulose can be observed, located at 3296 cm^{-1}, 1369 cm^{-1}, 1060 cm^{-1}, and 691 cm^{-1}, respectively. This also confirms the presence of microcrystalline cellulose in the composite. The Ti—O bond in anatase makes the infrared spectrum have a characteristic absorption peak at 691 cm^{-1}. The infrared spectrum of the cellulose/titanium dioxide composite prepared by dissolving cellulose in lithium chloride/DMAc solution also shows absorption peaks at similar positions, namely 3344 cm^{-1}, 1420 cm^{-1}, 1115 cm^{-1}, and 619 cm^{-1}. There are slight differences in the absorption peak intensity of the infrared spectra of cellulose-based composites obtained from two different solution systems of dissolved cellulose. Based on the aforementioned X-ray diffraction and Fourier transform infrared spectroscopy analysis results, the cellulose/titanium dioxide composite was synthesized using the hydrothermal method.

The SEM images of the cellulose/titanium dioxide composite prepared by dissolving cellulose in the NaOH/urea system observed that when the volume ratio of tetrabutyl titanate to cellulose solution is 1:2, titanium dioxide nanoparticles are uniformly distributed on the surface of cellulose. When the volume ratio of the two is 1:14, similar fibrous and granular shapes can be observed. However, using the LiCl/DMAc system to dissolve cellulose to prepare cellulose/titanium dioxide nanocomposites, the SEM images indicated that only a small amount of particles existed on the surface of cellulose. In addition, when the volume of tetrabutyl titanate decreased to 1 mL, the particle size on the surface of cellulose actually increased. Therefore, the concentration of tetrabutyl titanate and different solvent systems for dissolving cellulose not only had a certain impact on the phase of cellulose-based nanocomposites but also had a certain degree of influence on the particle size and morphology of cellulose-based nanocomposites. The thermogravimetric curve of the cellulose/titanium dioxide composite prepared by dissolving cellulose in the sodium hydroxide/urea system indicated that the total mass loss of the sample was observed to be 17.0%, and three exothermic absorption peaks were observed in the DTA curve at temperatures of 299.8, 460.0, and 527.7°C, respectively. However, when LiCl/DMAc solutions were used, the total mass loss of the sample was 55.8%, and three exothermic absorption peaks appeared on the corresponding DTA curve at temperatures of 242.5, 480.0, and 560.6°C. The thermogravimetric curves of cellulose/titanium dioxide composites prepared by the two different solvent systems for dissolving cellulose mentioned before indicated that the cellulose dissolved in the sodium hydroxide/urea system was more advantageous for the preparation of cellulose/titanium dioxide composites, as the products prepared in this system had higher thermal stability.

The antimicrobial activity of cellulose/titanium dioxide composite is shown in Fig. 3.12. Among them, microcrystalline cellulose is used as a blank reference sample. It observed that no antibacterial ring phenomenon was found

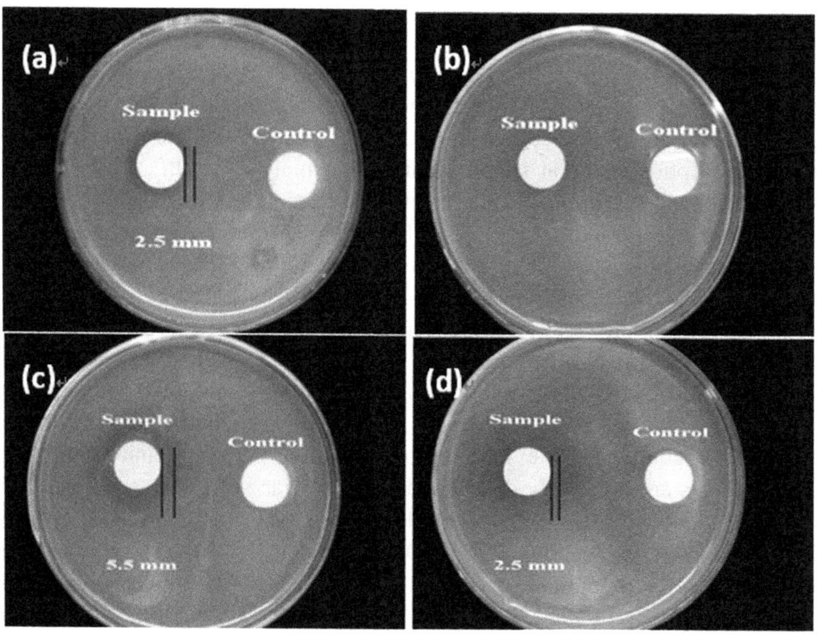

FIGURE 3.12 Antimicrobial activities of the cellulose/TiO_2 composites: (a and b) using LiCl/DMAc solution; (c and d) using NaOH/Urea solution. *(a and c) is for ***Escherichia coli***; (b and d) is for ***Staphylococcus aureus.*** *Reproduced from Li et al. (2013) with permission from Elsevier.*

around the blank reference sample. Therefore, it has been concluded that pure cellulose did not have any antibacterial properties. In the process of preparing composites, cellulose existed as the substrate for the diffusion of titanium dioxide crystals. Due to the introduction of cellulose in the composite, the cost of antibacterial materials was greatly reduced. In Fig. 3.12a, the size of bacteriostatic circle of cellulose/titanium dioxide nanocomposite prepared from cellulose dissolved in LiCl/DMAc solution against *E. coli* is 2.5 mm, which also confirms that titanium dioxide has good antibacterial properties. However, in Fig. 3.12b, the mixture prepared by dissolving cellulose in lithium chloride/DMAc solution showed no antibacterial effect on *S. aureus*, indicating that no antibacterial ring was observed around the sample. In the bacteriostasis experiment, the samples of the two kinds of bacteria to be tested have the same quality and size, but under the same conditions, the antibacterial effect on *E. coli* is significantly better than that of *S. aureus*, which may also be due to the difference between the cell walls of gram-negative bacteria and gram-positive bacteria (Feng et al., 2000). In the system of sodium hydroxide/urea dissolving cellulose, the size of inhibition zone of cellulose/titanium dioxide nanocomposites against *E. coli* and *S. aureus* is 5.5 and 2.5 mm, respectively. Based on the aforementioned results of antimicrobial activity, compared with the LiCl/DMAc solution system, microcrystalline cellulose dissolved in NaOH/urea solution has better antibacterial effect in preparing the mixture, which also indicates that this solution system has better optimized the antibacterial performance of the product.

Based on the analysis of the aforementioned thermogravimetric results, to achieve complete degradation of cellulose, 600°C is chosen as the calcination temperature for the composite. After calcination of the previously prepared cellulose/titanium dioxide composite, titanium dioxide crystals are finally obtained. The X-ray diffraction pattern of the sample was obtained by the hydrothermal synthesis reaction of cellulose dissolved in the LiCl/DMAc system. No obvious characteristic diffraction peak of titanium dioxide was observed on it. However, the X-ray diffraction curve of the sample obtained from the hydrothermal synthesis reaction of cellulose dissolved in the sodium hydroxide/urea system indicated that the characteristic peak of the X-ray diffraction of the calcined sample belongs to rutile with a quadrangular structure (JCPDS 21-1276). Compared with anatase, rutile is a thermally stable form of titanium dioxide. After calcination at 600°C, most cellulose has been degraded into carbon dioxide and water vapor, and anatase has also been transformed into rutile at high temperature. Fourier transform infrared spectrum of the calcined product indicated that the infrared spectrum curves of the samples obtained from two different cellulose dissolution systems have the same absorption peaks at the same position, but there are slight differences in the intensity of the absorption peaks at certain wavelengths between the two. The characteristic absorption peak below 800 cm^{-1} is caused by the vibration of Ti—O bond groups. Due to the presence of water molecules, a weak absorption peak appeared at 1741 cm^{-1}. The absorption peaks at 1599 and 1507 cm^{-1} are caused by the C—H bond of hydrocarbons, while the absorption peaks at 1000—1200 cm^{-1} belong to the vibration of the C—O bond. The scanning electron microscopy images of the calcined sample indicated that a large number of particles are attached to the surface of the fibrous object. Fibrous substances may be carbon materials formed from cellulose after high-temperature calcination at 600°C. Rutile nanoparticles are dispersed on the surface of the irregular matrix. It has been observed that there are highly agglomerated titanium dioxide nanoparticles dispersed on the surface of the matrix.

In this section, cellulose is dissolved using a "nonaqueous" solvent system-LiCl/DMAc homogenous solution system and a new generation solvent-NaOH/urea solution system. In the LiCl/DMAc solution system, LiCl and DMAc form a complex. Firstly, the nitrogen atoms and carbonyl groups in DMAc undergo a complex reaction with Li^{+} on LiCl. The hydroxyl groups on the cellulose molecular chain then interact with the free Cl^{-} in LiCl, leading to the destruction of hydrogen bonds within and between cellulose molecules, ultimately leading to the dissolution of cellulose. As is well known, the NaOH/urea solution system is a nonderivative solvent for dissolving cellulose to prepare cellulose solution. The dissolution of cellulose in this system can be divided into three stages: the first stage is to immerse cellulose in a precooled NaOH/urea system, surrounded by alkali hydrates, urea hydrates, and free water molecules around the cellulose macromolecular chains. In the second stage, at lower temperatures, NaOH, urea, and free water can penetrate into the macromolecular chains of cellulose, disrupting the hydrogen bonding between molecules, causing the cellulose chains to dissolve. In this stage, the hydrates in NaOH and urea enter the cellulose macromolecular chains, and the system further develops into a swollen colloidal form. In the third stage, the outer layer is wrapped with cellulose macromolecular chains of alkali hydrates, urea hydrates, and free water as "protective coatings." During this process, cellulose self-assembles with small molecules in the solvent through hydrogen bonding to form an inclusion complex. Finally, cellulose enters the aqueous solution, is dispersed, and forms a transparent cellulose solution. At low temperature, the presence of alkali hydrates and urea hydrates in the solution system is the key to cellulose dissolution, and this system is one of the good solvent systems for cellulose not to undergo derivatization. Under the dissolution of the system, small molecules such as NaOH and urea in the solvent form a highly stable hydrogen bond network with cellulose. A certain amount of NaOH hydrates are connected to cellulose, while urea hydrates in the solution act as hydrogen bond donors and acceptors to prevent cellulose macromolecules from getting close to each other, ultimately leading to better dispersion of cellulose in the solution. Corresponding to the hydrothermal synthesis reaction to prepare cellulose/TiO_2 nanocomposites, we speculate that the surface of the matrix phase is rich in a large number of hydroxyl groups, which undergo hydrothermal reactions with the titanium salt solution to form surface covalent titanium salts, resulting in nanoscale titanium dioxide. Therefore, by comparing the dissolution mechanisms of the two different cellulose dissolution systems mentioned above, it is speculated that in the solvent system where NaOH/urea solution dissolves cellulose, cellulose forms a relatively stable pipeline inclusion complex with NaOH hydrate and urea at low temperature. Compared with the complex obtained by dissolving cellulose in LiCl/DMAc system, this cellulose NaOH urea pipeline inclusion complex may be more stable. Therefore, for cellulose/TiO_2 nanocomposites prepared using two different cellulose solvent systems, the products prepared by dissolving cellulose in the NaOH/urea system are superior to those prepared by predissolving cellulose in the LiCl/DMAc system in terms of thermal stability, phase, and antibacterial properties.

In summary, this section explores a preparation method for hydrothermal synthesis of cellulose/titanium dioxide composites using LiCl/DMAc solution system and NaOH/urea solution system to predissolve cellulose, respectively. Compared with the LiCl/DMAc system, the composite prepared by pretreating cellulose with NaOH/urea system is superior to the composite prepared by predissolving cellulose with LiCl/DMAc system in terms of phase, thermal stability, and antibacterial properties. The composites synthesized by the two systems all have good antimicrobial activity. Among them, the size of the inhibition zone of the composites prepared by predissolving cellulose in LiCl/DMAc system against *E. coli* and *S. aureus* is 2.5 mm and no obvious inhibition zone, respectively. The size of the inhibition zone of the composites prepared by predissolving cellulose in NaOH/urea system against *E. coli* and *S. aureus* is 5.5 and 2.5 mm, respectively. This section used the calcining composites method to remove cellulose and obtain TiO_2, which is also a new approach to prepare titanium dioxide nanomaterials. This composite is a potentially valuable antibacterial material in the biomedical field, which will open up a new path for the high-value application of biomass cellulose.

3.3 Chapter summary

In summary, this chapter introduced antibacterial cellulose-based nanocomposites, such as cotton stalk cellulose/Ag, cellulose/Ag/AgCl, cellulose/AgCl, Ag@Fe_3O_4@nanocellulose, and cellulose/TiO_2 nanocomposites. The inorganic particles showed uniform dispersion in the cellulose matrix. The physicochemical properties, cytocompatibility, and antibacterial properties of the composites were also summarized. There are some problems with cellulose composites for antibacterial application, such as complex preparation method, poor dispersion of reinforcing phase in cellulose matrix, the low antibacterial performance, and biocompatibility and safety issues. Currently, the preparation method of cellulose-based antibacterial nanocomposites is relatively complex, which may limit their large-scale production and application. In cellulose-based antibacterial nanocomposites, inorganic antibacterial materials (such as silver, silver chloride, etc.) as reinforcing phases may have poor dispersion in the cellulose matrix, which can affect the performance and stability of the composites. Although cellulose antibacterial composites have certain antibacterial effects, their antibacterial performance still needs further improvement to meet higher standards of antibacterial needs. For cellulose antibacterial composites in the biomedical field, their biocompatibility and safety are crucial. At present, research on this aspect is not in-depth enough, and further exploration and verification are needed. Due to the complexity of preparation methods and the cost of raw materials, the production cost of cellulose antibacterial composites may be higher, which may limit their application in antibacterial fields. To solve the aforementioned problems, researchers are constantly exploring new preparation methods and processes to improve the performance of cellulose composites, reduce costs, and promote their applications in biomedical, tissue engineering, and other fields. At the same time, research on biocompatibility and safety is constantly deepening to ensure the safety and effectiveness of such cellulose composites in practical applications.

Acknowledgments

The financial support from the National Natural Science Foundation of China (22478036) and Beijing Nova Program (20230484431) is gratefully acknowledged.

References

Benaissi, K., Johnson, L., Walsh, D.A., Thielemans, W., 2010. Synthesis of platinum nanoparticles using cellulosic reducing agents. Green Chem. 12 (2), 220–222. https://doi.org/10.1039/b913218j.

Bian, J., Peng, P., Peng, F., Xiao, X., Xu, F., Sun, R.C., 2014. Microwave-assisted acid hydrolysis to produce xylooligosaccharides from sugarcane bagasse hemicelluloses. Food Chem. 156, 7–13. https://doi.org/10.1016/j.foodchem.2014.01.112.

Biçer, M., Şişman, I., 2010. Controlled synthesis of copper nano/microstructures using ascorbic acid in aqueous CTAB solution. Powder Technol. 198 (2), 279–284. https://doi.org/10.1016/j.powtec.2009.11.022.

Cao, S.W., Zhu, Y.J., Cheng, G.F., Huang, Y.H., 2009. ZnFe2O4 nanoparticles: microwave-hydrothermal ionic liquid synthesis and photocatalytic property over phenol. J. Hazard Mater. 171 (1–3), 431–435. https://doi.org/10.1016/j.jhazmat.2009.06.019.

Chi, Y., Zhao, L., Yuan, Q., Li, Y., Zhang, J., Tu, J., Li, N., Li, X., 2012. Facile encapsulation of monodispersed silver nanoparticles in mesoporous compounds. Chem. Eng. J. 195–196, 254–260. https://doi.org/10.1016/j.cej.2012.05.010.

Dallinger, D., Kappe, C.O., 2007. Microwave-assisted synthesis in water as solvent. Chem. Rev. 107 (6), 2563–2591. https://doi.org/10.1021/cr0509410.

Das, D., Kar, T., Das, P.K., 2012. Gel-nanocomposites: materials with promising applications. Soft Matter 8, 2348–2365.

Davies, R.L., Etris, S.F., 1997. The development and functions of silver in water purification and disease control. Catal. Today 36, 107–114.

Huang, H.Z., Yang, X.R., 2004. Synthesis of polysaccharide stabilized gold and silver nanoparticles: a green method. Carbohydr. Res. 15, 2627–2631.

Dong, Y.Y., Li, S.M., Ma, M.G., Yao, K., Sun, R.C., 2014. Compare study cellulose/Ag hybrids using fructose and glucose as reducing reagents by hydrothermal method. Carbohydr. Polym. 106 (1), 14–21. https://doi.org/10.1016/j.carbpol.2014.02.023.

Dong, Y.Y., Fu, L.H., Liu, S., Ma, M.G., Wang, B., 2015. Silver-reinforced cellulose hybrids with enhanced antibacterial activity: synthesis, characterization, and mechanism. RSC Adv. 5 (118), 97359–97366.

Dong, Y.Y., Liu, S., Liu, Y.J., Meng, L.Y., Ma, M.G., 2017. Ag@Fe_3O_4@cellulose nanocrystals nanocomposites: microwave-assisted hydrothermal synthesis, antimicrobial properties and good adsorption of dye solution. J. Mater. Sci. 52, 8219–8230.

Eichhorn, S.J., Dufresne, A., Aranguren, M., Marcovich, N.E., Capadona, J.R., Rowan, S.J., Weder, C., Thielemans, W., Roman, M., Renneckar, S., Gindl, W., Veigel, S., Keckes, J., Yano, H., Abe, K., Nogi, M., Nakagaito, A.N., Mangalam, A., Simonsen, J., Benight, A.S., Bismarck, A., Berglund, L.A., Peijs, T., 2010. Review: current international research into cellulose nanofibres and nanocomposites. J. Mater. Sci. 45 (1), 1–33. https://doi.org/10.1007/s10853-009-3874-0.

Feng, Q.L., Kim, T.N., Wu, J., Park, E.S., Kim, J.O., Lim, D.Y., Cui, F.Z., 1998. Antibacterial effects of Ag-HAp thin films on alumina substrates. Thin Solid Films 335 (1–2), 214–219. https://doi.org/10.1016/s0040-6090(98)00956-0.

Feng, Q.L., Wu, J., Chen, G.Q., Cui, F.Z., Kim, T.N., Kim, J.O., 2000. A mechanistic study of the antibacterial effect of silver ions onEscherichia coli andStaphylococcus aureus. J. Biomed. Mater. Res. 52 (4), 662–668. https://doi.org/10.1002/1097-4636(20001215)52:4<662::aid-jbm10>3.0.co;2-3.

Feng, X., Liu, Y., Lu, C., Hou, W., Zhu, J.J., 2006. One-step synthesis of AgCl/polyaniline core-shell composites with enhanced electroactivity. Nanotechnology 17 (14), 3578–3583. https://doi.org/10.1088/0957-4484/17/14/037.

Ferraria, A.M., Boufi, S., Battaglini, N., Botelho, do R.A.M., ReiVilar, M., 2009. Hybrid systems of silver nanoparticles generated on cellulose surfaces. Langmuir 26, 1996–2001.

Fu, L.H., Dong, Y.Y., Ma, M.G., Li, S.M., Sun, R.C., 2013. Compare study $CaCO_3$ crystals on the cellulose substrate by microwave-assisted method and ultrasound agitation method. Ultrason. Sonochem. 20, 839–845.

Fu, L.H., Deng, F., Ma, M.G., Yang, J., 2016. Green synthesis of silver nanoparticles with enhanced antibacterial activity using holocellulose as a substrate and reducing agent. RSC Adv. 6 (34), 28140–28148.

Fu, L.H., Qi, C., Ma, M.G., Li, J.F., 2018. Microwave-hydrothermal rapid synthesis of cellulose/Ag nanocomposites and their antibacterial activity. Nanomaterials 8, 978.

Habibi, Y., Lucia, L.A., Rojas, O.J., 2010. Cellulose nanocrystals: chemistry, self-assembly, and applications. Chem. Rev. 110, 3479–3500.

He, J.H., Kunitake, T., Nakao, A., 2003. Facile in situ synthesis of noble metal nanoparticles in porous cellulose fibers. Chem. Mater. 15, 4401–4406.

He, J.H., Kunitake, T., Watanabe, T., 2005. Porous and nonporous Ag nanostructures fabricated using cellulose fiber as a template. Chem. Commun. (6), 795–796.

Inoue, Y., Hoshino, M., Takahashi, H., Noguchi, T., Murata, T., Kanzaki, Y., Hamashima, H., Sasatsu, M., 2002. Bactericidal activity of Ag–zeolite mediated by reactive oxygen species under aerated conditions. J. Inorg. Biochem. 92 (1), 37–42. https://doi.org/10.1016/s0162-0134(02)00489-0.

Jia, N., Li, S.M., Zhu, J.F., Ma, M.G., Xu, F., Wang, B., Sun, R.C., 2010. Microwave-assisted synthesis and characterization of cellulose-carbonated hydroxyapatite nanocomposites in NaOH-urea aqueous solution. Mater. Lett. 64 (20), 2223–2225. https://doi.org/10.1016/j.matlet.2010.07.029.

Klemm, D., Heublein, B., Fink, H.P., Bohn, A., 2005. Cellulose: fascinating biopolymer and sustainable raw material. Angew. Chem. Int. Ed. 44, 3358.

Kokot, S., Anh, N., Rintoul, T.L., 1997. Discrimination of reactive dyes on cotton fabric by Raman spectroscopy and chemometrics. Appl. Spectrosc. 51, 387–395.

Lam, E., Male, K.B., Chong, J.H., Leung, A.C.W., Luong, J.H.T., 2012. Applications of functionalized and nanoparticle-modified nanocrystalline cellulose. Trends Biotechnol. 30 (5), 283–290. https://doi.org/10.1016/j.tibtech.2012.02.001.

Li, S.M., Jia, N., Zhu, J.F., Ma, M.G., Sun, R.C., 2010. Synthesis of cellulose-calcium silicate nanocomposites in ethanol/water mixed solvents and their characterization. Carbohydr. Polym. 80, 270–275.

Li, S.M., Jia, N., Zhu, J.F., Ma, M.G., Xu, F., Wang, B., Sun, R.C., 2011. Rapid microwave-assisted preparation and characterization of cellulose-silver nanocomposites. Carbohydr. Polym. 83, 422–429.

Li, S.M., Fu, L.H., Ma, M.G., Zhu, J.F., Sun, R.C., Xu, F., 2012. Simultaneous microwave-assisted synthesis, characterization, thermal stability, and antimicrobial activity of cellulose/AgCl nanocomposites. Biomass Bioenergy 47, 516–521.

Li, S.M., Dong, Y.Y., Ma, M.G., Fu, L.H., Sun, R.C., Xu, F., 2013. Hydrothermal synthesis, characterization, and bactericidal activities of hybrid from cellulose and TiO_2. Carbohydr. Polym. 96, 15–20.

Liau, S.Y., Read, D.C., Pugh, W.J., Furr, J.R., Russell, A.D., 1997. Interaction of silver nitrate with readily identifiable groups: relationship to the antibacterial action of silver ions. Lett. Appl. Microbiol. 25 (4), 279–283. https://doi.org/10.1046/j.1472-765X.1997.00219.x.

Liu, S., Zhang, L., Zhou, J., Wu, R., 2008. Structure and properties of cellulose/Fe2O3 nanocomposite fibers spun via an effective pathway. J. Phys. Chem. C 112 (12), 4538–4544. https://doi.org/10.1021/jp711431h.

Luo, X., Liu, S., Zhou, J., Zhang, L., 2009. In situ synthesis of Fe_3O_4/cellulose microspheres with magnetic-induced protein delivery. J. Mater. Chem. 19, 3538–3545.

Luo, Z.T., Zheng, K.Y., Xie, J.P., 2014. Engineering ultra-small water-soluble gold and silver nanoclusters for biomedical applications. Chem. Commun. (J. Chem. Soc. Sect. D) 50, 5143–5155.

Ma, M.G., Zhu, J.F., Jia, N., Li, S.M., Sun, R.C., Cao, S.W., Chen, F., 2010. Rapid microwave-assisted synthesis and characterization of cellulose-hydroxyapatite nanocomposites in N,N--dimethylacetamide solvent. Carbohydr. Res. 345, 1046–1050.

Maneerung, T., Tokura, S., Rujiravanit, R., 2008. Impregnation of silver nanoparticles into bacterial cellulose for antimicrobial wound dressing. Carbohydr. Polym. 72 (1), 43–51.

Maria, L.C.S., Santos, A.L.C., Oliveira, P.C., Valle, A.S.S., Barucl, H.S., Messaddeq, Y., Ribeiro, S.J.L., 2010. Preparation and antibacterial activity of silver nanoparticles impregnated in bacterial cellulose. Polimeros 20 (1), 72–77. https://doi.org/10.1590/S0104-14282010005000001.

Miranda, O.R., Dollahon, N.R., Ahmadi, T.S., 2006. Critical concentrations and role of ascorbic acid (vitamin C) in the crystallization of gold nanorods within hexadecyltrimethyl ammonium bromide (CTAB)/tetraoctyl ammonium bromide (TOAB) micelles. Cryst. Growth Des. 6 (12), 2747–2753.

Nadagouda, M.N., Speth, T.F., Varma, R.S., 2011. Microwave-assisted green synthesis of silver nanostructures. Accounts Chem. Res. 44, 469–478.

Nagaoka, S., Arinaga, K., Kubo, H., Hamaoka, S., Sakurai, T., Takafuji, M., Ihara, H., 2005. Cellulose/TiO2 hybrid spherical microbeads prepared by a viscose phase separation method: control of the distribution of TiO2 particles in a sphering system. Polym. J. 37 (3), 186–191. https://doi.org/10.1295/polymj.37.186.

Natile, M.M., Boccaletti, G., Glisenti, A., 2005. Properties and reactivity of nanostructured CeO_2 powders: comparison among two synthesis procedures. Chem. Mater. 17, 6272–6286.

Padalkar, S., Capadona, J.R., Rowan, S.J., Weder, C., Won, Y.H., Stanciu, L.A., Moon, R.J., 2010. Natural biopolymers: novel templates for the synthesis of nanostructures. Langmuir 26 (11), 8497–8502. https://doi.org/10.1021/la904439p.

Pappas, C., Tarantilis, P.A., Daliani, I., Mavromoustakos, T., Polissiou, M., 2002. Comparison of classical and ultrasound-assisted isolation procedures of cellulose from kenaf (Hibiscus cannabinus L.) and eucalyptus (Eucalyptus rodustrus Sm.). Ultrason. Sonochem. 9 (1), 19–23. https://doi.org/10.1016/s1350-4177(01)00095-5.

Patel, A.C., Li, S., Wang, C., Zhang, W., Wei, Y., 2007. Electrospinning of porous silica nanofibers containing silver nanoparticles for catalytic applications. Chem. Mater. 19 (6), 1231–1238. https://doi.org/10.1021/cm061331z.

Peter, A., Mihaly-Cozmuta, L., Mihaly-Cozmuta, A., Nicula, C., Ziemkowska, W., Basiak, D., Danciu, V., Vulpoi, A., Baia, L., Falup, A., Craciun, G., Ciric, A., Begea, M., Kiss, C., Vatuiu, D., 2016. Changes in the microbiological and chemical characteristics of white bread during storage in paper packages modified with Ag/TiO2–SiO2, Ag/N–TiO2 or Au/TiO2. Food Chem. 197, 790–798. https://doi.org/10.1016/j.foodchem.2015.11.048.

Phuruangrat, A., Kuntalue, B., Thongtem, S., Thongtem, T., 2016. Synthesis of cubic CuFe2O4 nanoparticles by microwave-hydrothermal method and their magnetic properties. Mater. Lett. 167, 65–68. https://doi.org/10.1016/j.matlet.2016.01.005.

Pinto, R.J.B., Marques, P.A.A.P., Neto, C.P., Trindade, T., Daina, S., Sadocco, P., 2009. Antibacterial activity of nanocomposites of silver and bacterial or vegetable cellulosic fibers. Acta Biomater. 5 (6), 2279–2289. https://doi.org/10.1016/j.actbio.2009.02.003.

Rai, M., Yadav, A., Gade, A., 2009. Silver nanoparticles as a new generation of antimicrobials. Biotechnol. Adv. 27 (1), 76–83.

Ruan, D., Huang, Q., Zhang, L., 2005. Structure and properties of CdS/regenerated cellulose nanocomposites. Macromol. Mater. Eng. 290, 1017–1024.

Sambhy, V., MacBride, M.M., Peterson, B.R., Sen, A., 2006. Silver bromidenanoparticle/polymer composites: dual action tunable antimicrobial materials. J. Am. Chem. Soc. 128, 9798–9808.

Sharma, V.K., Yngard, R.A., Lin, Y., 2009. Silver nanoparticles: green synthesis and their antimicrobial activities. Adv. Colloid Interface Sci. 145, 83–96.

Shin, Y., Bae, I.T., Arey, B.W., Exarhos, G.J., 2008. Facile stabilization of gold-silver alloy nanoparticles on cellulose nanocrystal. J. Phys. Chem. C 112 (13), 4844–4848.

Sondi, I., Goia, D.V., Matijević, E., 2003. Preparation of highly concentrated stable dispersions of uniform silver nanoparticles. J. Colloid Interface Sci. 260, 75–81.

Song, D., Yang, R., Wang, C., Xiao, R., Long, F., 2016. Reusable nanosilver-coated magnetic particles for ultrasensitive SERS-based detection of malachite green in water samples. Sci. Rep. 6. https://doi.org/10.1038/srep22870.

Stenstad, P., Andresen, M., Tanem, B.S., Stenius, P., 2008. Chemical surface modifications of microfibrillated cellulose. Cellulose 15 (1), 35–45.

Sun, R.C., Sun, X.F., Liu, G.Q., Fowler, P., Tomkinson, J., 2002. Structural and physicochemical characterization of hemicelluloses isolated by alkaline peroxide from barley straw. Polym. Int. 51 (2), 117–124. https://doi.org/10.1002/pi.815.

Sun, Y.C., Wen, J.L., Xu, F., Sun, R.C., 2011. Structural and thermal characterization of hemicelluloses isolated by organic solvents and alkaline solutions from Tamarix austromongolica. Bioresour. Technol. 102 (10), 5947–5951.

Sureshkumar, M., Siswanto, D.Y., Lee, C.K., 2010. Magnetic antimicrobial nanocomposite based on bacterial cellulose and silver nanoparticles. J. Mater. Chem. 20 (33), 6948–6955.

Terbojevich, M., Cosani, A., Conio, G., Ciferri, A., Bianchi, E., 1985. Mesophase formation and chain rigidity in cellulose and derivatives. 3. Aggregation of cellulose in N,N--dimethylacetamide-lithium chloride. Macromolecules 18 (4), 640–646. https://doi.org/10.1021/ma00146a011.

Uddin, M.J., Cesano, F., Bonino, F., Bordiga, S., Spoto, G., Scarano, D., Zecchina, A., 2007. Photoactive TiO2 films on cellulose fibres: synthesis and characterization. J. Photochem. Photobiol. Chem. 189 (2–3), 286–294. https://doi.org/10.1016/j.jphotochem.2007.02.015.

Venkateswarlu, S., Natesh Kumar, B., Prathima, B., Anitha, K., Jyothi, N.V.V., 2015. A novel green synthesis of Fe3O4-Ag core shell recyclable nanoparticles using Vitis vinifera stem extract and its enhanced antibacterial performance. Phys. B Condens. Matter 457, 30–35. https://doi.org/10.1016/j.physb.2014.09.007.

Wang, Y., Wang, S., Niu, H., Ma, Y., Zeng, T., Cai, Y., Meng, Z., 2013. Preparation of polydopamine coated Fe3O4 nanoparticles and their application for enrichment of polycyclic aromatic hydrocarbons from environmental water samples. J. Chromatogr. A 1283, 20–26. https://doi.org/10.1016/j.chroma.2013.01.110.

Wu, M.S., Chiang, P.C.J., Lee, J.T., Lin, J.C., 2005. Synthesis of manganese oxide electrodes with interconnected nanowire structure as an anode material for rechargeable lithium ion batteries. J. Phys. Chem. B 109 (49), 23279–23284. https://doi.org/10.1021/jp054740b.

Xiong, R., Lu, C., Wang, Y., Zhou, Z., Zhang, X., 2013. Nanofibrillated cellulose as the support and reductant for the facile synthesis of Fe3O4/Ag nanocomposites with catalytic and antibacterial activity. J. Mater. Chem. A 1 (47), 14910–14918. https://doi.org/10.1039/c3ta13314a.

Yao, K., Liu, S., Dong, Y.Y., Wang, B., Bian, J., Ma, M.G., 2016. Comparative study of CuO crystals on the cellulose substrate by the microwave-assisted hydrothermal method and hydrothermal method. Mater. Des. 90, 129–136. https://doi.org/10.1016/j.matdes.2015.10.121.

Yuan, X., Setyawati, M.I., Leong, D.T., Xie, J., 2014. Ultrasmall Ag^+-rich nanoclusters as highly efficient nanoreservoirs for bacterial killing. Nano Res. 7 (3), 301–307.

Zawadzki, J., Wisniewski, M., 2002. 13C NMR study of cellulose thermal treatment. J. Anal. Appl. Pyrol. 62, 111–121.

Zeng, J., Liu, S., Cai, J., Zhang, L., 2010. TiO2 immobilized in cellulose matrix for photocatalytic degradation of phenol under weak UV light irradiation. J. Phys. Chem. C 114 (17), 7806–7811. https://doi.org/10.1021/jp1005617.

Zhang, H., Wang, Z.G., Zhang, Z.N., Wu, J., Zhang, J., He, J.S., 2007. Regenerated-cellulose/multiwalled- carbon-nanotube composite fibers with enhanced mechanical properties prepared with the ionic liquid 1-Allyl-3-methylimidazolium chloride. Adv. Mater. 19 (5), 698–704. https://doi.org/10.1002/adma.200600442.

Zhang, T., Wang, W., Zhang, D., Zhang, X., Yurong, M., Zhou, Y., Qi, L., 2010. Biotemplated synthesis of cold nanoparticle-bacteria cellulose nanofiber nanocomposites and their application in biosensing. Adv. Funct. Mater. 20 (7), 1152–1160. https://doi.org/10.1002/adfm.200902104.

Zhao, Y., Tao, C., Xiao, G., Wei, G., Li, L., Liu, C., Su, H., 2016. Controlled synthesis and photocatalysis of sea urchin-like Fe3O4@TiO2@Ag nanocomposites. Nanoscale 8 (9), 5313–5326. https://doi.org/10.1039/c5nr08624h.

Zhu, Y.J., Chen, F., 2014. Microwave-assisted preparation of inorganic nanostructures in liquid phase. Chem. Rev. 114 (12), 6462–6555. https://doi.org/10.1021/cr400366s.

Zhu, C., Xue, J., He, J., 2009. Controlled in-situ synthesis of silver nanoparticles in natural cellulose fibers toward highly efficient antimicrobial materials. J. Nanosci. Nanotechnol. 9 (5), 3067–3074.

Chapter 4

Cellulose-based composites for water treatment

Peilin Wang, Tian Mal and Mingguo Ma*

MOE Engineering Research Center of Forestry Biomass Materials and Bioenergy, Research Center of Biomass Clean Utilization, Beijing Key Laboratory of Lignocellulosic Chemistry, College of Materials Science and Technology, Beijing Forestry University, Beijing, PR China

Corresponding author: E-mail: mg_ma@bjfu.edu.cn

Water remediation has been applied to cellulose-based materials as advanced technologies. The most abundant biopolymer in nature, high biocompatibility, and several functional groups are some of the advantages exhibited by them. Several derivatives including nanomaterials such as cellulose nanocrystals (CNCs), cellulose nanofibers (CNFs), and TEMPO (2,2,6,6-tetramethylpiperidine-1-oxyl radical)-mediated oxidized cellulose nanofibrils (TOCNF) can be prepared from cellulose. Extra functional groups can be offered by modifying or grafting the functional groups such as carboxylic and hydroxyls groups with organic moieties, customizing for specific applications. The capability of cellulose biopolymers to be modified with nanoparticles such as metal−organic frameworks (MOFs), graphene oxide (GO), silver (Ag) nanoparticles, and zinc oxide (ZnO) nanoparticles is ensured by these functional groups. Thus, water pollutants including heavy metal ions, organic dyes, drugs, and microbial species can be removed by applying them for water remediation. Microorganisms can be also removed by using cellulose-based materials as membranes or antibacterial agents. Various forms such as membranes, sheets, papers, foams, aerogels, and filters can be proceeded by them. This chapter summarized the methods such as adsorption, catalysis, and antifouling for water remediation via the applications of cellulose-based materials. The high performance of cellulose-based materials and their simple processing methods ensure the high potential for water remediation.

Surface Science and Advanced Functional Applications of Cellulose Composites
https://doi.org/10.1016/B978-0-443-13675-7.00004-X

4.1 Introduction

Water is essential for human life, and all living organisms depend on it. Safety and healthy life require clean drinking water. However, water pollution has emerged as a global problem in the past 20 years, owing to the development of economy, industry, urbanization, and agriculture (Ali and Gupta, 2006). By 2025, half of the world's population will face water-stressed demand, as stated by WHO. A basic drinking water service is lacked by 785 million people (WHO, 2019). Pollution decreases clean drinking water over time. The rain-water is rendered acidic by the air pollution with toxic gases such as sulfur oxides and nitrogen oxide. Contaminants or pollutants, such as toxic inorganic anions, synthetic dyes, pesticides, heavy metals, micropollutants, and organics, were reported into natural waterbodies (Beyki et al., 2016).

WHO reported that clean drinking water requires at least a basic service for 90% of the global population (6.8 billion people). Water treatment aims to remove or mitigate the toxicity of pollutants that could be metal ions, organic dyes, drugs, or microorganism species. Various sources of water pollution exist, such as oil–water pollution, which is caused by the discharge of oil-containing wastewater from domestic, petrochemical, and food industries into natural water bodies (Wang et al., 2015a). Industrial effluents are a major source of water pollution that affects the health and well-being of humans, plants, animals, and ecosystems. These effluents contain contaminants such as pathogens, heavy metals, dyes, and pigments that can infiltrate the soil or enter the water system directly, altering the composition and quality of the groundwater and surface water. The adverse effects of these untreated effluents are significant and alarming, stimulating research and development activities to address the water problems and comply with the regulatory standards for toxic metals and dyes in wastewater, such as heavy metal ion pollution, which poses a serious threat to human health and the environment due to its non-degradability, carcinogenicity, and high toxicity (Islam et al., 2015; Schwarzenbach et al., 2006; Scheuhammer et al., 2015). Moreover, dye pollution, which originates from different industrial sectors such as paint, plastics, textiles, cosmetics, and paper industries (Pang et al., 2019; Akar et al., 2009; Mahmoodi et al., 2019), affects the quality and aesthetics of water.

Some of these toxic metals and dyes have high industrial demand for various applications, but they can cause severe health problems when present in domestic water, even at trace levels. Several highly toxic pollutants such as heavy metals of arsenic (As), lead (Pb), cadmium (Cd), and Mercury (Hg) are present. Dyes can be classified as cationic or anionic, depending on their charge and affinity for different substrates. They are composed of persistent organic molecules that are resistant to various degradation processes. Phenol red, violet crystals, safranin, basic fuchsine, and methylene blue (MB) are examples of cationic dyes (Mu et al., 2019). Anionic dyes include Remazol, Reactive Black and Red (azo dyes), Nigrosin, and Eosin. Both types of dyes have deleterious effects on aquatic organisms, terrestrial ecosystems, and human health when bioavailable. Therefore, dyes are considered as organic water pollutants, while heavy metals are classified as inorganic water pollutants. The treatment of wastewater in an economical and environmentally friendly manner is a major challenge that needs to be addressed. Various methods have been employed for the treatment of water contaminated with toxic metals and dyes. Among these methods, adsorption and photocatalysis are some of the most promising and widely used techniques due to their simplicity, eco-friendliness, and applicability to different water streams.

Various techniques such as adsorption, filtration, ion exchange, coagulation, precipitation, electrolysis, electrodialysis, and reverse osmosis are being employed to eliminate these pollutants in wastewater (Almeida et al., 2020; Noor et al., 2022; Batool and Valiyaveettil, 2021). Among these methods, adsorption and filtration have gained popularity due to their ease of use, high economic value, removal of pollutants at low concentrations, and high efficiency (Khulbe and Matsuura, 2018). However, most of these techniques require further treatment because of their high cost and low removal efficiencies (Tshikovhi et al., 2020). Hence, the exploration of materials that enhance high separation efficiency, low energy consumption, affordable cost, simple operation, and environmentally friendly is highly needed. In recent years, nanotechnology has introduced a number of innovative nanocellulose-based materials that have shown promise in wastewater treatment (Tshikovhi et al., 2020; Dutt et al., 2020). Some materials have been well researched for their high efficiency and short time requirement in water treatment, but they are not cheap and green. Therefore, they have not been successfully applied to large-scale wastewater treatment. The development of green and efficient water treatment materials is urgently needed, which should have high activity/efficiency, environmentally friendliness, and easy treatment. Biodegradable biomass materials are potential candidates for practical water treatment systems, but their applications need to be further investigated. The use of biodegradable biomass as raw material is a rapidly growing area of research, which offers many attractive alternatives for wastewater treatment. Biomass material refers to materials derived directly or indirectly from plants, such as crop wastes, forestry biomass, aquatic plants, etc. (Jin et al., 2014; Cai et al., 2015; Annadurai et al., 2002). Nanocellulose is a cellulosic material that is recovered or separated from native cellulose found in plants, animals, or microbes that have at least one dimension in the nanoscale range with outstanding features such as availability, low cost, high hydrophilicity, surface area, aspect ratio, mechanical strength, and stability (Liu et al., 2021a; Mateo et al., 2021). Nanocellulose has emerged as a green potential adsorbent for removing various contaminants from wastewater due to worldwide environmental issues and its exceptional binding and adsorbing capacities (Trache et al., 2020). Nanocellulose can be categorized into three main types depending on its isolation technique, structure, size, and orientation (Dhali et al., 2021). Moreover, the surface of nanocellulose contains a large number of free hydroxyl groups, which can be easily modified by some functional groups to increase their performance (Gopakumar et al., 2018). The process of nanocellulose modification with various types of functional groups makes it one of the most widely used materials as an adsorptive and separative materials depending on the class of pollutants of wastewater (Mahfoudhi and Boufi, 2017). The development of novel biomaterial as dual-functional membranes for pollutants removal in wastewater was aided by the excellent adsorption and filtration properties of nanocellulose. Nanocellulose as adsorptive membrane has gained substantial attention due to its cost, biodegradability, eco-friendly, availability, and performance attributes (Qalyoubi et al., 2021).

Cellulose fiber, the most abundant natural biopolymer on earth, is a promising candidate for wastewater treatment owing to its high efficiency, low cost, and chemical free nature. Cellulose is a carbohydrate polymer composed of β-1,4-linked anhydro-D-glucose units with three hydroxyl groups per anhydroglucose unit (AGU), and it is widely distributed in plants and some marine animals. Cellulose-based materials have been employed in various fields such as food, industrial, pharmaceutical, paper, and textile production, as well as in wastewater treatment applications, because of their low cost, renewability, biodegradability, and nontoxicity. In the oil and gas industry, cellulose-based materials can serve as adsorbents, flocculants, and oil/water separation membranes for water treatment.

The first extraction of cellulose from plants was performed by Anselme Payen, in 1839, and various modifications have been studied since then (Peng et al., 2020). Later, cellulose was extracted from algae, tunicates, and bacteria, and the extraction methods were systematically investigated (Moon et al., 2011). Cellulose is a linear polysaccharide composed of β-D-glucopyranose units linked by β-1,4 glycosidic bonds in a repeating unit (Rojas and Rojas, 2016; Olivera et al., 2016). Cellulose is insoluble in common organic solvents and water due to the large number of hydrogen bonds between cellulose chains. Moreover, the unmodified cellulose has poor adsorption capacity and mechanical properties. However, cellulose contains many polar groups that can be modified physically or chemically, and the appropriate modification techniques can greatly enhance the application potential of cellulose (Habibi, 2014). Chemical modification is the most common technique used by researchers, which mainly relies on the hydroxyl groups in the cellulose chain. This hydroxyl functional modification mainly involves oxidation reaction (Cao et al., 2020a), etherification reaction (Aguir and M'Henni, 2006), esterification reaction (Li et al., 2009), and grafting copolymerization reaction (Cai et al., 2013). Nanocellulose has better mechanical properties than natural cellulose by eliminating the macroscopic defects of natural cellulose. Nanocellulose is generally classified into four categories based on their origin and production methods, namely, microfibrillated cellulose (MFC), CNCs, CNFs, and bacterial nanocellulose (BNC) (Carpenter et al., 2015).

The removal of contaminants from wastewater by CNCs is a recent technology that relies on the ability of the constituent and functional groups present in cellulose. Cellulose can be prepared in powder and suspension form, which enables its application in various water treatment techniques such as adsorption, coagulation/flocculation, nanocatalysis, and membrane synthesis. However, surface modifications are often required to tailor the surface chemistry according to the contaminant to be removed. Cationic surfactant introduction onto cellulose has proven to be highly effective in wastewater treatment. CNCs are highly electronegative and have very low affinity to remove some contaminants such as suspended solids. Surface modification provides an alternative way to improve and partially convert the surface of the cellulose to positively charged ions (Barobay and Guyen, 2019). The surface of cellulose has been reengineered by many researchers for efficient application in wastewater treatment (Badshah et al., 2018; Ahsan et al., 2019).

Cost-effective and environmentally sustainable solutions for industrial scalability can be developed by using nanocellulose (Peng et al., 2020). A new frontier for the scientists has been created by the prospect of transforming cellulose into nanoscale materials by different physicochemical approaches, to pursue the features and potential scope of this nanomaterial in wastewater treatment. Effective removal of pollutants is preferred by nanocellulose over microdimensional materials, owing to their high surface area and nanodimension (Carpenter et al., 2015; Reshmy et al., 2020). Surface modifications can be enabled by the convenience and diversity of functional groups, which can be incorporated into the nanocellulose framework, to adapt and refine the backbone of cellulose for a specific purpose (Bethke et al., 2018). Furthermore, nanocellulose can be more economical over others nanomaterials such as graphene and ceramics, due to the affordable cost of the raw material and easy processability. Suitable cost-effective adsorbents can be made from industrial and agricultural derivatives with CNC, which display excellent adsorption capacity with the significant option of substituting activated carbons (Jamshaid et al., 2017; Mautner, 2020).

The use of cellulose-based materials for the treatment of various water pollutants is an emerging research area (Fig. 4.1). Among the conventional technologies for pollution control, different options are available for efficient treatment of different pollutants. Adsorption technology, which has advantages such as ease of operation, high efficiency, and simple and optimal performance, has been widely applied in the treatment of dye pollutants. Biomass adsorption method utilizes the chemical structure and composition of biomass to adsorb chemical pollutants dissolved in water through solid–liquid two-phase separation. Besides chemisorption, van der Waals forces, hydrogen bonding, hydrophobicity, polar and spatial interactions, dipole induction, etc. are the main physical forces governing the adsorption of cellulose-based adsorbents.

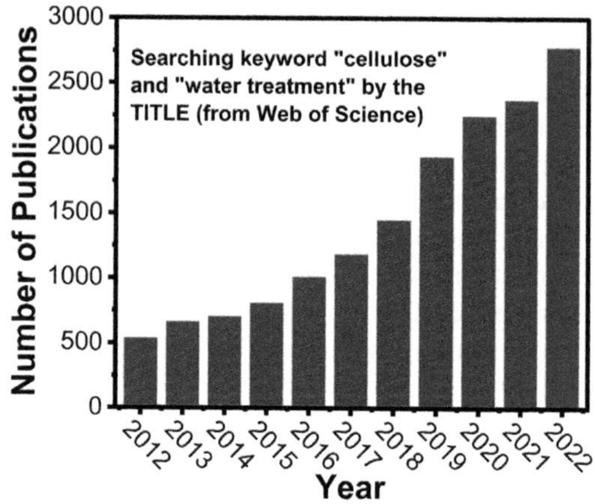

FIGURE 4.1 Numbers of the published articles related to cellulose-based materials for water treatment indexed in Web of Science by the title of "cellulose" and "water treatment."

The water matrix to be treated determines the modification routes used in the development of adsorbents. Therefore, the chemistry of adsorbate and adsorbent is important. Cellulose is negatively charged and requires cationic chemicals for modification to adsorb negatively charged adsorbate. Adsorbent is usually in solid or powdered form and needs a cost-effective drying route without affecting the cellulose structure and efficiency. In some cases, modification of adsorbent is also needed because of its low adsorption capacity even if the charges differ. Heavy metals in aqueous solution are mostly positively charged and attract cellulose. However, cellulose might have low affinity for certain metals at its natural state, which implies that modification of cellulose is also required in such cases. Wang et al. (2019a) developed a facile route for the preparation of cellulose-based adsorbent, carboxymethylated cellulose fiber (CMF), for water purification. The as-prepared CMF bioadsorbent was synthesized via a controllable carboxymethylation modification while maintaining inherent fibrous framework. The CMF bioadsorbent efficiently adsorbed heavy metal ions, such as Cu (II) and Ni (II) ions, from aqueous solution. Macrocyclic pyridine pentamer and succinic acid anhydride were also used to modify cellulose in another similar study (Zhou et al., 2019). The modified cellulose had an adsorptive performance that removed about 75% of Ba^{2+} from simulated water. The efficacy of cellulose paper consolidated with cellulose acetate and mixed adsorbents (bentonite: activated clay: celite = 37.5: 50: 12.5 g with 1 g citric acid/100 g mixed adsorbents) was also investigated (Udomkun et al., 2018). Cellulose fiber and cellulose acetate were produced from waxy corn husks, modified and utilized for the reduction of polyunsaturated fatty acids.

Metal−oxide cellulose nanocomposites can be prepared by different methods, such as hydrothermal or wet chemical, solvothermal, wet spinning, sol−gel, coprecipitation, and microwave synthesis methods (Fu et al., 2015;

Ramesh et al., 2018). The sol–gel method, which involves the use of polar and hydrophilic polymers such as polyvinyl alcohol in the synthesis of organic–inorganic nanocomposite, offers advantages such as biocompatibility, easy processability, and high water swelling ability. A sol–gel method of preparing cellulose-PVA-TiO_2 nanocomposites was reported by Ramesh, Kim, and Kim (Ramesh et al., 2018). In this method, a solution containing cellulose/LiCl/*N,N*-dimethylacetamide was heated and stirred, and then tetraethylene glycol and PVA were added in the cellulose sol. Titanium tetraisopropoxide was finally introduced, and the entire solution is heated for 10 h at 95 °C. The solution was washed with ethanol and oven-dried to obtain the composite. NFC/ZnO nanocomposites were prepared by Martins et al. (2013) by treating the nanofibrillated cellulose with poly(diallyldimethylammonium chloride) (PDDA) solution and later incorporating the ZnO nanoparticle solution. A composite of natural cellulose fibers and manganese dioxide nanosheets with high porosity was prepared by a simple hydrothermal method. In this method, aqueous potassium permanganate ($KMnO_4$) solution was impregnated into the cellulose matrix and with the introduction of oleic acid, which acted as a reducing agent (Biliuta and Coseri, 2019). The reaction conditions could also be changed to obtain incorporated MnO_2 nanoparticles in different shapes. Cellulose acetate/Fe_2O_3 nanocomposite was prepared by a solution blending method in which the Fe_2O_3 nanoparticles were ultrasonically dispersed in cellulose acetate and stirred (Bakhsh et al., 2018). A two-step method was applied to synthesize cellulose–iron oxide composites: iron sulfate was ultrasonicated in an alkaline medium to give γ-Fe_2O_3 or α-FeOOH nanoparticles. The iron oxides (maghemite γ-Fe_2O_3) in the second step were thermally treated at a temperature of over 300 °C or at low temperature of 20–70°C for the goethite (α-FeOOH). A one-pot approach, which involved the use of ultrasonication in the synthesis of the metal–oxide cellulose nanocomposite, thereby avoiding the pretreatment step that involves energy, was also reported (Rotaru et al., 2018).

The cost-effectiveness and high efficiency of the adsorbent material are essential for an adsorption process. This process is regarded as one of the most economical techniques for pollutant removal in general. However, the cost of the final adsorbent material is seldom discussed in research papers. A range of 10 USD–200 USD per million liters was reported for the adsorption process of pollutants from wastewater depending on the type adsorbent (de Andrade et al., 2018). Other treatment technologies such as ion exchange or electrodialysis and reverse osmosis can cost up to 450 USD per million liters, making the removal process comparatively more expensive. Activated carbon, a very common choice for heavy metals and organic pollutants removal among the adsorbents, has a cost between 586 and 1171 USD/kg, as reported by Iakovleva and Sillanpää (Iakovleva et al., 2020). The regeneration costs of activated carbon are also very high, which is why low-cost adsorbents have been explored by researchers in the recent years. Cellulose, derived mostly from organic materials, is an eco-friendly and a biodegradable raw material that can reduce production cost significantly due to its abundance and low cost (Long et al., 2018). A total cost of only 4.50 USD was reported for preparing 25 g CNFs from jute bags by Ahuja et al. (2021). Native cellulose is generally modified by various chemicals to achieve the desired efficiency, durability, and physical characteristics of the aerogel. However, this also implies that more chemicals and treatment are required for a modified aerogel, which will increase the cost of the final material. For example, commercial cellulose costs about 259.5 USD/kg (Bello et al., 2019). Many researchers have also opted to use cellulose directly from nature or from manufacturers in bulk, which will reduce the cost. The cost of nanocellulose is very volatile and can be between 60 and 1180 USD/kg (Bulota et al., 2013). A CNC modified via TEMPO oxidation can be considered to have an idea about the costs of modified cellulose. In this case, the items required for the synthesis included 1000 g of CNC (300 \$/kg), 59 g of TEMPO reagent (570 \$/kg), 325 g NaBr (112 \$/kg), 7.1 L NaClO methanol (190 \$/L), 0.1–0.5 M NaOH (8 \$/L), and 0.1–1.0 M HCl (10 \$/L), which is a total of 2800 USD/kg, whereas an unmodified CNC would cost only 300 USD/kg (Batmaz et al., 2014; Hokkanen et al., 2016). However, modifying the adsorbent will cut down costs by being more efficient, having the ability to be reused multiple times and by being chemically and mechanically stable throughout its lifespan. The total cost is highly dependent on the type of modifications applied to the cellulose and the method of preparation of the aerogels.

Various isotherm and kinetic models have been revealed by literature based on distinctive assumptions. The assumption with respect to Langmuir isotherm was monolayer adsorption with a definite number of adsorption sites on the homogenous surface. In contrast, Freundlich model was more suited to heterogenous surfaces that could be applied to multilayer adsorption. These two (Langmuir and Freundlich) were the most popularly used models in depicting the performance of adsorbents. A decrease in adsorption heat to be linear was assumed by Tempkin isotherm. Dubinin–Radushkevich isotherm was commonly applied in determining whether the adsorption mechanism was chemical or physical with a heterogenous surface hosting free energy. The spontaneous and feasibility nature of the adsorption process was the major assumption associated with Flory–Huggins isotherm. Another important assumption related to Hill isotherm was how the binding sites of a ligand in a macromolecule could influence other binding sites within the same macromolecule. All the aforementioned were categorized as two-parameter isotherms. Redlich–Peterson isotherm model, on the other hand, was the most popular and commonly used three parameter isotherms. It was a combination of Langmuir and Freundlich models. It could be used over a vast range of concentrations and could be used in both homogenous and heterogenous systems. Other three parameter isotherms included Sips, Toth,

Koble–Corrigan, Khan, and Radke–Prausnitz isotherm (Foo and Hameed, 2010; Singh et al., 2018).

Adsorption kinetic studies were also very crucial to have a very good understanding of the dynamics of an adsorption process, with Lagergren pseudo-first order, pseudo-second order, intraparticle diffusion, and Elovich models being the most popular and widely used. It was always very crucial to establish the most suitable adsorption system for equilibrium data in an attempt to explore novel adsorbents for an efficient adsorption process, which should reliably predict the correlation of adsorption parameters and also quantitatively compare the adsorbents behavior for systems. The most suitable system was an equilibrium relationships (generally known as adsorption isotherms) because it perfectly described the interaction between pollutants and the adsorbent materials, thus becoming very vital in optimizing the mechanisms of adsorption processes, expressing the adsorbents capabilities and surface properties as well as effectively designing the adsorption systems. Isotherms were functions relating the adsorbate amount on the adsorbent with its pressure in the case of gas or its concentration if liquid at a fixed temperature. It could be used to study and explain the whole adsorption and desorption processes. Generally, adsorption isotherms comprised invaluable curves that depicted how the phenomenon of a substance was governed. Such phenomena included the mobility or retention of such substances from aqueous porous media/environments to a solid phase at a fixed pH and temperature. Adsorption isotherm models were established based on different hypothesis, and no particular isotherm model was universally applicable in all experimental data. Also, a situation whereby some hypotheses of a model were unable to describe experimental data absolutely demanded that more than one isotherm models be used for adsorption data analysis. Two adsorption isotherm models, namely, Langmuir and Freundlich were commonly used to describe the adsorption processes.

Water pollution has become increasingly severe with the development of industrialization. Water pollutants are generally divided into two categories: organic and inorganic (Hokkanen et al., 2016). This review mainly considers dyes, which belong to organic pollutants, and heavy metals, which belong to inorganic pollutants. Dyes are a significant source of organic pollution in water because of their obstinacy (Abouzeid et al., 2019). Heavy metal ion pollutants in water are highly migratory and accumulate along the food chain (O'Connell et al., 2008). Heavy metal ions primarily include Pb (II), Cd (II), Cu (II), Fe (III), Cr (VI), Hg (II), and Co (II). The removal methods of dyes and heavy metal ions can usually be divided into chemical precipitation, coagulation/flocculation, ion exchange, reverse osmosis, electrochemical treatment, adsorption, etc. (Liu et al., 2015a; Unuabonah and Taubert, 2014). Chemical precipitation or electrochemical treatment is inefficient for removing metal ions when the ion concentration is below 100 mg/L or when the ion product is below K_{sp} (Liu et al., 2015a). Adsorption is a fast, easy to operate, and cost-effective method for treating low-concentration wastewater (Tshikovhi et al., 2020; Kose et al., 2020). Activated carbon is a widely used adsorbent at present, but its manufacturing cost is high, and the regeneration conditions are harsh (Kose et al., 2020; Maatar and Boufi, 2015). Natural biopolymer materials have gradually attracted attention because of their environmental friendliness, low cost, reproducibility, and good adsorption performance (Bhatnagar et al., 2015). Cellulose has the advantages of a high yield (1011−12 t/y), nontoxicity, and a high content of hydroxyl groups, which are conducive to further modification (Tshikovhi et al., 2020; Kose et al., 2020). Cellulose is assembled from long single chains, each of which is composed of β-D-glucopyranosyl units linked by β-1,4-glycosidic bonds (Hokkanen et al., 2016; Lavoine et al., 2012). Many hydrogen bonds are formed by cellulose, resulting in a small specific area (1−4 m^2/g) with less hydroxyl exposure. This leads to poor hydrophilicity and adsorption (Anirudhan and Shainy, 2015; Voisin et al., 2017). The specific surface area of nanocellulose obtained by hydrolysis or homogenization treatment of cellulose has increased dramatically (Voisin et al., 2017). Generally, the specific surface area of cellulose nanoparticles is between 50 and 200 m^2/g, and the specific surface area of nanocellulose aerogels is between 250 and 350 m^2/g (Mishra et al., 2018).

Small molecules that link the desired functional groups to the hydroxyl groups of nanocellulose are modified by various reactions, such as esterification, etherification, halogenation, and oxidation, for surface functionalization of nanocellulose (Hokkanen et al., 2016; Voisin et al., 2017). The surface functionalization of nanocellulose include carboxylation, amination, quaternization, phosphorylation, sulfhydrylation, grafting of polymers, grafted with carboxylated polymers (Qiao et al., 2021). By reacting with various reagents, such as TEMPO, maleic anhydride, $NaIO_4/NaClO_2$, succinic anhydride, carboxymethylated cellulose, ammonium persulfate (APS), citric acid/hydrochloric acid, and others (Sehaqui et al., 2011; He et al., 2013; Sheikhi et al., 2015; Wang et al., 2018; Chen et al., 2009; Stephen et al., 2011), the surface carboxylation modification of nanocellulose can be achieved. Positively charged cationic dyes and metal ions can be adsorbed by surface carboxylation−modified nanocellulose, and the suitable pH value is usually 4−10. The surface amino modification of nanocellulose can be achieved by reacting with various reagents, such as diethylenetriamine, 3-aminopropyl-triethoxysilane (APTES), acrylamide, ethylenediamine (EDA), and others (Shen et al., 2009; Hokkanen et al., 2014; Jin et al., 2015; Singh et al., 2014). Negatively charged anionic dyes and metal ions can be adsorbed by the surface amino-modified nanocellulose, which is positively charged at the appropriate pH value of 2.5−5. Surface-aminated modified nanocellulose can also be used to adsorb positively charged metal ions at the suitable pH value of 5−7, where nanocellulose is almost uncharged and the effect is a nonelectrostatic interaction. Various reagents, such as imidazolium, glycidyl trimethylammonium chloride, and Girard's reagent T (Eyley and Thielemans, 2011; Pei et al.,

2013), are reacted with nanocellulose for the surface quaternization modification of nanocellulose. Anionic dyes and negatively charged metal ions can be absorbed by the surface-quaternized nanocellulose, which is positively charged over a wide pH range. Mono-anionic dyes, such as orange II (O), can be adsorbed by the azidated CNC surface, which was grafted with imidazolium bromide salt. By reacting with adenosine-5′-triphosphate (ATP) and sodium alendronate (Liu et al., 2015a; Sirvio et al., 2016), nanocellulose can be phosphorylated on the surface. Positively charged metal ions can be adsorbed by surface-phosphorylated nanocellulose at a pH range of 2–7, under which the phosphorylated nanocellulose is negatively charged. Reagents such as 3-mercaptopropyl-trimethoxysilane, L-cysteine, and 3-mercaptopropionic acid can be used to sulfhydrylate nanocellulose on the surface (Eyley and Thielemans, 2011; Pei et al., 2013; Li et al., 2019). Surface thiol–modified nanocellulose can adsorb positively charged metal ions, mainly for the selective adsorption of Hg(II), at a pH range of 5–7. Adsorbents rich in carboxyl and sulfhydryl groups were obtained by freeze-drying a TEMPO-oxidized CNF suspension in the presence of hydrolyzed 3-mercaptopropyl-trimethoxysilane sols. A variety of specific groups are connected to the main chain of cellulose by free radical polymerization, ring-opening polymerization (ROP), and controlled radical polymerization (CRP) to enable nanocellulose to adsorb pollutants, as referred to by polymer grafting of nanocellulose (Hokkanen et al., 2016; Kedzior et al., 2019). Nanocellulose grafted with poly(acrylic acid), poly(methacrylic acid-co-maleic acid), methacrylic acid (MAA), itaconic acid (IA), and other reagents can realize nanocellulose grafted with a carboxylated polymer (Maatar and Boufi, 2015; Anirudhan et al., 2016; Zhang et al., 2014). Positively charged metal ions can be adsorbed by nanocellulose grafted with carboxylic polymers at a pH range of 4.5–6. The adsorption rates of Pb (II), Cd (II), Zn (II), and Ni (II) by the poly(methacrylic acid-co-maleic acid)-grafted cellulose nanofibril aerogel were more than 95% for concentrations lower than 10 ppm, and 90%–60% for concentrations above 10 ppm (Maatar and Boufi, 2015).

By grafting nanocellulose with polyethyleneimine (PEI), polyetherimide, and other reagents (Jin et al., 2017; Tang et al., 2020), nanocellulose grafted with aminated polymer is realized. Positively charged metal ions and cationic dyes can be absorbed by nanocellulose grafted with aminated polymers, with the cooperation of carboxyl groups, at a pH range of 4.5–6. By grafting nanocellulose with cysteine, 2-mercapto-benzamide-modified itaconic acid (MB-IA), and other reagents (Anirudhan et al., 2016; Yang et al., 2014), nanocellulose grafted with sulfhydrylated polymer is realized. Positively or negatively charged metal ions can be adsorbed by nanocellulose grafted with sulfhydrylated polymer, with the cooperation of amine groups and carboxyl groups. Thiol-functionalized polycysteine-grafted CNFs were prepared by grafting cysteine onto the surface of CNFs in one pot using the amide bond formed between the cysteine amine group and the cellulose carboxyl group, and used to adsorb Cr(VI) and Pb(II) (Yang et al., 2014).

4.2 Adsorption mechanisms

4.2.1 Mechanism of oils adsorption

Oil in water can be treated by various methods, such as combustion, oil skimming, microbial degradation, the use of dispersants or curing agents, and adsorption (Angelova et al., 2011). Among these methods, the production of oil sorbents has attracted much attention due to their relatively high efficiency, low cost, and absence of secondary contamination. The oil absorption performance of sorbents is influenced by not only the viscosity and density of the oil but also the morphological parameters of the sorbent itself (e.g., total pore volume, pore structure and surface wettability) and the hydrophobic interaction between the oil and the sorbent (Jiang and Hsieh, 2014; Wang et al., 2015b). Generally, porous adsorbent materials are more efficiently penetrated by lower viscosity oils. Cellulose, which has hydroxyl groups on the surface that facilitate its modification, and long, thin, and numerous fibers that contribute to its mesh structure and porosity, is a promising material for oil absorption. Besides designing structures for absorbing oil droplets in water, many researchers have exploited the inherent porosity of biomass materials (e.g., wood-based materials (Zhao et al., 2020)) for oil adsorption. To achieve higher separation efficiency and prevent secondary contamination of water bodies, materials with special infiltration properties have been developed. A membrane with both suboil and subwater superhydrophobic wettability was prepared by Cao et al. (2020b). Such a membrane requires surface roughness and low surface energy to attain superhydrophobicity. When the membrane is immersed in water, either water or oil can fully enter the layered microstructure. However, when another immiscible liquid enters, the oil–water interface immersed in the microstructure remains intact, repelling the second liquid and preventing it from entering the structure. Therefore, the customized wettability of the membrane can be applied to filter multiphase emulsion mixtures of water in light oil and water in heavy oil, when the membrane is double prewettable by oil and water.

4.2.2 Mechanism of heavy metal ions adsorption

Complexation, ion exchange, and adsorption on the surface and internal pores of the adsorbent are generally the adsorption mechanisms for low-valent heavy metal ions in water, such as Pb^{2+}, Ni^{2+}, Cu^{2+}, Cd^{2+}, Zn^{2+}, Co^{2+}, and Fe^{3+} (Abu-Danso et al., 2018; Sadeek et al., 2015; Mo et al., 2021; Xiong et al., 2017). Due to its high porous nature, large specific surface area, and hydrophilic hydroxyl groups on the surface, good adsorption capacity for heavy metal ions is exhibited by cellulose.

The form of metal ions in water is also affected by the pH of the solution, which can influence REDOX reactions, the hydrolysis, coordination and polymerization of metal ions, and the ionic state of the surface functional groups of the adsorbent. The adsorption capacity of the biomaterial decreases sharply when the pH value of the solution increases, indicating that electrostatic interaction plays a dominant role in the adsorption process. The surface charge of the adsorbent affects the adsorption of pollutants. The zero charge point (PZC) indicates the pH value of the chargeless adsorbent surface (Melo et al., 2018). Cationic molecules adsorb better when the solution pH is higher than the pH_{PZC} of the adsorbent, while anionic molecules adsorb better at a lower pH (Ghorai et al., 2014). Carboxyl groups, which adsorb cationic pollutants, are deprotonated at high pH. For instance, carboxylated CNFs via the citric acid/hydrochloric acid hydrolysis method had the highest adsorption capacity for Cu(II) and the cationic dye MB when they had the highest carboxyl content (Wang et al., 2018). The adsorption capacity of the adsorbent for MB increased with pH in the range of 4–10, due to the increased negative charge of CNFs from more deprotonated carboxyl groups. Likewise, Hg(II) removal by the pure carboxylated CNFs film increased from 5% at pH 4 to 54% at pH 11 for a solution with 50 μg/L Hg(II), which was related to the decreased zeta potential of CNFs from -6.6 ± 1.1 mV at pH 4 to -17.6 ± 0.6 mV at pH 11 (Silva et al., 2020). On the contrary, amine groups, which adsorb anionic pollutants, are protonated at low pH (2–3). Singh et al. (2015) showed that the zeta potential of dialdehyde CNCs grafted with diethylenetriamine decreased with pH, and the pH_{PZC} was 8.5. At pH = 2.5, the nanocellulose adsorbent had the maximum adsorption of As(V), because As(V) was mainly H_2AsO^{4-} and the amino groups were protonated. The adsorption efficiency can be affected by the surface structure of the adsorbent. A report used the template method to prepare macroporous fiber beads, and tentacle-type polyisohydroxamic acid was fixed on the fiber beads as adsorption ligand (Liu et al., 2021b). The mass transfer resistance was reduced by the large pores of the cellulose beads, thereby increasing the mass transfer rate of heavy metal ions within the adsorbent. At the same time, an accessible active site for polymer grafting was provided by the pores, thus increasing the adsorption capacity.

A process involving NC as a stationary surface and some specific solvents as a mobile phase containing dissolved heavy metal ions to be sorbed might have been the mechanism for heavy metal removal from aqueous solutions. Chemical adsorption, adsorption by physical forces onto surface and pores, chelation reaction, ion exchange, entrapment in intra- and interfibrillar cavities of the NC framework because of their concentration gradient while passing through the membrane are some of the major pathways involved in heavy metal removal by NC (Chitpong and Husson, 2016; Zhan et al., 2020). The adsorption process is regulated by the interaction between the sorbate and the sorbent, which is commonly described using adsorption isotherms, which represent the relationship between adsorbate surface concentration and bulk concentration of the adsorbent at a constant temperature. The Langmuir adsorption isotherm, which is the most extensively utilized isotherm for sorption on cellulose, can be utilized to evaluate the performance of different adsorbents for specific adsorbates in this case by using the maximal uptake and adsorption energy. The Freundlich isotherm, an empirical model that is frequently utilized for modeling adsorption behavior, takes into consideration multiple adsorption layers as well as heterogenous site energies. The Langmuir and Freundlich models can be combined if there is competition between multiple adsorbates, as is often the case in wastewaters (Voisin et al., 2017).

4.2.3 Mechanism of dyes adsorption

The physical and/or chemical interactions between the adsorbent and the dye and physical properties of the adsorbent determine the efficiency with which dye pollutants are attracted by adsorbents. Hydrophobicity, which is influenced by the size and polarity of the dyes, as well as the protonation state of the ionizable molecules, is the main factor that drives the adsorption of dyes from water to the surface of the adsorbent. The adsorptive capacity of the material is also enhanced by physical and chemical attraction and π-π interactions (especially for aromatic adsorbents). Electrostatic interactions and hydrogen bonding can occur via functional groups such as charged functional groups and hydroxyl.

The use of biomass materials as natural and modified forms of adsorbents has been investigated. The thermodynamic parameters indicated that the adsorption was an endothermic and entropy increasing process. The chemisorption of dyes on biomass adsorbents was responsible for the removal of organic pollutants from the aqueous environment. The electrostatic interaction between biomass adsorbents and Congo red molecules was the main factor for the adsorption process. Sugarcane-derived biological sorbents, which have various functional groups, such as $-OH$, $-NH_2$, $-COOH$, $-OCH_3$, $-CONH_2$, and $-SH$, were reported to be used to remove contaminants by chelation, complexation, coordination, and hydrogen bonding.

4.3 Cellulose–biomass materials for wastewater treatment

The use of green biomass materials in wastewater treatment to reduce or eliminate pollutants from aqueous environment has been widely investigated. However, the challenges of large-scale production, structural stability, and efficiency still hinder the commercial applications of biomass materials for wastewater treatment. Cellulose–biomass materials have been recognized by researchers for their oil, heavy metals and organic dyes removal, and adsorption capabilities due to their environmentally friendly and renewable and biodegradable properties. The modification of cellulose, pyrolysis, grafting of one or more suitable functional groups, and the preparation of aerogels by freeze-drying were the general methods for the preparation of cellulose-biomass materials.

The type of modification required to obtain an adsorbent that can act on specific or multiple adsorbates is usually determined by the target of adsorption (Kose et al., 2020). Nanocellulose, for example, must carry a charge opposite to that of the heavy metal ions for their adsorption through electrostatic interactions. The adsorption capacity of the adsorbent can be further improved by involving the complexation of functional groups to metal ions and other forms of interaction. Dyes can be adsorbed by electrostatic interactions, complexation, ion exchange, as well as van der Waals forces, hydrogen bonds, and π-π interactions. The corresponding adsorbent can therefore introduce not only a charge opposite to that of the dye molecules but also aromatic ring structures. Surface functionalization, grafting of polymers, and synthesis of composites of nanomaterials are some of the modification methods of nanocellulose (Abouzeid et al., 2019; Wang et al., 2019b). Carboxylated, aminated, quaternized, phosphorylated, sulfhydrylated, and other functionalized nanocellulose adsorbents are obtained as a result of the modification of nanocellulose.

4.3.1 Adsorption of heavy metals

A severe public health concern is posed by heavy metal ions found in water, such as (V), As (III), Cr (VI), Pb (II), and Cu (II) (Kong et al., 2020; Qiu et al., 2022). Cr (VI) is the reported ion for its oxidizing characteristics, carcinogenic, and toxic nature, all of which induce cell tissue irritation. Wastewater from many industrial operations, such as the textile industry, leather tanning, metal polishing, and electroplating, releases the Cr (VI) content (Zeng et al., 2021). Some suitable reducing agent is applied to effectively adsorb Cr (VI) or the pH is adjusted to precipitate as Cr (III) in the form of oxide and hydroxide. However, this process has problems of hazardous sludge formation and disposal to the environment (Marghaki et al., 2022). A highly toxic inorganic element available in the form of As (III) and As (V) in water is arsenic, causing cardiovascular, respiratory issues, immune deficiency, nervous disorder, and cancer (Marghaki et al., 2022). The ability of commercially available active carbon to adsorb arsenic is limited, particularly for As (V) (Nqombolo et al., 2021). Numerous biochemical, physiological, and behavioral abnormalities are caused by Pb (II) (Minh Trang et al., 2020). Protein synthesis and zinc enzymes can be impeded by Pb (II) by interacting with key components. Fertilizers, energy storage materials, the dye industry, cable coatings, plumbing fittings, and soldering are the principal sources of Pb (II) in water (Zou et al., 2021). Heart disorders, skin problems, memory disorders, and serious problems in the pancreas are caused by Cu (II) accumulation in the body. Waste streams from industrial activities such as mining, the dye industry, semiconductor manufacturing, and fungicides contain Cu (II) (Minh Trang et al., 2020).

Heavy metal contaminants are typically removed from the environment by sorption, chemical precipitation, membrane separation, and electrochemical treatment. Among these methods, sorption is one of the most effective and efficient processes, with activated carbon being the main sorptive medium. Cellulose nanomaterials (CNs) are a promising alternative adsorbent due to their high surface area-to-volume ratio, low cost, high natural abundance, and inherent environmental inertness. Furthermore, the surface of CNs can be easily functionalized with chemical moieties that may increase the binding efficiency of pollutants to the CNs. Carboxylation of CNs is the most studied method for enhancing their sorptive capacity. It was found by Yu et al. (2013) that the binding efficiency to Pb^{2+} and Cd^{2+} from aqueous solutions was significantly increased by the incorporation of succinic acid groups onto CNCs. The conversion of the carboxylic acid groups to sodiated carboxylates further improved their ability to remove these toxic metal ions from solution. The advantage of incorporating carboxylate groups was also demonstrated on CNFs. For example, Srivastava et al. (2012) showed that COO^{-}-modified CNFs could sorb Ni^{2+}, Cr^{3+}, Cd^{2+}, and Pb^{2+} with efficiencies 3%−10% higher than unmodified CNFs. Ma et al. (2012a) further supported the remediative utility of CNFs by illustrating their ability to remove radioactive uranyl ions ($UO_2{}^{2+}$) from solution. The $UO_2{}^{2+}$ ions coordinated to the carboxylate groups of TEMPO-oxidized CNFs up to 167 mg/g, which is two to three times greater than that achieved with traditional adsorbents (i.e., montmorillonite, polymer particles, silica particles, and hydrogels).

Nanocellulose-based adsorbents for heavy metal removal have been explored extensively in recent research. The removal of Cu(II), Pb(II), Mg(II), and Fe(II) by adsorption process onto alginate/nanocellulose beads was investigated by Abou-Zeid et al. (2021). Three types of nanocellulose (CNC, CNF, and tricarboxylate cellulose nanofibers, TPC-CNF) were combined with alginate to eliminate the metal cations. The modified biopolymeric beads (Alg-CNF, Alg-CNC, and Alg-TPC-CNF) showed high efficiency in removing the metal cations Cu^{2+}, Pb^{2+}, Mg^{2+}, and Fe^{2+}. The synthesis, characterization, and application of the polypyrrole-functionalized nanocellulose for the removal of Cr(VI) from aqueous solution were reported by Alsaiari et al. (2021). The functionalized nanocellulose exhibited better performance for Cr(VI) absorption compared with previous studies, with a maximum adsorption capacity of 560 mg/g according to the Langmuir isotherm. The results indicated that the functionalized nanocellulose enhanced the adsorbent's effectiveness of Cr(VI) by two mechanisms: adsorption and reduction. A high equilibrium Cu^{2+} removal capacity of 115.3 mg/g at pH 5.0 was achieved by the modified CNFs according to Qin et al. (2019). The feasibility of using carboxymethyl nanocellulose (CMNc) stabilized nZVI for effective Cr(VI) reduction in aqueous medium was examined by Kumar et al. (2019). The findings suggested that adsorption efficiency was about 100% with an initial concentration of 6 mg/L and only 0.015 g of the produced material. The CNFs adsorbent had maximum adsorption capacities of 571 mg/g, 462 mg/g, 361 mg/g, 263 mg/g, and 208 mg/g for Pb(II), Cu(II), Zn (II), Cd(II), and Mn(II), respectively (Mo et al., 2021). The adsorbent was able to rapidly remove Pb(II) species (87% and 100% of its equilibrium uptake in 2 and 10 min, respectively) (see Fig. 4.2).

The ability of hydrogel composites based on chitosan-g-poly(acrylic acid) matrices filled with cellulose nanowhiskers (CNWs) to adsorb Pb (II) and Cu (II) ions from water was investigated by Rodrigues et al. (2019). The highest

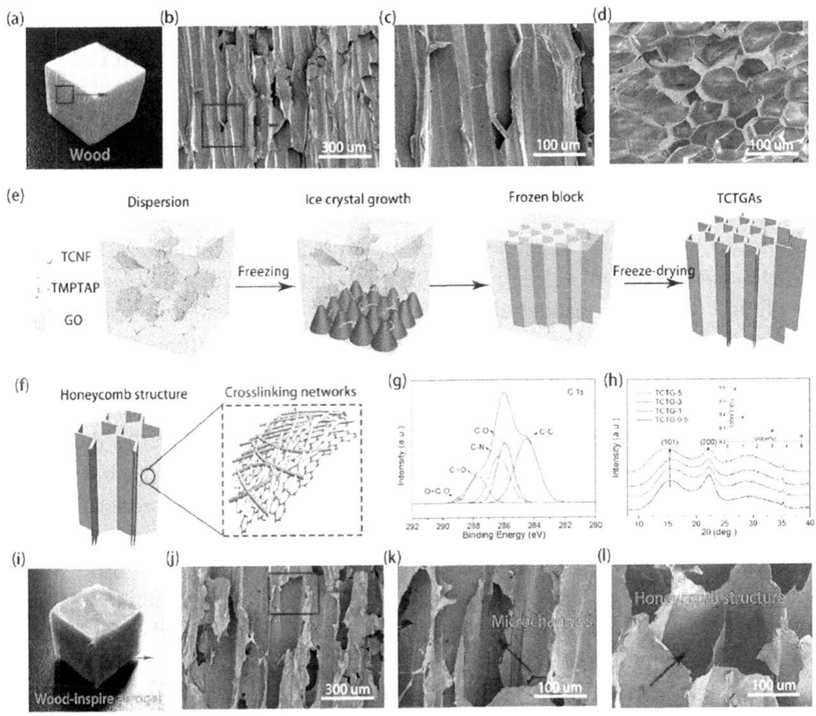

FIGURE 4.2 (a) Optical image and microscopic structure (b–d) of wood at different magnifications. (e) Schematic illustration of the synthesis of wood-inspired highly elastic TCTGAs with honeycomb-like macroporous architecture via a directional freezing method. (f) The cross-linking networks among TCNF, TMPTAP, and GO of TCTGAs channel walls. (g) The XPS C 1s spectra for TCTGAs suggesting the presence of covalent bond. (h) XRD patterns of TCTGAs with different mass fraction of GO (0.5 wt%, 1 wt%, 3 wt%, and 5 wt%, respectively), suggesting that 0.5 wt% of GO relative to TCNF was optimum. (i–l) Morphology of TCTGAs at various magnifications demonstrating the wood-inspired honeycomb macropore structure with aligned channels. *Reproduced from Mo et al. (2021) with permission from Elsevier.*

adsorption of Pb (II) (818.4 mg/g) and Cu (II) (325.5 mg/g) was obtained within 30 min, at pH 4.0, using 20 mg of the hydrogel composite containing 10% w/w of CNWs. Silva et al. (2020) synthesized dual nanofibrillar-based biosorbent films composed of CNFs and lysozyme nanofibrils (LNFs) for the removal of mercury (II) from aqueous solutions. This material exhibited excellent mechanical properties with a Young's modulus of more than 5 GPa and thermal stability up to 250°C. The Hg (II) removal efficiency in ultrapure and natural spring waters contaminated with ambient realistic amounts of mercury was 50 g/L. The removal efficiency of Hg (II) was pH-dependent, reaching at 98% after 24 h. A magnetic hybrid aerogel was developed by Wei et al. (2019a) by combining nanocellulose and ferroferric oxide (Fe_3O_4) nanoparticles for efficient heavy metal ion adsorption from water. The hybrid aerogel showed similar adsorption behavior on plumbum (Pb) (II) and copper (Cu) (II) ions as well as a high adsorption efficiency on the Cr(VI) ion of 2.2 mg/g. Zhang et al. (2021) reported the incorporation of NFC to the vinyl imidazole (VIM) monomer in the development of poly(VIM)/MFC cryogel composite for continuous and efficient heavy metal removal.

Palliparambil et al. (2016) fabricated cellulose acetate membranes with and without zero valent iron (ZVI) nanoparticles by phase inversion method and evaluated their performance for nickel removal from aqueous solutions. The incorporation of ZVI enhanced the nickel removal efficiency, which was influenced by various parameters such as contact time, initial pH, adsorbent dosage, and initial ion concentration. The maximum nickel removal efficiency was found to be 96%. Bai et al. (2020) reported that the addition of cellulose nanocrystals (CNCs-M) into nanocomposite polyethersulfone (PES) membranes increased their porosity, hydrophilicity, elasticity, and pure water flux compared with carbon nanotubes (CNTs-M) neat (see Fig. 4.3). They also demonstrated that nanocellulose is a potential alternative to CNTs due to its advantages such as abundance, eco-friendliness, biocompatibility, biodegradability, renewability, high hydrophilicity, strong mechanical properties, and low cost. Bai et al. further developed CNC-blended PES membranes for improved natural organic matter (NOM) removal and fouling mitigation. Surface morphology analysis revealed that the CNC-containing nanocomposite membranes had homogenous surfaces. The results showed that the nanocomposite membranes had higher porosity, zeta potentials, and pure water flux than the conventional PES membrane. The CNC-PES membranes also exhibited enhanced anti-NOM fouling properties, better cleaning performance, and effective fouling control.

The adsorption performance of cellulose-based materials is influenced by various factors such as pH, temperature, interaction time, reaction conditions, and ionic nature (Pourfadakari et al., 2017). Three types of synthesis routes have been reported for the development of CNC-based adsorbents: modification, grafting, and polymerization. Modification involves adapting a functional structure to cellulose (Dufresne, 2013). Grafting involves attaching other polymeric materials

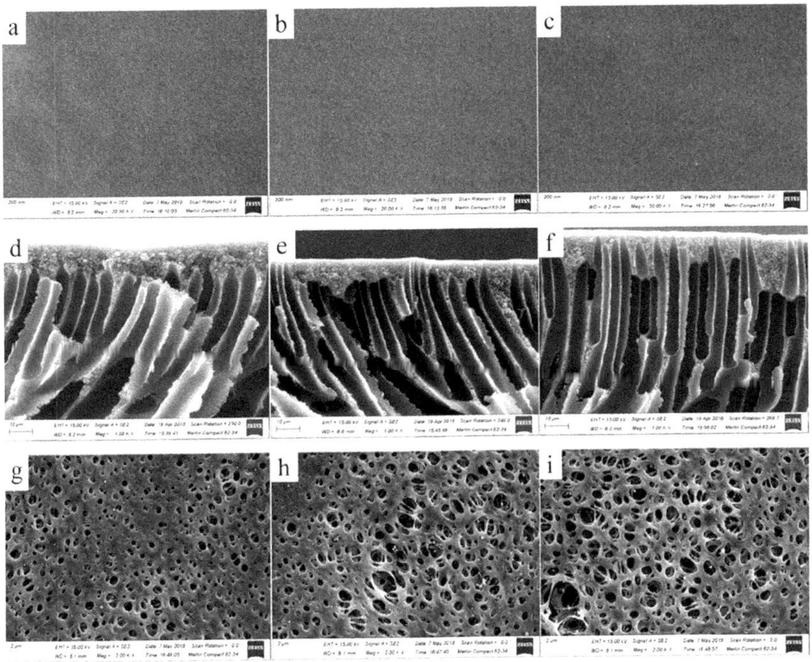

FIGURE 4.3 SEM images of the top surface: (a) neat PES; (b) CNC-M3; (c) CNC-M5; cross section of membranes: (d) neat PES; (e) CNC-M3; (f) CNC-M5; and bottom surface: (g) neat PES; (h) CNC-M3; (i) CNC-M5. *Reproduced from Bai et al. (2020) with permission from Elsevier.*

directly onto the CNC surface by either "grafting to" or "grafting from" methods. "Grafting to" refers to the direct binding of polymer chains to the CNC surface, while "grafting from" refers to the in situ development of polymer chains on the CNC surface by the addition of an initiator (Kose et al., 2020; Putro et al., 2017). Polymerization involves the formation of various monomers in the presence of cellulose (Jamshaid et al., 2017). These synthesis routes can introduce different functional groups such as carboxy, amino, and hydroxyl groups to the CNC surface, which can enhance its adsorption efficiency for the removal of hazardous heavy metals from contaminated water. These functional groups can chelate with metal ions through electrostatic interactions, ion exchange, or covalent bonding. For example, Taleb et al. (2019) observed that amino-terminal functions improved the performance of CNC as arsenic adsorbents. Anirudhan and Deepa (2015) developed a nanobentonite/CNC adsorbent with various carboxy functional groups for the effective removal of Co^{2+} and U^{6+}. Anirudhan et al. (2016) reported a structural modification of CNC using poly(itaconic acid)—poly(methacrylic acid) as an effective adsorbent for removing Co^{2+} from industrial wastewater. Voicu and Thakur (2021) modified CNC with amino-propyltriethoxysilane and demonstrated its ability to adsorb various heavy metals, including Cd^{2+}, Cu^{2+}, and Ni^{2+}.

CNC biocomposites have a greater chance of succeeding as biosorbents than other biosorbents since they combine two or more unique components to create a new substance that performs better than their constituent components. The time-dependent properties such as pH, contact time, adsorption dosage, and temperature are considered as vital controlling parameters during the designing of better CNC-based biosorbents (Qasem et al., 2021). The proportional heavy metal sorption increases with increasing contact time and reaches an optimal level. After equilibrium was reached, the percentage of metal ions sorption did not vary with the progression of time (Kardam et al., 2014).

Aerogels are novel adsorbent materials that are porous, ultralow density and have favorable properties such as specific surface area and high pore volume (Vareda et al., 2019). They can be classified into organic and inorganic aerogels. Cellulose-based aerogels, also known as bioaerogels, are organic aerogels with superior adsorption properties. They have high strength and do not break under compression, unlike traditional silica aerogels. A plastic deformation of up to 80% strain before pore wall collapse has been reported (Budtova, 2019). Moreover, the abundant hydroxyl groups on native cellulose allow easy chemical modification to create improved versions of cellulose-based aerogels with enhanced adsorption ability, mechanical strength, and the ability to regenerate for repeated use. The production of aerogel from cellulose does not involve toxic compounds. Since the information available on cellulose-based aerogel for the removal of heavy metals is scattered and limited, this review paper aims to bridge the gaps in this research by bringing together recent literature in this topic.

The biosorbent is reused and recycled after metal ions are removed from the contaminated aqueous media, which makes the biosorption process economically viable. The desorbing media used for CNC regeneration is nontoxic to the biosorbent, which is a desired characteristic. Furthermore, the heavy metals deposited on the surface of the CNC are retrieved after biosorption. Chemicals including HCl, NaOH, and chelating agents such as EDTA have been employed to desorb metal ions while simultaneously regenerating the loaded biosorbent, according to the literature (Gupta et al., 2015). Although HCl is commonly employed as a desorbing agent that recovers more than 90% of metal ions, it also causes simultaneous hydrolysis of the functional groups present on the biomass, as shown by numerous studies. The biosorption efficiency of the regenerated biosorbent would suffer as a result. It should be emphasized that the mechanism behind metal ion biosorption is the same regardless of the origin, kind, or source of the biosorbent. The regeneration of CNC biosorbent loaded with Cr^{4+} using 2 M NaOH was reported by Lin et al. (2014). The biosorption method is cost-effective since it recovers more than 70% of metal ions and uses biosorbent for continuous sorption/desorption to remove metal ions from wastewaters. The regeneration process of fatigued CNC-based adsorbents is an essential element in determining the potential economic use of these biomaterials. Extra attention must be paid to this issue in CNC-based adsorbents due to greater cost of CNC synthesis and surface modification as compared with traditional adsorbents. The utilization of ethylenediaminetetraacetic acid as a regeneration reagent after fully recovering Cd^{2+} from the CNC membranes was reported by Chitpong et al. (Chitpong and Husson, 2016). A straightforward regeneration process that has no detrimental impact on adsorption capacity would definitely help to promote the usage of NC-based adsorbents in real-world applications. A variety of techniques have been proposed and tested for the regeneration of the adsorbent, based on the adsorbate (Mahfoudhi and Boufi, 2017). The recovery of the carboxymethyl cellulose adsorbent is critical. This is necessary to avoid secondary environmental contamination. Magnetic elements have been added into the structure of heavy metal impregnated carboxymethyl cellulose by certain researchers to improve recovery. For example, Zong et al. (2019) employed NC-supported magnetic graphene oxide biocomposites to remove uranium ions from polluted solutions. An external magnetic field was used to detach the biosorbent from the aqueous medium followed by acid pickling using 0.01 mol/L HNO_3 to remove uranium ions from the biosorbent. Dahlan et al. also reported the utilization of external magnetic field for regenerating the biosorbent carboxymethyl cellulose grafted with Fe_3O_4 and carboxymethyl polyvinyl alcohol (Dahlan et al., 2017). A carboxymethyl cellulose/Fe_3O_4 biocomposite was also employed by Cai et al. (2017) as a europium (III) ion adsorbent from radioactive wastewater. The removal capacity was found to be 2.78×10^{-4} mol/g. Furthermore, an external magnetic field might be used to readily separate the biosorbent.

Several studies have shown that surface modifications to NC-based biosorbents can enhance their metal binding characteristics, but they also increase the total cost of the procedure, making it comparable with that of commercial ion-exchange resins. However, increased biosorption is not always observed as a result of the addition of functional groups due to chemical modification, which may be attributed to conformational, steric, or other factors. The complexity in obtaining reliable and inexpensive raw biomass, the lower robustness, and nonselectivity of the biosorbents to metal mixture solutions, the difficulty in regeneration, and the adverse effects of coexisting heavy metal ions on biosorptive capacity are some of the factors that hinder the widespread industrial application of biosorption technology. Before industrial use, the following parameters need to be evaluated: pH, coexisting ions, ionic strength, and suspended particles.

Cellulose is a popular material due to its abundant availability, renewable nature, flexibility, biodegradability, and ease of structural reformation. It can be used as an adsorbent for heavy metal removal. However, its applications in this field are limited by its larger size, lower inner porosities, reduced surface volume ratio, and presence of internal diffusion. Therefore, nanosized cellulose has emerged as the material of choice for wastewater remediation due to its superior adsorption properties compared to microcellulose. This chapter summarizes the most recent and advanced information of nanocellulose in the field of heavy metal wastewater remediation. Novel synthetic methods coupled with feasible functionalization are used to produce nanoscale adsorbents that can efficiently remove pollutants from wastewater. Nanocellulose-based biosorbents have been reported to have several additional advantages in most publications, besides the enhanced heavy metal absorption capacity, such as faster kinetics, efficiency across a wide pH and temperature range, and low operating costs. Nanocellulose-based biosorbents have been widely promoted as a cost-effective, environmentally friendly, and efficient technology that can be used as superabsorbent aerogels, impregnated composites, and forward osmosis membranes. Despite the market demand for cheaper and greener remediation technologies and the significant benefits of biosorption technology, it has received very limited industrial acceptance, especially when used in conjunction with other conventional treatment methods. Therefore, this chapter will contribute to a better understanding of the potential of using nanocellulose as biosorbents to remove various contaminants, especially heavy metal ions.

4.3.2 Oil/water separation

The production of refractory oily wastewater, which poses a serious threat to the environment and humans, has increased due to various factors, such as rapid industrial development, urbanization, crude oil exploitation and transportation, oil spill accidents, industrial production, and daily life (Deng et al., 2020). Oily wastewater comprises immiscible oil/water mixtures ($>20\ \mu m$) and emulsified oil/water mixtures ($<20\ \mu m$) (oil-in-water and water-in-oil emulsions) (Junaidi et al., 2021). The separation of emulsified droplets is more challenging than that of immiscible mixtures because of their higher stability and complex pollution mechanism (Zheng et al., 2021). Therefore, effective methods for controlling oily wastewater pollution are urgently needed. Several approaches have been proposed for treating oily wastewater, including gravity separation, mechanical recovery, centrifugation, in situ combustion, chemical solidification, electrolysis, and bioremediation (Xiang et al., 2022). However, these methods suffer from limitations such as low separation efficiency, secondary pollution, high energy consumption, and high cost, which hinder their practical large-scale applications (Baig et al., 2021). Moreover, these methods often fail to meet the reuse requirements because of their limited oil/water separation performance (Fu and Guo, 2022). Furthermore, the materials used in these methods, such as graphene–polyurethane sponges, superoleophobic stainless steel mesh, superhydrophobic sepiolite/silica, TiO_2 nanowire membranes (Zhou et al., 2016; Zhu et al., 2021; Pang et al., 2022; Pan et al., 2019), etc., are usually not recyclable and eco-friendly. Hence, there is a need to design and develop oil–water separation materials and technologies that are ecofriendly, economical, convenient, and efficient for oily wastewater treatment. Inspired by natural phenomena such as the superhydrophobicity of lotus leaves, researchers have developed superhydrophobic surfaces for oil/water separation by modifying the surface energy and creating micronanostructures on the surface of materials (Yang et al., 2022).

The use of superwetting porous natural materials for oil/water separation has been considered the best candidate to replace artificial complex materials. The natural superwetting materials have the advantages of being low-cost, ecofriendly, and easily available, which make them promising candidates for tackling the global challenge of oil contamination. Cellulose-based materials are promising candidates for this purpose, owing to their low cost, nontoxicity, and biodegradability. Cellulose-based aerogels, which have attracted considerable attention in recent years (Xiao et al., 2018), can be applied for oil/water separation. Their excellent performance also provides new ideas and directions for the manufacture of efficient, green, and cheap oil–water separation materials (see Fig. 4.4). However, the hydrophilicity of cellulose, which is inherent to its structure, often hinders its performance in this application (Mulyadi et al., 2016).

The wettability of the surface, which is closely related to the adsorption of oils, is determined by the surface chemistry and morphology. The water contact angle is a measure of wettability. A hydrophilic surface is characterized by a water contact angle on a flat surface of less than 90 degrees, a hydrophobic surface by a water contact angle between 90 degrees and 150 degrees, and a superhydrophobic surface by a water contact angle over 150 degrees. Similar measurements can be performed for different kinds of oils, and the solid material can be characterized as oleophilic, oleophobic, or superoleophobic, respectively (Doshi et al., 2018). The oil sorption capacity is enhanced by the hydrophobicity of the sorbent, which prevents water sorption and therefore reduces the competition between water and oil molecules. Superhydrophobicity can be found in natural materials such as lotus leaves, but it can also be introduced into an otherwise nonsuperhydrophobic material by increasing the surface roughness and decreasing the surface energy. A high porosity and surface area, and a selectivity to oil are desirable characteristics of an ideal sorbent for oil sorption. For oil spill treatment applications, the sorbent should also have mechanical and chemical stability, recyclability, environmental friendliness, low density, and good buoyancy properties.

Polysaccharide aerogels with different shapes and properties, which are synthesized by chemical or physical methods from cellulose raw materials of varying compositions, are usually referred to as cellulose-based aerogels. Three types of cellulose aerogels can be distinguished: natural cellulose aerogels, regenerated cellulose aerogels, and aerogels made from cellulose derivatives (Long et al., 2018). The structural properties of cellulose aerogels

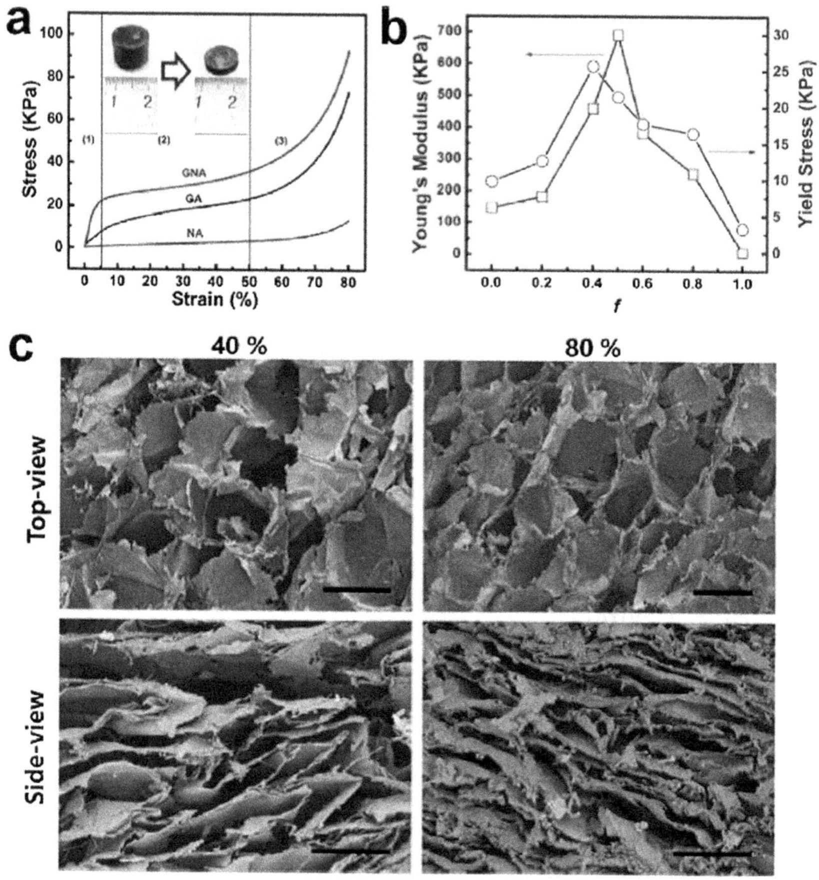

FIGURE 4.4 (a) Compressive stress−strain curves of GNA (ρ = 18.4 mg/cm3, f = 0.5), GA and NA. (b) Young's modulus and yield stress of GNA as a function of f. (c) Top- and side-view SEM images of GNA compressed to 40% and 80% strains. The inset in (a) shows photographs of GNA before and after compression to 80% strain, and the scale bar in (c) represents 100 μm. *Reproduced from Xiao et al. (2018) with permission from Elsevier.*

largely depend on the properties and concentration of the raw cellulose; most of them exhibit good reproducibility, biocompatibility, and degradability. (1) Nanocellulose aerogels and bacterial cellulose aerogels are the main types of natural cellulose aerogels. The raw materials typically display a cellulose I-type structure (Wan et al., 2019). The drying method of natural cellulose aerogel prepared from raw materials such as bleached cellulose fibers, cellulose whisker, and CNC was summarized by Yun-Xuan Weng et al., who also characterized the properties of the samples such as density, porosity, specific surface area, and modulus (Long et al., 2018). In 2014, Zhang et al. prepared for the first time a bacterial cellulose (BC)/graphene oxide (GO) ultralight nanocomposite aerogel, which can effectively adsorb common organic substances (cyclohexane, DMF) in water, indicating its potential prospects in future water pollution treatment (Wang et al., 2014). (2) Regenerated cellulose aerogel (RCA) is a cellulose aerogel prepared through four steps: dissolution, regeneration, solvent exchange, and drying. The process of regeneration and solvent exchange destroys and reorganizes the cellulose crystal structure, resulting in the conversion of cellulose I to cellulose II (Wan et al., 2019). Recently, a large number of researches on regenerated cellulose aerogels have been carried out; Li et al. (Wan et al., 2015) used wheat straw, bamboo fiber, filter paper, and cotton as raw materials to construct regenerated cellulose aerogels and characterized and compared them. The characteristics of large pore size, large pore volume, and high porosity were obtained. The regenerated cellulose aerogel has a simple preparation process, is low cost, and has a good application prospect in the field of oil/water separation. This type of aerogel is extensively researched and developed by scientists in the field of oil/water separation. (3) Cellulose derivative aerogels mainly refer to two types of aerogels: cellulose-derived aerogels and nanocellulose-derived aerogels.

Cellulose fibers can be chemically or mechanically processed to obtain pure cellulose microfibril units (5–100 nm diameter) with several micrometer lengths, which are referred to as nanofibrillated cellulose (NFC) and CNCs (Qua et al., 2011; Chen et al., 2011). The high crystallinity and large aspect ratio of nanocellulose enable the fabrication of aerogels with a very low shrinkage rate (<10%) and no cracking after drying, preserving a good internal structure compared with traditional aerogels.

Bacterial cellulose (BC) is a microbial nanofiber biological material that differs from NFC and CNCs. It is produced by the metabolism of various bacteria, with *Acetobacter xylinum* being the main producer due to its high yield and economic benefits (Foresti et al., 2017). BC has a high chemical purity (close to 100% cellulose content), high tensile strength, high degree of polymerization, and high crystallinity (Zhang et al., 2011; Qiu and Netravali, 2014). It also has a natural 3D structure network, high adsorption, and high wet strength (Sai et al., 2014). Moreover, it exhibits good biocompatibility, toughness, and nontoxicity, making it a promising candidate for the development of green and environmentally friendly aerogels. Bacterial cellulose aerogel (BCA) can be obtained by removing bacteria and other impurities and applying a drying process. Further modification of BCA by chemical or physical methods (such as esterification, oxidation, etherification, and ultrasound) can enhance its performance and broaden its applications in oil–water separation, drug delivery, and other fields (Foresti et al., 2017; Paximada et al., 2016; Sai et al., 2015).

The hydrophobic interaction of cellulose aerogel and its morphological parameters greatly influence its oil/water separation properties; however, pure cellulose aerogels usually exhibit a hydrophilic wettability during preparation, which can be converted into a superhydrophobic material by appropriate modifications, while retaining the original characteristics (Dilamian and Noroozi, 2021). These modifications enable cellulose aerogels to act as effective oil adsorbents or oil—water separators (see Fig. 4.5). The main driving force for oil absorption by hydrophobic cellulose aerogel is the hydrophobic interaction between modified cellulose and oil (or organic solvent) and the capillary effect of the pores. Therefore, the adsorption process is mostly physical adsorption.

A natural cellulose derivative, CNC, can also be used to fabricate superhydrophobic and lipophilic separation materials. The properties of raw materials can be improved by mixing them with some inorganic substances. A high-porosity cellulose composite aerogel was prepared by Lin et al. (Zhu et al., 2018) using an organic—inorganic hybrid strategy, in which rod-shaped CNC (extracted from biomass waste) and red mud (RM, extracted from industrial waste) were cross-linked. The hydrophilic hydroxyl groups of CNC were converted to hydrophobic groups by silanization, resulting in a water contact angle (WCA) of 134.5 degrees. The aerogel exhibited high adsorption

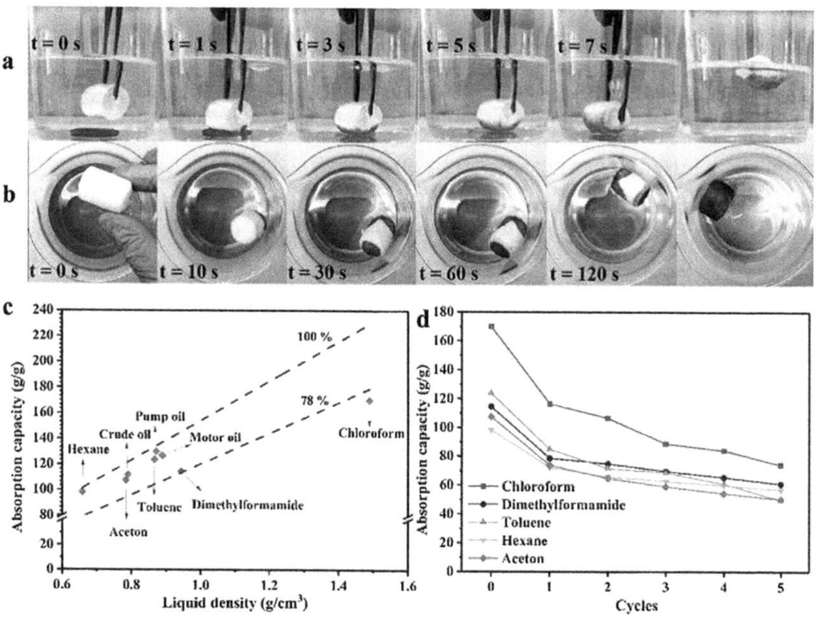

FIGURE 4.5 Oil and organic solvents—water separation performance of s-MNCAs. Removing (a) chloroform at the bottom of the water (b) and toluene layer on the surface of the water (both stained with Sudan Red 5B). (c) A mass-based adsorption capacity of s-MNCA (density and porosity of 6.5 mg/cm^3 and 99.6%, respectively) with dashed lines representing 100% and 78% volume-based adsorption capacities, and (d) reusability of the s-MNCA (density and porosity of 9.3 mg/cm^3 and 99.4%, respectively). *Reproduced from Dilamian and Noroozi (2021) with permission from Elsevier.*

capacity for oils and organic molecules, up to 30 times its own weight, owing to its hydrophobicity and high porosity. It could be used as an adsorbent for oil–water separation.

Basic research on a new type of cellulose aerogel based on chemically modified CNF for oil–water separation was conducted by Korhonen et al. (2011). A high permeability nanocellulose aerogel with a density of 20–30 mg/cm^3 and a porosity of >98% was prepared by using microfibrillated cellulose hydrogel as the raw material and the vacuum freeze-drying method. A hydrophobic but oleophilic coating was formed by covering the natural cellulose nanofiber aerogel with titanium dioxide to achieve functionalization. The aerogel exhibited highly selective absorption of nonpolar liquids and oils, as shown by adsorption experiments. The oil in the aerogel remained trapped when the aerogel was floating on water or taken out of water, after the adsorption. The total amount of adsorption was close to 80%–90% of the total volume of the aerogel, and the adsorption capacity for various organic liquids ranged from 20 to 40 (wt/wt). Ethanol could be used to remove oil from the aerogel after adsorption. The adsorption capacity of the aerogel for toluene did not significantly decrease after repeated soaking and drying 10 times, indicating good reusability. This work provided a foundation for the sustainable application of nanocellulose aerogel in the field of oil–water separation.

Various oils or organic solvents can be adsorbed effectively by cellulose-based aerogels that have been chemically modified by methods such as surface coverage, simple immersion, chemical vapor deposition (CVD), silicidation, and in situ polymerization. This enhances the applicability of these aerogels in oil–water separation scenarios. Moreover, their mechanical properties are also improved to some extent. However, the mechanical performance of polysaccharide-based aerogels remains a challenge that needs to be addressed. To overcome this issue, researchers have carbonized cellulose aerogels or incorporated carbon components into them. The adsorption performance and mechanical properties of CNF are enhanced by carbonization. A superhydrophobic, superelastic aerogel (CNGGK) with excellent fatigue resistance was prepared by Long et al. by mixing CNF with various polymers (GO, glucose, and kaolin) (Long et al., 2020). The aerogel exhibited good lipophilicity due to the graphene component, enabling it to float on the water surface. The absorption efficiency of the aerogel for organic solvents and oils was 75–255 times its weight. The aerogel retained almost 100% of its initial height and wavy layer structure after being subjected to 90% strain for 100 cycles, demonstrating excellent fatigue resistance and structural robustness.

Cellulose-based aerogels are a new generation of aerogels that have been proposed as the most sustainable oil–water separation materials owing to their sustainability, low toxicity, biocompatibility, and renewability. Since 2014, the applications of cellulose-based aerogels in supercapacitors, catalysis, active agent carriers, and oil–water separation engineering have increased significantly. The researchers anticipate that more novel polysaccharide-based aerogels will be further developed in the next few years, especially composite aerogels that can modify properties and add new functions according to the target application, which will enable applications that were not possible in oil–water separation before (for example, due to issues such as poor mechanical properties, sensitivity to moisture, or low adsorption capacity). By reviewing the recent literature on cellulose aerogels in oil—-water separation, the researchers identify that optimizing the performance of polysaccharide aerogels currently used in this field will be one of the future research hot spots in aerogel materials. With the current environmental problems caused by oily wastewater, polysaccharide aerogels can play their unique advantages and replace traditional materials. However, further research on cellulose aerogels still has room for improvement. Based on the aforementioned literature, several suggestions are made to guide the future development of cellulose aerogels:

(1) Natural waste as a source of cellulose-based aerogels, such as straw, waste wood, and waste cotton, is rich in cellulose and can be used as raw materials for the preparation of cellulose-based aerogels. These raw materials have the advantages of high yield, wide availability, low cost, and environmentally friendliness. By combining them with conventional inorganic or organic materials, nano/micron-scale raw materials can be obtained that possess the properties of both natural polysaccharides and synthetic materials.

(2) Freeze-drying methods and scaling-up: Freeze-drying methods are commonly used to prepare polysaccharide-based aerogels in the laboratory, but they pose challenges for industrial large-scale production. For instance, the development of large-scale supercritical drying equipment is still problematic, and safety issues may arise at the production scale. Therefore, more reasonable drying equipment, parameter optimization, and drying methods are needed to achieve large-scale preparation of polysaccharide-based aerogels.

(3) Hydrophobic modifications for oil–water separation applications: Pure cellulose aerogels are usually hydrophilic and may be damaged or degraded by adverse environments in practical oil–water separation applications. Therefore, hydrophobicity is essential for polysaccharide-based oil–water separation materials. Chemical modification, carbonization, or addition of carbon components have been attempted in recent years to obtain hydrophobic/lipophilic cellulose-based aerogels. However, most of these methods may cause some environmental damage after achieving satisfactory separation performance. Hence, finding a more suitable method to obtain hydrophobic cellulose-based aerogels is a major challenge for the future.

At every stage of oil production, water can be polluted by various sources, such as oil spills in offshore oil exploration and during oil transportation;

contamination of produced water, which is a mixture of formation water, reinjected water, and treatment chemicals used for drilling or oil recovery; and contamination of process water generated during oil refining (Munirasu et al., 2016). Produced water contains various organic and inorganic components, and its discharge can pollute soil and surface/underground water. The adoption of technologies for the treatment of contaminated water that is economically feasible is a challenge for the oil industry. Mechanical collection (skimming), incineration, enzymatic or dispersant treatment, membrane filtration, and the use of absorbent materials are some of the current strategies for the collection and removal of spilled oil (Mullin and Champ, 2003; Bayat et al., 2005). In situ combustion, which is the most economical and efficient method and removes large amounts of oil pollution quickly, is still opted for large areas of oil pollution at sea in practice; however, a large number of harmful smoke and sticky residues that cause environmental and safety problems are created by the combustion. Oil skimmers, which are too costly in terms of labor and very time-consuming, can be used on a small scale. Organisms in the water can be affected by the health impact of dispersants and fixatives. Therefore, economical, effective, and environmentally friendly treatment methods still need to be studied. One of the most economical and efficient methods is to use a sorption material, which is a broader class of materials that exhibit absorption, adsorption, and sometimes ion-exchange capabilities (Li et al., 2013).

Cellulose, a low-cost and abundant adsorbent, is ideal for removing oil and organic solvents. However, unmodified cellulose has limitations such as variable physical stabilities and low heavy metal adsorption capacities (O'Connell et al., 2008), among others. Therefore, surface engineering via chemical modification has been explored in recent years to enhance the properties of cellulose nanomaterials for the sorption of organic contaminants. Both organic and inorganic groups can be inserted on the surface of cellulosic nanomaterials, as demonstrated by the atomic layer placement of titanium dioxide (TiO_2) nanoparticles (NPs) onto cellulose nanomaterial aerogels (Korhonen et al., 2011). The TiO_2 veneer created a low-energy surface on the fibers, resulting in a nanocellulose-based material with oleophilic and hydrophobic properties that could absorb oil and various organic solvents from the water's surface with 20–40 g/g and 80%–90% vol/vol capacity.

The self-cleaning surface of cellulose membranes can enable their reuse. However, cellulose aerogels can only be reused by extrusion, which leaves some oil residue and limits their reusability. High compressibility is not essential for desorbing the oil adsorbed in the adsorption material. If the adsorption material has high heat resistance, the adsorbed oil can be desorbed by in situ combustion, and the heat and CO_2 generated by burning can be further utilized. The surface of biomass materials containing cellulose has certain hydrophobicity, lipophilicity, and 3D porous structure, which can be designed as cellulose-based carbon aerogel for oil/water separation (Jing et al.,

2019). Cellulose carbon aerogel is a bulk material with low density, high specific surface area, high porosity, amorphous characteristic, and controllable 3D continuous porous structure (Li et al., 2017a). Cellulose-based carbon aerogels are not obtained by simple carbonization of cellulose-based aerogels (Sankaranarayanan et al., 2021). There are two main types of cellulose-based carbon aerogels used for oil/water separation: one is biochar obtained by direct pyrolysis and carbonization and the other is activated carbon material obtained by adding an activation process during carbonization. Both the direct carbonization and activated carbonization processes can independently design the density, elasticity, porosity, heat resistance, wettability, and other characteristics of a cellulose-based carbon aerogel to achieve oil/water separation. The unmodified cellulose-based carbon aerogels have poor hydrophobicity and low adsorption capacity. The modification of cellulose-based carbon aerogels can be divided into precarbonization modification and postcarbonization modification. The precarbonization modification involves doping of silane reagent, micro–nanoparticles, and polymer in the cellulose dispersion to realize the in situ construction of micro–nanostructures during carbonization. The postcarbonization modification involves depositing silane reagents in aerogels by vapor deposition to improve the wettability of carbon aerogels. For the design and preparation of cellulose-based carbon aerogels, preparation costs, recovery methods, oil–water separation scenarios, and other aspects must be considered to achieve inexpensive and efficient oil–water separation. Cellulose-based carbon aerogels prepared from cellulose separated from agricultural and forestry waste have attracted extensive attention because of their attractive characteristics such as low density, large specific surface area, and rich porous structure. Cellulose is dispersed and freeze-dried to obtain cellulose aerogels, which can be used for oil/water separation after direct pyrolysis and carbonization without oxygen. Cellulose-based carbon aerogels prepared directly from cellulose extracted from agricultural and forestry waste often require higher carbonization temperature to obtain better oil/water separation performance. For example, Lingxiao Li et al. prepared pressure-sensitive and conductive carbon aerogels with tubular catkin cellulose in an N_2 atmosphere at 1000°C (Li et al., 2017b). The porous 3D network and tubular structure of cellulose-based carbon aerogel provide sufficient environment for the adsorption of oil and organic liquid. Moreover, it can be reused after combustion, and the carbon skeleton remains intact after combustion.

A top-down method can be used to directly modify wood, a degradable, renewable, and sustainable natural polymer material with a hierarchical porous structure and rich active functional groups, into oil/water separation materials (Wang et al., 2021). The porous structures in wood, such as ducts, pits, tracheids, and other micronanopores facilitate the adsorption of pollutants and the filtration of wastewater containing pollutants by providing a channel for water transmission in wood (Fang et al., 2021). Moreover, cellulose in wood provides the skeleton structure, and lignin and hemicellulose act as the binder. They form a reinforced concrete-like structure on wood. The pore structure of wood can be adjusted and controlled by various means, which can improve the chemical accessibility of wood and endow it with special functions (Zhu et al., 2019). Inspired by wood structures, anisotropic aerogels based on cellulose, lignin, and chitosan were designed and prepared using directional freezing technology. Depending on the application scenarios of the materials in wood-based oil/water separation, they are mainly classified into filtration-type and adsorption-type materials, corresponding to the membrane and aerogel materials of the three materials mentioned before (Wang et al., 2019c).

Wood can be sawn into various sizes and styles, which enables wood-based oil/water separation materials to switch between filtering and adsorption functions without being limited by material shape or usage scenarios. The wettability of the wood surface plays a major role in the oil/water separation process. The porous structure is conducive to adsorbing and containing oil pollutants and to enhancing the demulsification performance. The main design and preparation direction that imparts special wettability to wood oil/water separation materials is to construct micro—nanostructures on the surface or inside of wood and modify them with low/high surface energy materials.

The application potential of wood in the adsorption of oil pollutants is great, owing to its honeycomb structure and hollow fiber arranged along the growth direction (Zhu et al., 2019). However, the chemical accessibility, permeability, and porosity of natural wood are poor, which limit its application in oil/water separation. Wood-based aerogels are obtained by partially or completely removing lignin from natural wood, which can retain the regularly arranged skeleton structure of cellulose in wood and can produce nanopores in the wood cell wall, improving the porosity and specific surface area of wood aerogels (Li et al., 2023). Oil water selectivity is crucial for improving the oil/water separation performance of materials. For adsorption-type wood-based oil/water separation materials, the oil water mixing situation should be considered for corresponding modifications. Delignified wood aerogel has excellent hydrophilicity, which enables it to separate water from water-in-oil without complicated chemical modification. A superior hydrophilicity cellulose aerogel for separating aqueous oils was prepared by Junwei Wu et al. using balsa wood as the natural polymer composite with cells (Wu et al., 2023). However, in oil-in-water, delignified wood aerogels need to be modified to fully develop the adsorption characteristics of wood aerogels. Generally, liquid deposition or vapor deposition are used to modify wood-based aerogels. The silanization modification can form hydrophobic polysiloxane in situ in wood-based aerogels to enable them to be used for oil/water separation. A wood-based sponge was obtained by Ying Liu et al. using a two-step chemical treatment to remove lignin from Balsa wood and then modified it with methyl trimethoxysilane by liquid-phase deposition (Liu et al., 2021c). In addition, some porous nanoparticles with large specific surface area, high porosity, and low density, such as MOFs, COFs, and MXenes (Wang et al., 2022), can impart oil/water separation properties to wood after in situ growth (see Fig. 4.6).

The problems of easy loss and difficult recovery of nanoparticles can be overcome by growing nanoparticles in situ in wood. Wood/COF composite with a unique porous structure was formed by Yaning Xu et al. using an in situ growth strategy to uniformly deposit COFs on wood microchannels, which shows fast and effective characteristics in oil–water separation, dye, and iodine adsorption (Xu et al., 2022). After delignification, the hydroxyl group of cellulose in the wood is almost completely exposed, which facilitates the proper modification to achieve the in situ construction of micro–nanostructures in the wood-based aerogel. The low surface energy and rough surface structure contribute to good absorption and help in maintaining oil. The mechanical properties of wood will be reduced by the complete removal of lignin and hemicellulose, which cannot meet long-term recycling

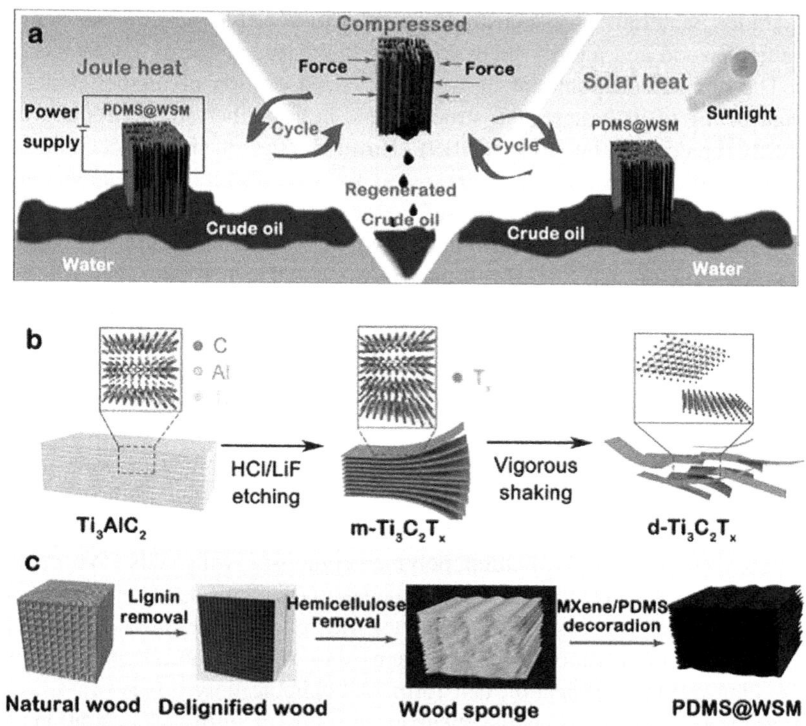

FIGURE 4.6 (a) Schematic illustration of Joule-heating and solar-heating PDMS@WSM to clean up and recovery a viscous crude oil spill. (b) Schematic showing the synthesis process of MXene nanosheet. (c) Schematic illustration of the fabrication of wood sponge and PDMS@WSM. *Reproduced from Wang et al. (2022) with permission from Elsevier.*

requirements. The mechanical properties of delignified wood-based aerogels can be improved by polymer impregnation, electroless metal plating, in situ grafting, and other methods. A magnetic wood-based aerogel (PDMS-Ni-PW) was obtained by Zhu Chu et al. by modifying the delignificated porous wood (PW) aerogel by amination, and then chemically depositing Ni and polydimethylsiloxane (PDMS) into aerogel (Chu et al., 2022). The aerogel can not only adsorb the oil pollution underwater but also adsorb and treat the oil pollution area by moving toward the polluted area on the water under the control of the magnetic field. Wood has abundant organic carbon, which can generate carbon materials under conditions of isolation from air during high-temperature thermal decomposition. In addition, during high-temperature carbonization, wood can maintain its natural cell structure, pore orientation, and anisotropy (Bai et al., 2023).

Various types of oil/water mixtures can be effectively separated by cellulose, lignin, chitosan, and wood-based oil/water separation materials in biomass materials. Moreover, using this biomass is beneficial because of its rich source, low cost, easy modification, environmentally friendliness, and low carbon, and it has considerable application prospects in oil/water separation. The design of better microporous structures and surface chemistry is extremely important to obtain biomass-based oil/water separation materials with higher performance and more functions. A number of issues still require to be addressed in the process of material design and preparation and the actual oil/water separation application, although considerable progress has been made in biobased materials of oil/water separation:

(1) The original advantages of biomass materials should be given full play. Biomass materials are wastes in the agricultural, forestry, and fishery industries, which retain some of the original biological characteristics. The use and modification of biomass materials should be based on the biological characteristics of biomass and the nature of the materials to improve the application range of biomass materials in the oil/water separation field, when biomass materials are used as raw materials in the preparation of oil/water separation. Cost savings (such as various natural wastes, waste fibers, recycled polymers, etc.) should be continued to explore and their stability under extreme conditions in complex water bodies should be furtherly improved in the future.

(2) The oil−water separation effect of the lipophilic−hydrophobic nanofiber-based aerogels is conspicuous, but the majority of them lack long-term durability and antipollution performance. Siloxane is usually adopted to modify nanofiber-based aerogels to obtain hydrophobic and lipophilic, but the method has inherent drawbacks, including high energy consumption, uniform modified layers, and harm to the environment. Therefore, emerging approaches, green, effective and low energy consumption, to obtain hydrophobic aerogels should be explored.

(3) Promising advanced composite materials could be obtained by the combination of nanofiber-based aerogels with other materials, and this strategy can improve the porosities of the aerogels, such as the specific surface area and mechanical properties, and provide novel functional nanofiber-based aerogels with flame retardancy, response performance, and magnetic properties. However, there are few studies on the electrical and thermal properties (Joule heat or photothermal) of nanofiber-based aerogels for oil–water separation, and the material–property relationships are not well established. Therefore, the relationship between the materials and properties of nanofiber-based aerogels should be built, the initial properties of materials should be used to build new functional nanofiber-based aerogels, and the application depth of nanofiber-based aerogels in oil–water separation should be continuously expanded.

(4) Multifunctional biobased oil/water separation materials should be designed and prepared: Various functions of biobased oil/water separation materials tend to look flashy and impractical and are inadequate for practical oil/water separation applications, despite the development of biobased oil/water separation materials. The water content of oily wastewater is relatively miscellaneous in actual production and life, and it is often mixed with organic dyes, antibiotics, heavy metals, bacteria, and other traditional and emerging pollutants, in addition to oil pollution. Therefore, it is of great practical significance for practical industrial applications to develop stable, efficient, and low-cost multifunctional biobased oil/water separation materials and realize the all-around removal of pollutants.

(5) Intelligent biobased oil/water separation materials based on intelligent response should be designed and prepared: Today, intelligently responsive biobased oil/water separation materials often respond to relatively simple and harsh trigger conditions. The components and structures of materials with response functions are easily damaged in the actual application process, thus losing the response function, greatly limiting the development and application of intelligently responsive biobased oil/water separation materials.

4.3.3 Dye removal

Surface and groundwater systems are seriously polluted by widespread use and discharge of water-soluble dyes in recent years, which has posed a serious environmental challenge and drawn global attention (Bhatnagar and Sillanpää, 2010). Among different types of pollution, dyes are a major category of pollutants that impair the quality and usability of water (Chequer et al., 2009; Dawood and Sen, 2014). Dyes are colored compounds that are widely used in cosmetics, printing, and plastic industries and generate a large amount of colored wastewater. Many dyes have synthetic origins and complex molecular structures, which make them highly stable and difficult to degrade naturally (Gupta, 2009). It has been estimated that 2% of dyes are discharged by manufacturing operations, while 10% of dyes are discharged by textile and related industries in the effluent (Allen et al., 2004). The emergence of dyes as harmful pollutants has raised major environmental concerns and necessitated their removal from effluents. Adsorption method, which has good selectivity and high reuse rate, is the best method for dye treatment (Ozacar and Sengil, 2005). Low-cost, nontoxic, and nonpolluting biomass is increasingly being developed as dye adsorbents (Luo et al., 2018; Yan et al., 2011; Wang and Li, 2013; Yamaki et al., 2005). In fact, effective adsorbents are being made by modifying biomass materials, which is the focus of wide range of research. Cellulosic materials, which are biodegradable, cost-effective, and renewable polymers, have been used as low-cost adsorbents. However, their solubility in most solvents is limited by their crystalline nature and hydrogen bonding. Therefore, modified cellulosic materials have been developed for enhanced dye removal from wastewater. These materials offer the advantages of pollution control, economic viability, and availability.

4.4 Effects of pH and temperature on adsorption

Solution pH is a crucial factor that influences adsorbent capacity in wastewater treatment. The adsorption efficiency depends on the solution pH, as it affects the ionization degree of the adsorbed molecule and the surface charge of the adsorbent (Saleem et al., 2022). Therefore, the adsorption rate varies significantly with the solution pH. For anionic dyes, higher dye removal percentages are observed at low pH, while lower dye removal percentages are observed at high pH. The opposite trend is found for cationic dyes. This is because hydrogen ions compete with cationic ions for adsorption sites at low pH, while functional groups on the adsorbent surface are ionized and attract cationic ions at high pH. For anionic dyes, low pH enhances adsorption because the adsorbent surface is positively charged and can easily adsorb anionic dyes (Noreen et al., 2020). The point of zero charge (pHpzc) is a key parameter for understanding the adsorption mechanism. The pHpzc indicates the adsorption ability and the type of active sites of the adsorbents. When pH > pHpzc, functional groups such as carboxyl (COO−) and hydroxyl (OH−) exist on the adsorbent surface and cationic dye adsorption occurs. When pH < pHpzc, the adsorbent surface has a positive charge and anionic dye adsorption occurs (Noreen et al., 2020).

The adsorption isotherm is essential for understanding the adsorption nature of the adsorbate onto the respective adsorbents in equilibrium study. The concentration of adsorbate in solution reaches dynamic equilibrium with the adsorbent interface. Mathematical models can be developed from the analysis of equilibrium data, which can help to quantify the results and predict the adsorption of ions based on the assumptions of such models (Babarinde et al., 2016). The equilibrium concentration is temperature-dependent, so the experiments are performed at a specific temperature. Several isotherms, such as Langmuir, Freundlich, Dubinin—Radushkevich (DR), Redlich—Paterson, Sips and Halsey isotherms, are widely used in adsorption studies. Different adsorption characteristics are explained by each equilibrium model, but the most common ones are the Langmuir and the Freundlich. The Langmuir isotherm (Langmuir, 1918) is an ideal model of a monolayer, which assumes that all adsorbed molecules have the same adsorption energy and form a monolayer on the adsorbate ion. The Freundlich isotherm (Freundlich, 1906) represents multilayer adsorption and is usually applied to the adsorption of interactive species or organic components of porous materials and activated carbons (Foo and Hameed, 2010). The AR-42 dye adsorption onto composite showed heterogenous adsorption, suggesting that the Freundlich model is more suitable (Leshaf et al., 2019). Composite organophilic montmorillonite-cellulose acetate was used to remove Eosin Y, and the Freundlich model did not fit well because of small 1/n values. The Langmuir model fitted better based on the correlation coefficient value for both isotherm models (Goswami and Das, 2019). LNC/MMT nanocomposite was used for CR dye removal, and the Langmuir model was more suitable than other models applied (Zhao et al., 2012). Similarly, composite CNC—alginate hydrogel beads were used for MB dye removal, and the results revealed that the Langmuir isotherm model fitted better than the others. The values of R2 from the linear form of the Langmuir model were close to unity, compared with those from the Freundlich model, indicating that the Langmuir model describes the adsorption process very well. A comparison of adsorption isotherms showed that the Langmuir isotherm was more suitable than the Freundlich isotherm for describing the adsorption process (Mohammed et al., 2015).

The dyeing industry is a major source of water pollution due to the large volume of effluents discharged into the water bodies. The composition and toxicity of dye effluents vary depending on the type of raw materials used in the industry. A wide variety of dye substances are available, such as direct, acid, basic, reactive, mordant, metal complex, vat, sulfur, and disperse dyes. Some of these dyes are synthetic and derived from organic or inorganic compounds. They are commonly used in industries such as textile and paper. Some dye effluents contain high suspended solids and toxic chemicals (Hethnawi et al., 2017), while others contain high concentration of residual dye type, formaldehyde-based dye fixing agents, hydrocarbon-based softeners, and nonbiodegradable chemicals. The effluents are also released at high temperature and pH, which are extremely damaging to the environment. Textile industry is reported to be the second highest cause of water pollution worldwide (Kant, 2011). About 72 toxic chemicals have been identified to reach water bodies through textile dyeing. The workers involved in the manufacturing of textile or paper may be exposed to harmful chemicals and their health hazards due to the toxicity of dyes.

Significant water pollution has been created by the chemicals coming from textile factories, which are highly toxic. Severe environmental and health problems can be led to by the contamination of water sources with dyes. Breathing problems and irritation of the skin are induced by prolonged exposure to the dye. Many of the dyes in fact are mutagenic and carcinogenic and increase cancer risk. However, hard to treat is dye-contaminated water, due to the nonbiodegradable behavior of the dyes and their defiance to oxidizing agents, light, and heat. A complex aromatic structure is displayed by these dyes, and cationic, anionic, and nonionic character may be displayed. Functionalized NCs with anionic moieties have been used to extract cationic dyes. The possibility of carboxylated CNC as a potential sorbent for cationic dyes was explored by Batmaz and colleagues. The carboxylic group is introduced using the TEMPO-facilitated oxidation. They observed that 769 mg/g at pH 9.0 for MB is the carboxylated CNC's adsorption (Batmaz et al., 2014). A method to functionalize lignocellulosic fibers by an emulsion from neem oil phenolic resin was documented by Manna et al. Rapid MB adsorption over a wide pH range was shown by the treated lignocellulosic fibers. The chemical characterization showed that largely a consequence of hydrogen bonding interactions, electrostatic interactions, and van der Waals forces between the dye and the adsorbent is the quick adsorption of MB (Manna et al., 2017).

Organic dyes are complex organic pollutants with cationic, anionic, and/or nonionic properties that originate from various industrial sources such as textile, printing, pulp and paper, rubber leather tanning, and cosmetic industries for coloring different products; their elimination is required as they are notable environmental afflictions in various parts of the world (Rafatullah et al., 2010; Nasrollahzadeh et al., 2018). The 1.6 million tons of toxic dyes are generated each year, and $\sim$10%–15% of this volume is released as wastewater, as estimated. Various cationic dyes are eliminated by using CNCs and CNFs modified with anionic moieties as adsorbent materials or catalysts. Carboxylation of cellulose-based nanomaterials is one of the most investigated procedures for enhancing their sorption capacity. Carboxylated nanocelluloses fabricated via a single-extraction step hydrolysis process were investigated for their adsorption characteristics (He et al., 2013). Carboxylated or COO-modified CNCs were prepared by ammonium persulfate (APS) hydrolysis of microcrystalline cellulose, in which carboxyl groups were introduced on their surface during the cellulose hydrolysis. Positively charged dyes were bound by the carboxylate groups; the adsorption capacity of MB onto CNCs reached a balance (0.32 mmol/g) at 22°C after 10 min, as confirmed by adsorption studies for a cationic dye (e.g., MB). The MB desorption from CNCs by ethanol was highly efficient with more than 90% dye removal up to seven desorption cycles. Similarly, carboxylated CNCs were prepared by a facile and single-step method using HCl/citric acid hydrolysis of the microcrystalline cellulose, which was utilized for the adsorption of MB (Yu et al., 2016); nearly complete UV degradation of methyl blue by the COO-modified CNCs was observed after 4 h, and with an increased rate compared with other types of CNCs prepared using acids such as sulfuric acid and formic acid. This result was attributed to the surface modification of carboxylated CNCs wherein additional carboxyl groups could effectively serve as a binding site for the dyes.

Cellulose-based porous biosorbent was used by Liu et al. (2015b) to remove cationic dye MB and anionic dye acid blue 93 (AB93) from single and double dye solutions in the natural water environment, which may contain ingredients such as salt and surfactants. The adsorption behavior of the biosorbent was determined by the electrostatic interaction between the biosorbent and the dye molecules, as reported by the authors. The adsorbent could still remove the dyes after three cycles of the adsorbent use in a complex environment containing two kinds of dyes, as well as salts and surfactants. The adsorbent derived from biomass materials has also been reported to remove dyes in sewage. The two major categories of biomass material processing, namely, modification and carbonization are generally used. The type of dye, the water environment where the dye is located, and the functionalization of the adsorbent are the important factors affecting the dye adsorption by biomass, as stated by the authors. Therefore, the adsorption capacity can be significantly improved by the high affinity between the modified biomass material and the dye.

Water treatment techniques based on advanced oxidation processes (AOPs) are well known for their high efficiency in dye removal. AOPs degrade the contaminants to less harmful products without producing voluminous secondary waste, and the reduced product can be recovered and reused efficiently in another application, thereby improving the economic values of this method of water treatment (Nkalane et al., 2019; Rueda-Marquez et al., 2020). In photocatalysis reactions, catalysts speed up the photoinduced reactions and promote an electron from the valence band to the conduction band, resulting in the generation of hole and electron pairs. The holes are highly oxidative and quickly react with organic molecules adsorbed on the catalyst surface, leading to their degradation. Excited electrons also dissolve O_2, producing $\cdot$OH and O_2. The success of any photocatalysis depends on how low the catalysts can prolong the recombination of holes and electrons (Dong et al., 2014). Au–ZnO core–shell nanoparticles with cubic Au core were used as catalysts for photocatalytic degradation of MB and methyl orange (MO) dyes. The effects of process parameters such as the type of UV lamp, dye concentration, nanoparticle loading, and solution pH on dye degradation were also investigated. The results showed that hydroxyl radical and the use of two semiconductors as catalysts delayed the recombination of holes and electrons (Verma et al., 2019). Adsorption and ion exchange are related and very common in the removal of heavy metals and dyes. Ion exchange can be done using either cationic or anionic resins depending on the contaminant or ion to be exchanged.

Adsorption and ion exchange are related and very common in the removal of heavy metals and dyes. Ion exchange can be done using either cationic or anionic resins depending on the contaminant or ion to be exchanged. A colorimetric paper sensor that adopted a cation–exchange strategy for the visual determination of heavy metals was reported by Dong et al. (2020). The results revealed that the final concentration of Cu^{2+} was 1 ppm under visually detected concentration. Adsorption techniques in water treatment are based on surface chemistry involving adsorbate and adsorbent. They have been proven to be highly efficient in heavy metals and dyes removal from water even at trace levels. The main advantage of this process is the reversibility and reuse of spent adsorbent by desorption using a suitable eluent. In fact, carbon desorbed from drinking water treatment plant was successfully reused for pharmaceutical compounds removal up to their allowable concentration (Lujan-Facundo et al., 2019). Various adsorbents have been used for the removal of heavy metals and dyes such as activated carbon, chemically synthesized adsorbents, and naturally synthesized adsorbents. However, the choice and the properties of adsorbent used are crucial for the success of any adsorption process. Surface area, porosity, and particle size play a major role in the adsorbent efficiency (Sekar et al., 2004).

The techniques and the composition of wastewater determine the choice of water treatment products. Heavy metals removal and semiconductor photocatalysts for the degradation of dyes are common applications of activated carbon. However, some of these materials have high cost and limited performance, and efforts to improve their efficiency and sustainability are ongoing. Chemically synthesized materials are mostly used as water treatment products nowadays. Although they are efficient in water treatment, they pose environmental and ecological risks by leaching other hazardous chemicals into the water system (Sekar et al., 2004; DeMessie, 2015; Mellah et al., 2006). Therefore, the investigation of environmentally friendly and highly efficient water treatment products is necessary. Green synthesized CNCs are a relatively novel solution for water quality and water reuse issues. The presence of flavonoids in some agricultural wastes enhances their ability to coordinate to metal ions and their catalytic activity for degradation of dyes in aqueous solution (Vilet et al., 2020). In some cases, polysaccharides and flavonoids bind during the cellulose extraction, which makes the estimation of flavonoid percentage in cellulose difficult (Guo et al., 2018). However, a flavonoid yield as high as 18.75 mg/g was reported when cellulase was used as base material under the optimum conditions (Wei et al., 2019b). Metal oxides are also effective, eco-friendly, and highly efficient in water treatment, especially in the degradation of dyes. The combination of these water treatment products is expected to be highly efficient, effective, and eco-friendly for the treatment of industrial effluents.

Thin porous membranes are nowadays commonly applied for wastewater treatments to remove water contaminants. These membranes are engineered to obtain well-organized pore size with tunable surface to eliminate pollutants through processes such as size-dependent elimination using chemical interactions. One of the simple and cost-effective methods for the production of porous membrane is electrospinning (Shirazi et al., 2020). Two different aspects are possessed by the uses of NCs for membrane filtration: firstly, as nanopapers using pristine NCs only, or as NC/polymer composites for improving membrane characteristics (Mohammed et al., 2018). The development of membranes that provide finest access to functional sites for better mechanical characteristics lies in the key challenges while NC is used for fabricating membranes (Shak et al., 2018). The most important processing systems for developing NC membranes are (1) impregnating various electrospun mats with NC, (2) vacuum filtration followed by dip coating of NC, and (3) freezing with or without matrix phases. The possibility of processing charged microporous membranes with anisotropic or isotropic structures was confirmed by the advantages of impregnating electrospun mats with NC (Choi et al., 2020; Li et al., 2020). Highly stable strong composite membrane was produced by heating the mixture at 100°C for 10 min followed by electrospinning, which can easily be achieved by the incorporation of CNC into nanofibrous polyacrylonitrile (PAN). This composite membrane was supported on polyethylene terephthalate (PET) to develop nonwoven substrates with pore diameter in 50 nm range (Ma et al., 2012b). A high density of negative surface charge was possessed by the CNC composite membranes, and therefore, strong adsorption of positively charged dye, namely, crystal violet was possessed. The study showed that higher adsorption efficiency to dye compared to a commercial MF membrane was exhibited by the NC-incorporated MF membrane (Gopakumar et al., 2017). The multilayered NC membranes were produced using CNF suspension vacuum filtration followed by sulphate or carboxyl surface group dip coating with CNC. By controlling drying conditions and treating acetone, it was likely to tune the aspect ratio, pore size, and water flux and improve wetting of the membrane. The highest tensile strength was shown by CNF coated with carboxyl surface groups with CNC (Karim et al., 2016). A novel membrane for process modification, the Janus cellulose membrane (JCM), was developed and demonstrated to have effective moisture ability and antibacterial properties (Lv et al., 2019). By immobilizing silver nanoparticles (AgNPs) on one surface of the cellulose membranes and modifying the other side with hydrophobic moieties, the JCM exhibited a remarkable hydrophobic/hydrophilic dual functionality. The JCM achieved a high capacity of more than 96.0% and showed excellent reusability and recyclability, with only minor differences in effectiveness and flux after 10 separation cycles (Duolikun et al., 2020). The JCM could be further modified to produce separation materials with industrial strategies for possible commercialization (Wu et al., 2019). Another hybrid cellulosic asymmetric membrane (CAM) with ZnO/MnO_2 was also reported to have effective removal performance, which included anisotropic moisture capability, effective separation, low cost, and ecofriendly features. This hybrid CAM was a promising candidate for a wide range of potential applications, especially for industrial wastewater treatment. The CAM had moisture and other functions and exhibited antibacterial properties suitable for realistic environments. The CAMs were also functionalized to improve their hydrophilicity for the management of wastewater from fuel industries, which contained various contaminants such as dissolved solids, petroleum–organic content, and oily substances. The modified CAMs had better antifouling efficiency without compromising their contaminant rejection (Duolikun et al., 2020).

An advanced filtration technique to purify municipal and industrial wastewater, the membrane bioreactors (MBs), was developed and involved a combination of processes such as nanofiltration and biological treatments. Compared with traditional common methods, MBs had many excellent contributions, particularly on environmental consequences, reductions in sludge output, the finest effluent, and complete biomass retention. In these situations, various strategies were employed to alleviate biofouling on the membrane, such as aeration, physiochemical cleaning, and surface improvement. For example, a modified CAM was fabricated by a conventional phase inversion strategy consisting of composites of functionalized graphene oxide/polyvinylidene fluoride (GO/PVDF). The cellulosic composite had improved durability, permeability, hydrophilicity, and antifouling characteristics when modified with PVDF membranes. The required performance of modified NC compositions was achieved by evolving targeted membranes through charging modifications, structural remodeling, and porosity adjustment. The modified CAM with AgNPs was obtained using surface chemistry that was stimulated by the bacteria's affinity with Ag. This promised a significant reduction of the accumulation of bacteria due to the dispersed AgNPs on the surface of the improved CAM. A polyethersulfone-based CAM for removal of *Staphylococcus aureus* was optimized for its selectivity for protein-in-water and fouling removal with effective antibacterial characteristics (Duolikun et al., 2020).

The environmental and health issues caused by dye contamination of water sources have been widely recognized. The dyes can cause respiratory issues and skin irritation as well as increase the risk of cancer due to their mutagenic and carcinogenic properties. The treatment of dye contaminated water is challenging because of the nonbiodegradable behavior and resistance of the colors to oxidizing chemicals, light, and heat. Adsorption is a simple and cost-effective method that is commonly employed for large and complex processes to remove colors from solutions (Muthu and Khadir, 2022). Functionalized nanocellulose has been investigated as a potential dye sorbent in many studies.

The removal of MB by a cellulose nanofibril-based aerogel generated from sago pith waste was investigated by Beh et al. (2020). The SPCNF aerogel, with a density of 2.1 mg/cm^3, showed high efficiency in removing MB, with a maximum MB adsorption of 222.2 mg/g at 20°C. More than 90% and almost 99% of MB were removed by 5 and 20 mg of SPCNF, respectively. A green manufacturing of porous microspheres incorporating CNC/MnO_2 nanohybrid for effective dye removal was established by Li et al. (2021). Porous microspheres based on CNC/MnO_2 were manufactured by freeze-drying the air bubble–templated emulsion in which sodium alginate (SA) was used as the cross-linked matrix. The color removal ratio of MB (800 mg/mL) was up to 95.4% in 10 min, and the equilibrium decolorization was 114.5 mg/g. Nguyen et al. (2022) reported another study on MB removal from an aqueous environment using CNFs derived from the petioles of the nipa palm tree and graphene oxide (GO). The aerogel was stable in the aquatic environment and could be reused five times with an adsorption capacity of more than 90%. The aerogel also exhibited a fast adsorption rate for MB adsorption, absorbing more than 99% of water in less than 20 min.

The removal of MB by nanocrystalline cellulose (NCC) flakes was developed by Tan et al. (2018). NCC flakes exhibited a high adsorption capacity (188.7 mg/g fixed at 0.7 g/L adsorbent dose, 25°C and pH 6). NCC flakes were easily removed from MB-containing effluent, unlike NCC powder. The optimal adsorbent dose of NCC flakes was 0.7 g/L, which aligned well with the Langmuir isotherm. The removal of two water-soluble model dyes, i.e., MB (cationic) and MO (anionic), from water was evaluated, and the results showed that both dyes were successfully removed under the studied conditions, with a maximum of ionic dye adsorption of approximately 4.4–4.5 mg/g. A new approach based on periodate oxidation of Cladophora nanocellulose resulted in micrometer-sized 2,3-dialdehyde cellulose (DAC) beads for Congo red adsorption was established by Ruan et al. (2018). The quick and high adsorption capacity of DACeCS beads for Congo red at pH 2 and good desorption characteristics at pH 12 suggested potential applications as a suitable material in chemical waste adsorption. A biobased and robust polydopamine-coated nanocellulose/amyloid composite aerogel for crystal violet removal was developed by Sorriaux et al. (2021). The biobased aerogel displayed high adsorption efficiencies of 93.1% in 30 min.

Pure biobased composite membranes for water purification were fabricated by freeze-drying and compacting CNCs in chitosan matrix by Karim et al. (2014). The CNCs were bonded in a stable and nano-porous membrane structure with a thickness of 250–270 μm by chitosan (10 wt%), which was further stabilized by cross-linking with gluteraldehyde vapors. The membranes efficiently eliminated 98%, 84%, and 70% of positively charged dyes such as Victoria Blue 2B, Methyl Violet 2B, and Rhodamine 6G during a 24 h contact period, despite the modest water flux (64 L m^2/h). A CNF membrane with an ultrathin graphene oxide (GO) layer was coated by two successive vacuum filtering stages by Liu et al. (2019), resulting in negative charge content for dye solution filtration. The standing GO/CNF membrane surface was found at low GO concentration, and they produced multiple mass transport nanochannels to achieve the outperformed water flux of 18,123 × 574 L m^2/h bar. The retention of three dyes (Victoria Blue 2B, Methyl Violet 2B, and Rhodamine 6G) of more than 90% at 0.9 bar was attributed to the combined effects of adsorption and size exclusion. A CNF polyelectrolyte filtration membrane with superior underwater superoleophobicity and antifouling performance for dye removal was developed by Yin et al. (2020). The membrane removed positively charged dyes, including a series of MB and gentian violet (GV), with high penetration fluxes (>10,000 L m^2/h) and rejection ratios (>98%). Nanocellulose-based membranes can be used for a range of applications other than adsorption and water filtrations, such as gas separation, photocatalysis, pollution sensors, and energy devices. Gas separation is another important application of nanocellulose-based membranes. Membranes allow CO_2 to pass from the feed side to the permeate side in a selective manner (Jaffar et al., 2022). Bacterial cellulose (BC) was used as an active material for membrane fabrication of CO_2 capture by Hosakun et al. (2017). The results showed that CO_2 has a strong interaction with the BC membrane, which is further enhanced by its modification with silk fibroin and ZnO nanoparticles. According to the CO_2 permeation experiment, the basic BC membrane, silk fibroin–modified BC membrane, and ZnO nanoparticles–modified BC membrane had CO_2 permeabilities of 2.73, 2.69, and 2.66 Barrer, respectively. Sensor applications are another potential use of nanocellulose-based membranes. The recent developments of nanocellulose-based sensors including electrochemical, optical, colorimetric, fluorescent, and biosensors for the detection of various types of pollutants such as heavy metal ions, water-soluble gases, minerals, and salts were summarized by Jaffar et al. (2022) and Fang et al. (2019). According to Jaffar et al. (2022), nanocellulose contains a number of hydroxyl groups that can be used to accommodate binding sites for selective analyte species adsorption, increasing electrochemical sensor selectivity, sensitivity, and durability. The presence of a large number of carboxyl and hydroxyl groups caused the sensors' ionic conductivity to adjust and adapt to gas exposure. Furthermore, nanocellulose-based materials could offer great potential in the development of renewable energy devices such as fuel cells, solar cells, and generators to tackle the real challenges of environmental sustainability.

The discharge of polluted wastewater containing hazardous compounds into surface or groundwater is the main cause of water pollution, which mainly depends on wastewater treatment. Adsorption is a cost-effective and efficient method for removing pollutants from wastewater and is widely used (Reshmy et al., 2022). The use of green materials derived from natural sources as biosorbents for wastewater pollutant removal is attracting attention. Nanomaterials-based adsorbents have been used to improve adsorption capacity, selectivity, and regeneration (Khulbe and Matsuura, 2018). Cellulose, which can be modified into derivatives such as nanocellulose, has advantages over other materials, such as low fabrication cost (Chen et al., 2019), superior mechanical properties, high surface area and pore volume (Reshmy et al., 2021), excellent biocompatibility, and adaptable surface chemistry (Yuan et al., 2020). Moreover, nanocellulose can be easily surface-modified to create novel binding sites and provide specific features for adsorbing various types of contaminants. However, significant modifications are required to overcome the limitations of nanocellulose, such as low thermal stability and incompatibility with many nonpolar thermoplastics, which are costly, time-consuming, and risky (Bierhalz, 2021). The distribution of hydrophilic nanocellulose in hydrophobic polymer matrices is challenging and requires surface grafting with low molecular weight polymers (Jaffar et al., 2022).

Several critical challenges for future growth in integrating cellulose or its derivatives as blending agents to attain sustainable water purification exist. The conversion of cellulose to nanocellulose to meet the high standard demands from industrial and commercial sectors is a foremost challenge due to its high production cost, which may be resolved in the future with thorough findings. Moreover, the inherent nanostructure of nanocellulose may result in aggregation due to the hydrogen bonding between the nanofibers. This issue can be overcome by preparing the cellulose-based beads in the form of aerogel beads, which involves freeze-drying to limit the self-aggregation of the fibers. Future studies may also focus on introducing target chemical moieties on the surface of cellulose to prevent its susceptibility for biological degradation while offering high adsorption capacity at the same time. Biological degradation limits its applicability in wastewater treatment. The elution process would be incomplete if a proper eluent is not employed to eliminate the adsorbed material from the bioadsorbents. Furthermore, the active sites of the adsorbent may be destroyed by the use of inappropriate eluent during the desorption treatment, which substantially reduces its adsorption performance. Therefore, future research should develop a more effective and affordable eluent to achieve highly recyclable cellulose-based beads for the sequestration of contaminants from aqueous media. Another important aspect where many research studies have not been concerned is the sustainable disposal of spent cellulose beads. Since the primary objective of fabricating cellulose-based beads is to eradicate wastewater effluents, the generation of toxic secondary materials through improper dumping of exhausted bioadsorbents is unacceptable. Therefore, the development of pertinent discarding procedures is essential to prevent any sort of consequential contamination (secondary contamination) to the environment with regard to wastewater remediation at a minor or large scale. As a recommendation, future studies could seek any possibilities of applying wasted cellulose-based beads containing heavy metal ions in photocatalysis to degrade organic contaminants. This approach might pave a novel strategy to engage depleted cellulose beads as photocatalysts, swelling their possible suitability at an industrial scale. Consequently, in attaining sustainable adsorbent materials, expanding the lifespan of these materials is desired before converting them into catalysts. Future research should also meet the criteria of synthesizing biodegradable cellulose-based beads through eco-friendly methodology and impart fewer adverse effects to nature. As a result, exploring green chemicals and materials to develop cellulose-based beads is a defying task. In particular, revalorization of agricultural wastes to obtain ecological cellulose-based beads would harness a viable industry.

Two major perspectives that need attention are the production cost and the need for advanced technology and equipment. Wastewater pollution has been a significant threat to humans and the environment, especially in developing countries, where the lack of cutting-edge facilities makes the wastewater remediation process impractical. Thus, simple methodologies with low energy and inexpensive starting materials that do not involve complex infrastructure to fabricate and recycle cellulose-based beads are needed to enable countries with poverty to overcome water contamination. It is also a challenge to achieve cellulose-based beads with high regenerative capability and excellent adsorption capacity at the same time to reduce the production cost pertaining to economic feasibility. Moreover, most desorbing agents that are currently used to remove the toxins from cellulose-based beads often contaminate the clean water. Hence, great attention should be paid to exploring green desorbing solutions to carry out reusability studies. On the other hand, if the adsorbent can be reused many times with relatively high removal efficiency without any desorption process, it could be a viable option since the recycled adsorbent could be directly disposed, provided that the adsorbent itself is biodegradable. Thus, researching the regeneration aptitude using inexpensive approaches is vital from an economic point of view. Furthermore, hydrogel beads typically absorb a large amount of water due to their high swelling ratio. Consequently, a simple oven drying could substantially minimize the volume (size reduction) of the spent gel beads before the final disposal.

The effective and low-cost adsorbents fabricated from naturally derived raw materials have garnered enormous awareness to overcome the potential threat of freshwater contamination. In particular, cellulose-based gel beads, which comprise a three-dimensional polymeric network of hydrophilic groups, have established promising potential in detoxifying heavy metal ions, dyes, and other organic contaminants from wastewater. Even though synthetic polymers can be employed in synthesizing cellulose-based beads, beads prepared with natural polymeric materials, such as nanocellulosic materials, have presented great potential for the removal of a multitude of chemical species in waterbodies due to their noble properties, including biodegradability, biocompatibility, less biological toxicity, reusability, low cost, high abundance, and high adsorption capability. Furthermore, the facility to tailor surface criteria through surface functionalization paves broader possibilities in wastewater treatment application with augmented adsorption capacity. Surface tailoring also enables combining nanocellulose with other polyelectrolytes, including chitosan and alginate, among other polymers, to fabricate in the form of spherical beads with superior mechanical strength and thermal and chemical stability. The bead structure facilitates easy regeneration without a significant loss in adsorption capacity. If the beads possess high porosity, the diffusion of chemical effluent species would be fast and effective owing to the interconnected pores. In general, ideal cellulose-based beads for potential industrial wastewater remediation should possess a number of prerequisites, which are (1) cost-effective, (2) excellent adsorption capacity with a high rate of adsorption, hence greater specific surface area, (3) outstanding mechanical robustness with stable structural integrity, and (4) high reusability performance. The unique ability of cellulosic materials, especially nanocellulose-based beads, to decontaminate environmental toxins in the concentration range of microgram per liter makes them economically beneficial and environmentally complaisant. The preparation of cellulose-based beads has been conducted primarily through the dropping technology, especially with the application of the ionotropic gelation method. Interestingly, most research studies on adsorptive removal employing cellulose-based beads have been carried out predominantly to detoxify heavy metal ions and dyes in aqueous media. Very few studies have focused on the decontamination capabilities for removing pharmaceutical residues, agricultural toxins, organic pollutants, oils, and other undesired effluents. Based on previous studies, the adsorption process comprises the interaction between the effluents and the cellulose-based beads through distinct mechanisms, including electrostatic attraction, hydrogen bonding, van der Waals forces, and π-π interactions. To study the adsorption behavior, a multitude of studies has typically employed the Langmuir and Freundlich isotherm models, among various other isotherms. Adsorptive removal of industrial effluents in aqueous solutions by most of the cellulose-based beads was consistent with the Langmuir model. On the other hand, the investigation of the adsorption rate through kinetic modeling revealed that the pseudo—second-order model is the most commonly followed.

Another major advantage of cellulose-based beads is the prospect of regeneration, which pushes the limits to fully explore its applicability on an industrial scale. However, the regeneration process should be efficient without adverse effects on the adsorption performance, given the higher cost of producing and chemically functionalizing cellulose-based beads compared to conventional adsorbents. A variety of approaches have been implemented for the regeneration of beads. In most cases, an acidic, such as HCl and $-HNO_3$, or ethylenediaminetetraacetic acid treatment is commonly employed for heavy metal ion desorption. In contrast, acid or alkaline desorbing agents are used for dye elimination. Recent investigations affirmed that cellulose-based beads could be recycled at least up to four times before exhaustion. The beads should be first separated from the aqueous medium concerning regeneration.

Separation could be achieved with ease if the cellulose beads are prepared with some magnetic properties by applying a magnetic field. However, very few studies have inspected the possibility to incorporate magnetic nanoparticles, such as magnetite (Fe_3O_4) or ferrimagnetic maghemite (γ-Fe_2O_3), into cellulose beads during the fabrication.

4.5 Solar-driven water purification

The global water crisis is a pressing challenge in the 21st century, affecting the quality of life, economic prosperity, and social stability of many regions. A low-cost and environmentally friendly option for wastewater treatment is provided by biodegradable cellulose-biomass materials. Pollutants from water, especially oil, heavy metal ions, and dyes, can be effectively removed by the diversity and versatility of biomass materials. Furthermore, biomass materials are widely sourced and have excellent properties such as economical, environmentally friendly, renewable, and degradable, which have broad application prospects in wastewater treatment. However, three main problems are faced by water treatment technology now: firstly, the difficulty of scaling up current laboratory treatment processes to achieve industrialization; secondly, the high economic cost of raw materials and industrial preparation; and thirdly, the complexity of substances in real water bodies, where a single treatment technology may be influenced by the substances in the water. Therefore, some important future prospects need to be considered in the deployment of biomass materials for water treatment: sustainability and toxicity issues need to be assessed; the stability of the modified biomass adsorption material should be researched; load materials such as metal compounds, carbon nanotubes, and organic compounds, etc., which are easy to cause secondary pollution, and the problem of leaching should be addressed via leaching tests. Innovative multifunctional biomass materials to improve their efficiency need to be further researched. At the same time, the optimal biomass raw material need to be researched on. Cost-effectiveness research should be carried out. The production cost is inevitably increased by chemical modification. In future research and development, new, efficient, and environment friendly technologies such as microwave, UV, etc. can be considered to reduce the production cost of modified biomass. Efficiency issues and reuse performance assessments are usually designed on a laboratory scale to simulate the real-world exposure conditions, but it is critical to investigate and evaluate results under real-world environmental conditions.

The discharge of polluted wastewater containing dangerous compounds into surface or groundwater is the main cause of water pollution, which relies primarily on wastewater treatment. Adsorption is a cost-effective and efficient method for removing pollutants from wastewater and is widely used (Reshmy et al., 2022). The use of green materials derived from natural sources as biosorbents for wastewater pollutant removal has attracted attention.

Nanomaterials-based adsorbents have been employed to improve adsorption capacity, selectivity, and regeneration (Khulbe and Matsuura, 2018). Cellulose, which can be modified into derivatives such as nanocellulose, has advantages over other materials, such as low fabrication cost (Chen et al., 2019), superior mechanical properties, high surface area and pore volume (Reshmy et al., 2021), excellent biocompatibility, and adaptable surface chemistry (Yuan et al., 2020). Moreover, nanocellulose can be easily surface-modified to create novel binding sites and provide specialized features for adsorbing various types of contaminants. However, significant modifications are required to overcome the limitations of nanocellulose, such as low thermal stability and incompatibility with many nonpolar thermoplastics, which increase the cost, time, and risks (Bierhalz, 2021). The distribution of hydrophilic nanocellulose in hydrophobic polymer matrices is also challenging and necessitates surface grafting with low-molecular-weight polymers (Jaffar et al., 2022). Furthermore, nanoparticles cause a large pressure drop if packed in an adsorption column. Therefore, nanoparticles must be either immersed in or coated on the surface of larger particles, compromising their effectiveness (Khulbe and Matsuura, 2018). The use of large amounts of adsorbents and additional chemicals to maintain a pH that allows for adsorption are the challenges in removing heavy metals from wastewater (Renu et al., 2017). In addition, cost-effective industrial procedures for nanocellulose synthesis and scaling are essential for commercialization (Jaffar et al., 2022). Polymeric membranes for hazardous pollutant adsorption could potentially contribute to the development of next-generation reusable and portable water treatment systems, according to Khulbe and Matsuura (2018). The dual function of membrane filtration and adsorption is particularly effective in removing trace levels of contaminants such as cationic heavy metals, anionic phosphates, and nitrates. Nanocellulose-based filtration membrane has benefits such as high porosity and excellent hydrophilicity, good mechanical stability, great chemical inertness, and antifouling capability, making it useful in industrial wastewater treatment (Liu et al., 2018). Moreover, nanocellulose-based membranes are simple to manufacture and inexpensive (Yuan et al., 2020). However, one of the drawbacks of this material is that isolating cellulose components is energy-intensive and requires the use of hazardous chemicals that are harmful to humans and the environment (Tshikovhi et al., 2020; Phanthong et al., 2018). Furthermore, the technology for large-scale industrial manufacturing has been difficult to develop (Yuan et al., 2020). Therefore, novel methods capable of enhancing the current process or enabling large-scale artificial synthesis are being sought.

The elimination of the use of nonbiodegradable raw materials for aerogel production is one of the main purposes of considering cellulose and its derivatives as a raw material. This will reduce carbon footprint and secondary pollution in the environment, following the principles of "green chemistry." Nanocellulose, which has a high aspect ratio compared with other forms of cellulose, is generally selected as the preferred form of cellulose for the production of cellulose-based aerogels. This means that the rate of shrinkage of nanocellulose is low. However, the extraction of nanocellulose is tedious because of the presence of other compounds such as cellulose and hemicellulose. Chemical pretreatment such as bleaching, which generally requires the use of harmful chemicals such as chlorine dioxide, is required to produce nanocellulose. Therefore, although the use of nonbiodegradable raw material is reduced to obtain the nanocellulose material, harmful wastewater may be created. Less toxic methods such as oxygen bleaching for nanocellulose production should be explored. The toxic effect of chemicals being used in cellulose-based aerogel production on the environment should also be evaluated by studies.

The functionality of cellulose-based aerogels can be improved by combining the aerogels with different compounds, creating novel adsorbent materials. The detailed mechanisms of adsorption and the interactions between the aerogel and heavy metal ions should be explored.

The drying process is a critical step in the production of cellulose-based aerogels. Freeze-drying is the preferred method of choice for synthesizing cellulose-based aerogels according to the reviewed literature. However, complications such as aggregation due to the formation of ice crystals, which can tamper with the internal network, form cracks on the aerogel, and reduce the mechanical durability, are present in the current freeze-drying method. The standard methods of preparing aerogels such as freeze-drying and supercritical drying are expensive and dangerous. Alternative routes should be explored for aerogel production. Ambient pressure drying, which is a more economical and less dangerous aerogel production method, can be considered as an option. Additionally, very limited information on the cost—benefit analysis on using aerogel as an adsorbent for heavy metals is available.

The contact of heavy metal ions and the active sites can be improved by producing novel morphologies of the cellulose-based aerogels. The effect of the morphologies on the performance of cellulose-based aerogels is not discussed by many studies. Therefore, more studies investigating the effect of different morphologies of cellulose-based aerogels on the removal of heavy metals should be explored. Different types of reactors such as packed bed reactor, continuous flow reactor, etc., have not been used in many experiments. Additionally, real wastewater instead of synthetically prepared heavy metal wastewater should be used in experiments. Multimetal systems need to be tested to achieve similar scenarios to industrial wastewaters. The removal of arsenic using cellulose-based aerogels has not been widely explored according to the conducted reviews. Although arsenic is a common heavy metal in water, it appears that the adsorption of arsenic using cellulose aerogel is not well studied. Hence, more studies should be conducted in this area. The mechanical stability of aerogels is still not at par to industrial standards and more methods on how to make aerogels structurally strong should be explored. A possible solution could be creating hybrid aerogels by combining the cellulose-based aerogels with other organic or inorganic compounds that can increase strength and durability.

4.6 Chapter summary

A low-cost and environmentally friendly option for wastewater treatment is provided by biodegradable cellulose–biomass materials. Pollutants from water, especially oil, heavy metal ions, and dyes, can be effectively removed by the diversity and versatility of biomass materials. Moreover, biomass materials are widely sourced and have excellent properties such as economical, environmentally friendly, renewable, and degradable, which offer broad application prospects in wastewater treatment. However, three main problems are faced by the current water treatment technology: firstly, the difficulty of scaling up laboratory treatment processes to achieve industrialization; secondly, the high economic cost of raw materials and industrial preparation; and thirdly, the complexity of substances in real water bodies, where a single treatment technology may be influenced by the substances in the water. Therefore, some important future prospects need to be considered in the deployment of biomass materials for water treatment: (1) Sustainability and toxicity issues need to be assessed; the stability of the modified biomass adsorption material should be researched; load materials such as metal compounds, carbon nanotubes, and organic compounds, which are prone to cause secondary pollution, and the problem of leaching should be addressed via leaching tests. (2) Innovative multifunctional biomass materials need to be further researched to improve their efficiency. At the same time, the optimal biomass raw material need to be identified. (3) Cost-effectiveness research should be conducted. Chemical modification inevitably increases the production cost. In future research and development, new, efficient and environment-friendly technologies such as microwave, UV, etc. can be considered to reduce the production cost of modified biomass. (4) Efficiency issues and reuse performance assessments are usually designed on a laboratory scale to simulate the real-world exposure conditions, but it is critical to investigate and evaluate results under real-world environmental conditions.

The development of effective and recyclable wastewater treatment for future generations could be facilitated by the application of adsorptive membranes to harmful contaminants, due to the global environmental issues. Nanocellulose-based adsorbent has been widely used for removing various contaminants in wastewater such as dyes, heavy metals, organic compounds, and oils, because of its exceptional binding and adsorbing capacities, which can be modified by various methods to enhance their performance. The adsorption performance of contaminant removal was found to depend on optimal values of parameters such as the nature of the pollutants, the type of nanocellulose, and adsorption conditions (pH, contact time, particle size, temperature, initial concentration of pollutant, nanocellulose dosage). The thickness, pore size, water flux, and contaminants adsorption of the membranes were affected by the modification and dosage of nanocellulose for filtration process. Moreover, outstanding recyclability, mechanical performance, and acid/alkali resistance were demonstrated by the incorporation of adhesive and additive materials with nanocellulose-based membrane, enhancing multifunction membranes for treating wastewater. Novel biomaterials as dual-functional membranes for the effective removal of contaminants in wastewater were developed by the nanocellulose modification. This hybrid separation process improved the ability to remove pollutants quickly and efficiently. The nanocellulose isolation method, modification and properties as well as membrane adsorption were conducted for future work.

Dependable access to clean and safe water is hampered by the increasing population, the pollution of freshwater sources, and the effects of global warming. Therefore, environmentally friendly and sustainable solutions are needed to address this issue without imposing further stress on the water resources. Solar-driven water purification technology, which utilizes sunlight as the energy source, is a promising and green approach in this context (Zhang et al., 2019). The key component of this technology is the photothermal agent, which converts solar energy into heat for water evaporation. Various materials and device designs have been explored to enhance the energy conversion efficiency of the photothermal agents.

Acknowledgments

The financial support from National Natural Science Foundation of China (22478036) is gratefully acknowledged.

References

Abou-Zeid, R.E., Ali, K.A., Gawad, R.M.A., Kamal, K.H., Kamel, S., Khiari, R., 2021. Removal of Cu(II), Pb(II), Mg(II), and Fe(II) by adsorption onto alginate/nanocellulose beads as biosorbent. J. Renew. Mater. 9, 601–613.

Abouzeid, R.E., Khiari, R., El-Wakil, N., Dufresne, A., 2019. Current state and new trends in the use of cellulose nanomaterials for wastewater treatment. Biomacromolecules 20, 573–597.

Abu-Danso, E., Peraniemi, S., Leiviska, T., Bhatnagar, A., 2018. Synthesis of S-ligand tethered cellulose nanofibers for efficient removal of Pb(II) and Cd(II) ions from synthetic and industrial wastewater. Environ. Pollut. 242, 1988–1997.

Aguir, C., M'Henni, M.F., 2006. Experimental study on carboxymethylation of cellulose extracted from Posidonia oceanica. J. Appl. Polym. Sci. 99, 1808–1816.

Ahsan, H.M., Zhang, X., Li, Y., Li, B., Liu, S., 2019. Surface modification of microcrystalline cellulose: physicochemical characterization and applications in the Stabilization of Pickering emulsions. Int. J. Biol. Macromol. 132, 1176–1184.

Ahuja, D., Dhiman, S., Rattan, G., Monga, S., Singhal, S., Kaushik, A., 2021. Superhydrophobic modification of cellulose sponge fabricated from discarded jute bags for oil water separation. J. Environ. Chem. Eng. 9, 105063.

Akar, T., Anilan, B., Gorgulu, A., Akar, S.T., 2009. Assessment of cationic dye biosorption characteristics of untreated and non-conventional biomass: pyracantha coccinea berries. J. Hazard Mater. 168, 1302–1309.

Ali, I., Gupta, V.K., 2006. Advances in water treatment by adsorption technology. Nat. Protoc. 1, 2661–2667.

Allen, S.J., McKay, G., Porter, J.F., 2004. Adsorption isotherm models for basic dye adsorption by peat in single and binary component systems. J. Colloid Interface Sci. 280 (2), 322–333.

Almeida, A.P.C., Oliveira, J., Fernandes, S.N., Godinho, M.H., Canejo, J.P., 2020. All-cellulose composite membranes for oil microdroplet collection. Cellulose 27, 4665–4677.

Alsaiari, N.S., Katubi, K.M., Alzahrani, F.M., Amari, A., Osman, H., Ben Rebah, F., Tahoon, M.A., 2021. Synthesis, characterization and application of polypyrrole functionalized nanocellulose for the removal of Cr(VI) from aqueous solution. Polymers 13, 3691.

Angelova, D., Uzunov, I., Uzunova, S., Gigova, A., Minchev, L., 2011. Kinetics of oil and oil products adsorption by carbonized rice husks. Chem. Eng. J. 172, 306–311.

Anirudhan, T.S., Deepa, J.R., Binusreejayan, 2015. Synthesis and characterization of multi-carboxyl-functionalized nanocellulose/nanobentonite composite for the adsorption of uranium(VI) from aqueous solutions: kinetic and equilibrium profiles. Chem. Eng. J. 273, 390–400.

Anirudhan, T.S., Shainy, F., 2015. Effective removal of mercury(II) ions from chlor-alkali industrial wastewater using 2-mercaptobenzamide modified itaconic acid-grafted-magnetite nanocellulose composite. J. Colloid Interface Sci. 456, 22–31.

Anirudhan, T.S., Deepa, J.R., Christa, J., 2016. Nanocellulose/nanobentonite composite anchored with multi-carboxyl functional groups as an adsorbent for the effective removal of Cobalt(II) from nuclear industry wastewater samples. J. Colloid Interface Sci. 467, 307–320.

Annadurai, G., Juang, R.-S., Lee, D.-J., 2002. Use of cellulose-based wastes for adsorption of dyes from aqueous solutions. J. Hazard Mater. 92, 263–274.

Babarinde, A., Ogundipe, K., Sangosanya, K., Bdaoe, A.H., 2016. Comparative study on the biosorption of Pb (II), Cd (II) and Zn (II) using Lemon grass (Cymbopogon citratus): kinetics, isotherms and thermodynamics. Chem. Int. 2, 89–102.

Badshah, M., Ullah, H., Khan, A.R., Khan, S., Park, J.K., Khan, T., 2018. Surface modification and evaluation of bacterial cellulose for drug delivery. Int. J. Biol. Macromol. 113, 526–533.

Bai, L., Wu, H., Ding, J., Ding, A., Zhang, X., Ren, N., Li, G., Liang, H., 2020. Cellulose nanocrystal-blended polyethersulfone membranes for enhanced removal of natural organic matter and alleviation of membrane fouling. Chem. Eng. J. 382, 122919.

Bai, Y., Yu, M., Zhang, X.-F., Yao, J., 2023. Deep eutectic solvent assisted preparation of ZnO deposited carbonized wood for efficient CO_2 storage and oil absorption. J. Mol. Liq. 376, 121409.

Baig, U., Faizan, M., Dastageer, M.A., 2021. Polyimide based super-wettable membranes/materials for high performance oil/water mixture and emulsion separation: a review. Adv. Colloid Interface Sci. 297, 102525.

Bakhsh, E.M., Khan, S.A., Marwani, H.M., Danish, E.Y., Asiri, A.M., Khan, S.B., 2018. Performance of cellulose acetate-ferric oxide nanocomposite supported metal catalysts toward the reduction of environmental pollutants. Int. J. Biol. Macromol. 107, 668–677.

Barsbay, M., Guven, O., 2019. Surface modification of cellulose via conventional and controlled radiation-induced grafting. Radiat. Phys. Chem. 160, 1–8.

Batmaz, R., Mohammed, N., Zaman, M., Minhas, G., Berry, R.M., Tam, K.C., 2014. Cellulose nanocrystals as promising adsorbents for the removal of cationic dyes. Cellulose 21, 1655–1665.

Batool, A., Valiyaveettil, S., 2021. Surface functionalized cellulose fibers - a renewable adsorbent for removal of plastic nanoparticles from water. J. Hazard Mater. 413, 125301.

Bayat, A., Aghamiri, S.F., Moheb, A., Vakili-Nezhaad, G.R., 2005. Oil spill cleanup from sea water by sorbent materials. Chem. Eng. Technol. Industrial Chem. Plant Equipment Process Eng. Biotechnol. 28, 1525–1528.

Beh, J.H., Lim, T.H., Lew, J.H., Lai, J.C., 2020. Cellulose nanofibril-based aerogel derived from sago pith waste and its application on methylene blue removal. Int. J. Biol. Macromol. 160, 836–845.

Bethke, K., Palantoeken, S., Andrei, V., Ross, M., Raghuwanshi, V.S., Kettemann, F., Greis, K., Ingber, T.T.K., Stueckrath, J.B., Valiyaveettil, S., Rademann, K., 2018. Functionalized cellulose for water purification, antimicrobial applications, and sensors. Adv. Funct. Mater. 28, 1800409.

Beyki, M.H., Bayat, M., Shemirani, F., 2016. Fabrication of core–shell structured magnetic nanocellulose base polymeric ionic liquid for effective biosorption of Congo red dye. Bioresour. Technol. 218, 326–334.

Bhatnagar, A., Sillanpää, M., 2010. Utilization of agro-industrial and municipal waste materials as potential adsorbents for water treatment-A review. Chem. Eng. J. 157, 277–296.

Bhatnagar, A., Sillanpaa, M., Witek-Krowiak, A., 2015. Agricultural waste peels as versatile biomass for water purification - a review. Chem. Eng. J. 270, 244–271.

Bierhalz, A.C.K., 2021. Cellulose nanomaterials in textile applications. Cellul. Chem. Technol. 55, 725–741.

Biliuta, G., Coseri, S., 2019. Cellulose: a ubiquitous platform for ecofriendly metal nanoparticles preparation. Coord. Chem. Rev. 383, 155–173.

Budtova, T., 2019. Cellulose II aerogels: a review. Cellulose 26 (1), 81–121.

Cai, T., Yang, Z., Li, H., Yang, H., Li, A., Cheng, R., 2013. Effect of hydrolysis degree of hydrolyzed polyacrylamide grafted carboxymethyl cellulose on dye removal efficiency. Cellulose 20 (5), 2605–2614.

Cai, T., Li, H.J., Yang, R., Wang, Y.W., Li, R.H., Yang, H., Li, A.M., Cheng, R.S., 2015. Efficient flocculation of an anionic dye from aqueous solutions using a cellulose-based flocculant. Cellulose 22, 1439–1449.

Cai, Y., Yuan, F., Wang, X., Sun, Z., Chen, Y., Liu, Z., Wang, X., Yang, S., Wang, S., 2017. Synthesis of core–shell structured Fe_3O_4@carboxymethyl cellulose magnetic composite for highly efficient removal of Eu(III). Cellulose 24, 175–190.

Cao, Y., Hua, H., Yang, P., Chen, M., Chen, W., Wang, S., Zhou, X., 2020a. Investigation into the reaction mechanism underlying the atmospheric low-temperature plasma-induced oxidation of cellulose. Carbohydr. Polym. 233, 115632.

Cao, G., Li, Q., Wang, R., Wang, C., Duan, X., Ho, S.-H., 2020b. Simultaneous separation of multiphase emulsion mixture and catalytic degradation of BPA via microalgae residue membranes. Chem. Eng. J. 393, 124750.

Carpenter, A.W., de Lannoy, C.-F., Wiesner, M.R., 2015. Cellulose nanomaterials in water treatment technologies. Environ. Sci. Technol. 49, 5277–5287.

Chen, S.Y., Zou, Y., Yan, Z.Y., Shen, W., Shi, S.K., Zhang, X., Wang, H.P., 2009. Carboxymethylated-bacterial cellulose for copper and lead ion removal. J. Hazard Mater. 161, 1355–1359.

Chen, W., Yu, H., Liu, Y., Chen, P., Zhang, M., Hai, Y., 2011. Individualization of cellulose nanofibers from wood using high-intensity ultrasonication combined with chemical pretreatments. Carbohydr. Polym. 83, 1804–1811.

Chen, W., Chen, F., Zhang, G., Liu, X., Kong, S., Cai, W., Wang, J., Du, L., Wu, C., 2019. Fabrication of cellulose nanocrystal composite filter papers for rapid and highly efficient removal of bacteria from aqueous solutions. Cellulose 26, 7027–7035.

Chequer, F.M.D., Angeli, J.P.F., Ferraz, E.R.A., Tsuboy, M.S., Marcarini, J.C., Mantovani, M.S., de Oliveira, D.P., 2009. The azo dyes disperse red 1 and disperse orange 1 increase the micronuclei frequencies in human lymphocytes and in HepG2 cells. Mutation Res. Genet. Toxicol. Environ. Mutagenesis 676, 83–86.

Chitpong, N., Husson, S.M., 2016. Nanofiber ion-exchange membranes for the rapid uptake and recovery of heavy metals from water. Membranes 6, 59.

Choi, H.Y., Bae, J.H., Hasegawa, Y., An, S., Kim, I.S., Lee, H., Kim, M., 2020. Thiol-functionalized cellulose nanofiber membranes for the effective adsorption of heavy metal ions in water. Carbohydr. Polym. 234, 115881.

Chu, Z., Li, Y., Zhou, A., Zhang, L., Zhang, X., Yang, Y., Yang, Z., 2022. Polydimethylsiloxane-decorated magnetic cellulose nanofiber composite for highly efficient oil-water separation. Carbohydr. Polym. 277, 118787.

Dahlan, N.A., Veeramachineni, A.K., Langford, S.J., Pushpamalar, J., 2017. Developing of a magnetite film of carboxymethyl cellulose grafted carboxymethyl polyvinyl alcohol (CMC-g-CMPVA) for copper removal. Carbohydr. Polym. 173, 619–630.

Dawood, S., Sen, T., 2014. Review on dye removal from its aqueous solution into alternative cost effective and non-conventional adsorbents. J. Chem. Proc. Eng. 1, 1–11.

de Andrade, J.R., Oliveira, M.F., da Silva, M.G.C., Vieira, M.G.A., 2018. Adsorption of pharmaceuticals from water and wastewater using nonconventional low-cost materials: a review. Ind. Eng. Chem. Res. 57, 3103–3127.

DeMessie, B., 2015. Sustainable water cleaning system for point-of-use household application in developing countries to remove contaminants from drinking water. In: Water Challenges and Solutions on a Global Scale. ACS Publications, pp. 285–317.

Deng, Y., Peng, C., Dai, M., Lin, D., Ali, I., Alhewairini, S.S., Zheng, X., Chen, G., Li, J., Naz, I., 2020. Recent development of super-wettable materials and their applications in oil-water separation. J. Cleaner Prod. 266, 121624.

Dhali, K., Ghasemlou, M., Daver, F., Cass, P., Adhikari, B., 2021. A review of nanocellulose as a new material towards environmental sustainability. Sci. Total Environ. 775, 145871.

Dilamian, M., Noroozi, B., 2021. Rice straw agri-waste for water pollutant adsorption: relevant mesoporous super hydrophobic cellulose aerogel. Carbohydr. Polym. 251, 117016.

Dong, Y., Tang, D., Li, C., 2014. Photocatalytic oxidation of methyl orange in water phase by immobilized TiO_2-carbon nanotube nanocomposite photocatalyst. Appl. Surf. Sci. 296, 1–7.

Dong, R.E., Kang, P., Xu, X.L., Cai, L.X., Guo, Z., 2020. Cation-exchange strategy for a colorimetric paper sensor: belt-like ZnSe nanoframes toward visual determination of heavy metal ions. Sens. Actuat. B Chem. 312, 128013.

Doshi, B., Sillanpaa, M., Kalliola, S., 2018. A review of bio-based materials for oil spill treatment. Water Res. 135, 262–277.

Dufresne, A., 2013. Nanocellulose: a new ageless bionanomaterial. Mater. Today 16, 220–227.

Duolikun, T., Ghazali, N., Leo, B.F., Lee, H.V., Lai, C.W., Bin Johan, M.R., 2020. Asymmetric cellulosic membranes: current and future aspects. Symmetry 12, 1160.

Dutt, M.A., Hanif, M.A., Nadeem, F., Bhatti, H.N., 2020. A review of advances in engineered composite materials popular for wastewater treatment. J. Environ. Chem. Eng. 8, 104073.

Eyley, S., Thielemans, W., 2011. Imidazolium grafted cellulose nanocrystals for ion exchange applications. Chem. Commun. 47, 4177–4179.

Fang, Z., Hou, G., Chen, C., Hu, L., 2019. Nanocellulose-based films and their emerging applications. Curr. Opin. Solid State Mater. Sci. 23, 100764.

Fang, Y., Jing, C., Li, G., Ling, S., Wang, Z., Lu, P., Li, Q., Dai, C., Gao, S., Chen, B., 2021. Wood-derived systems for sustainable oil/water separation. Adv. Sustain. Syst. 5, 2100039.

Foo, K.Y., Hameed, B.H., 2010. Insights into the modeling of adsorption isotherm systems. Chem. Eng. J. 156, 2–10.

Foresti, M.L., Vázquez, A., Boury, B., 2017. Applications of bacterial cellulose as precursor of carbon and composites with metal oxide, metal sulfide and metal nanoparticles: a review of recent advances. Carbohydr. Polym. 157, 447–467.

Freundlich, H., 1906. Over the adsorption in solution. J. Phys. chem 57, 1100–1107.

Fu, Y., Guo, Z., 2022. Natural polysaccharide-based aerogels and their applications in oil-water separations: a review. J. Mater. Chem. A 10, 8129–8158.

Fu, F.Y., Li, L.Y., Liu, L.J., Cai, J., Zhang, Y.P., Zhou, J.P., Zhang, L.N., 2015. Construction of cellulose based ZnO nanocomposite films with antibacterial properties through one-step coagulation. ACS Appl. Mater. Interfaces 7, 2597–2606.

Ghorai, S., Sarkar, A., Raoufi, M., Panda, A.B., Schoenherr, H., Pal, S., 2014. Enhanced removal of methylene blue and methyl violet dyes from aqueous solution using a nanocomposite of hydrolyzed polyacrylamide grafted xanthan gum and incorporated nanosilica. ACS Appl. Mater. Interfaces 6, 4766–4777.

Gopakumar, D.A., Pasquini, D., Henrique, M.A., De Morais, L.C., Grohens, Y., Thomas, S., 2017. Meldrum's acid modified cellulose nanofiber-based polyvinylidene fluoride microfiltration membrane for dye water treatment and nanoparticle removal. ACS Sustain. Chem. Eng. 5, 2026–2033.

Gopakumar, D.A., Manna, S., Pasquini, D., Thomas, S., Grohens, Y., 2018. 19 - nanocellulose: Extraction and application as a sustainable material for wastewater purification. In: Hussain, C.M., Mishra, A.K. (Eds.), New Polymer Nanocomposites for Environmental Remediation. Elsevier, pp. 469–486.

Goswami, M., Das, A.M., 2019. Synthesis and characterization of a biodegradable Cellulose acetate-montmorillonite composite for effective adsorption of Eosin Y. Carbohydr. Polym. 206, 863–872.

Guo, Q., Ma, Q., Xue, Z., Gao, X., Chen, H., 2018. Studies on the binding characteristics of three polysaccharides with different molecular weight and flavonoids from corn silk (Maydis stigma). Carbohydr. Polym. 198, 581–588.

Gupta, V., 2009. Application of low-cost adsorbents for dye removal—a review. J. Environ. Manag. 90 (8), 2313–2342.

Gupta, V.K., Nayak, A., Agarwal, S., 2015. Bioadsorbents for remediation of heavy metals: current status and their future prospects. Environ. Eng. Res. 20, 1–18.

Habibi, Y., 2014. Key advances in the chemical modification of nanocelluloses. Chem. Soc. Rev. 43, 1519–1542.

He, X.Y., Male, K.B., Nesterenko, P.N., Brabazon, D., Paull, B., Luong, J.H.T., 2013. Adsorption and desorption of methylene blue on porous carbon monoliths and nanocrystalline cellulose. ACS Appl. Mater. Interfaces 5, 8796–8804.

Hethnawi, A., Nassar, N.N., Manasrah, A.D., Vitale, G., 2017. Polyethylenimine-functionalized pyroxene nanoparticles embedded on Diatomite for adsorptive removal of dye from textile wastewater in a fixedbed column. Chem. Eng. J. 320, 389–404.

Hokkanen, S., Repo, E., Suopajarvi, T., Liimatainen, H., Niinimaa, J., Sillanpaa, M., 2014. Adsorption of Ni(II), Cu(II) and Cd(II) from aqueous solutions by amino modified nano-structured microfibrillated cellulose. Cellulose 21, 1471–1487.

Hokkanen, S., Bhatnagar, A., Sillanpaa, M., 2016. A review on modification methods to cellulose-based adsorbents to improve adsorption capacity. Water Res. 91, 156–173.

Hosakun, Y., Halász, K., Horváth, M., Csóka, L., Djoković, V., 2017. ATR-FTIR study of the interaction of CO_2 with bacterial cellulose-based membranes. Chem. Eng. J. 324, 83–92.

Iakovleva, E., Sillanpää, M., 2020. Chapter 4 - Novel sorbents from low-cost materials for water treatment. In: Sillanpää, M. (Ed.), Advanced Water Treatment. Elsevier, pp. 265–359.

Islam, M.S., Ahmed, M.K., Raknuzzaman, M., Habibullah-Al-Mamun, M., Islam, M.K., 2015. Heavy metal pollution in surface water and sediment: a preliminary assessment of an urban river in a developing country. Ecol. Indicat. 48, 282–291.

Jaffar, S.S., Saallah, S., Misson, M., Siddiquee, S., Roslan, J., Saalah, S., Lenggoro, W., 2022. Recent development and environmental applications of nanocellulose-based membranes. Membranes 12, 287.

Jamshaid, A., Hamid, A., Muhammad, N., Naseer, A., Ghauri, M., Iqbal, J., Rafiq, S., Shah, N.S., 2017. Cellulose-based materials for the removal of heavy metals from wastewater - an overview. ChemBioEng Rev. 4, 240–256.

Jiang, F., Hsieh, Y.L., 2014. Amphiphilic superabsorbent cellulose nanofibril aerogels. J. Mater. Chem. A 2, 6337–6342.

Jin, H.M., Capareda, S., Chang, Z.Z., Gao, J., Xu, Y.D., Zhang, J.Y., 2014. Biochar pyrolytically produced from municipal solid wastes for aqueous As(V) removal: adsorption property and its improvement with KOH activation. Bioresour. Technol. 169, 622–629.

Jin, L.Q., Li, W.G., Xu, Q.H., Sun, Q.C., 2015. Amino-functionalized nanocrystalline cellulose as an adsorbent for anionic dyes. Cellulose 22, 2443–2456.

Jin, X., Xiang, Z., Liu, Q., Chen, Y., Lu, F., 2017. Polyethyleneimine-bacterial cellulose bio-adsorbent for effective removal of copper and lead ions from aqueous solution. Bioresour. Technol. 244, 844–849.

Jing, Z., Ding, J., Zhang, T., Yang, D., Qiu, F., Chen, Q., Xu, J., 2019. Flexible, versatility and superhydrophobic biomass carbon aerogels derived from corn bracts for efficient oil/water separation. Food Bioprod. Process. 115, 134–142.

Junaidi, N.F.D., Othman, N.H., Fuzil, N.S., Shayuti, M.S.M., Alias, N.H., Shahruddin, M.Z., Marpani, F., Lau, W.J., Ismail, A.F., Aba, N.D., 2021. Recent development of graphene oxide-based membranes for oil-water separation: a review. Sep. Purif. Technol. 258, 118000.

Kant, R., 2011. Textile dyeing industry an environmental hazard. Nat. Sci 2, 22–26.

Kardam, A., Raj, K.R., Srivastava, S., Srivastava, M.M., 2014. Nanocellulose fibers for biosorption of cadmium, nickel, and lead ions from aqueous solution. Clean Technol. Environ. Policy 16, 385–393.

Karim, Z., Mathew, A.P., Grahn, M., Mouzon, J., Oksman, K., 2014. Nanoporous membranes with cellulose nanocrystals as functional entity in chitosan: removal of dyes from water. Carbohydr. Polym. 112, 668–676.

Karim, Z., Claudpierre, S., Grahn, M., Oksman, K., Mathew, A.P., 2016. Nanocellulose based functional membranes for water cleaning: tailoring of mechanical properties, porosity and metal ion capture. J. Membr. Sci. 514, 418–428.

Kedzior, S.A., Zoppe, J.O., Berry, R.M., Cranston, E.D., 2019. Recent advances and an industrial perspective of cellulose nanocrystal functionalization through polymer grafting. Curr. Opin. Solid State Mater. Sci. 23, 74–91.

Khulbe, K.C., Matsuura, T., 2018. Removal of heavy metals and pollutants by membrane adsorption techniques. Appl. Water Sc. 8 (1), 19.

Kong, Q.P., Preis, S., Li, L.L., Luo, P., Hu, Y., Wei, C.H., 2020. Graphene oxide-terminated hyperbranched amino polymer-carboxymethyl cellulose ternary nanocomposite for efficient removal of heavy metals from aqueous solutions. Int. J. Biol. Macromol. 149, 581–592.

Korhonen, J.T., Kettunen, M., Ras, R.H., Ikkala, O., 2011. Hydrophobic nanocellulose aerogels as floating, sustainable, reusable, and recyclable oil absorbents. ACS Appl. Mater. Interfaces 3, 1813–1816.

Kose, K., Mavlan, M., Youngblood, J.P., 2020. Applications and impact of nanocellulose based adsorbents. Cellulose 27, 2967–2990.

Kumar, N., Kardam, A., Rajawat, D.S., Jain, V.K., 2019. Suman, Carboxymethyl nanocellulose stabilized nano zero-valent iron: an effective method for reduction of hexavalent chromium in wastewater. Mater. Res. Express 6, 115013.

Langmuir, I., 1918. The adsorption of gases on plane surfaces of glass, mica and platinum. J. Am. Chem. Soc. 40, 1361–1403.

Lavoine, N., Desloges, I., Dufresne, A., Bras, J., 2012. Microfibrillated cellulose - its barrier properties and applications in cellulosic materials: a review. Carbohydr. Polym. 90, 735–764.

Leshaf, A., Ziani Cherif, H., Benmansour, K., 2019. Adsorption of acidol red 2BE-NW dye from aqueous solutions on carboxymethyl cellulose/organo-bentonite composite: characterization, kinetic and thermodynamic studies. J. Polym. Environ. 27, 1054–1064.

Li, W.Y., Jin, A.X., Liu, C.F., Sun, R.C., Zhang, A.P., Kennedy, J.F., 2009. Homogeneous modification of cellulose with succinic anhydride in ionic liquid using 4-dimethylaminopyridine as a catalyst. Carbohydr. Polym. 78, 389–395.

Li, D., Zhu, F.Z., Li, J.Y., Na, P., Wang, N., 2013. Preparation and characterization of cellulose fibers from corn straw as natural oil sorbents. Ind. Eng. Chem. Res. 52, 516–524.

Li, L., Li, B., Sun, H., Zhang, J., 2017a. Compressible and conductive carbon aerogels from waste paper with exceptional performance for oil/water separation. J. Mater. Chem. A 5, 14858–14864.

Li, L., Hu, T., Sun, H., Zhang, J., Wang, A., 2017b. Pressure-sensitive and conductive carbonaerogels from poplars catkins for selective oil absorption and oil/water separation. ACS Appl. Mater. Interfaces 9, 37509.

Li, W.X., Ju, B.Z., Zhang, S.F., 2019. A green l-cysteine modified cellulose nanocrystals biosorbent for adsorption of mercury ions from aqueous solutions. RSC Adv. 9, 6986–6994.

Li, L., Shi, W., Yu, S., 2020. Research on forward osmosis membrane technology still needs improvement in water recovery and wastewater treatment. Water 12, 1–27.

Li, M., He, B., Chen, Y., Zhao, L., 2021. Physicochemical properties of nanocellulose isolated from cotton stalk waste. ACS Omega 6, 25162–25169.

Li, X., Gao, L., Wang, M., Lv, D., He, P., Xie, Y., Zhan, X., Li, J., Lin, Z., 2023. Recent development and emerging applications of robust biomimetic superhydrophobic wood. J. Mater. Chem. A 11, 6772–6795.

Lin, C., Qiao, S., Luo, W., Liu, Y., Liu, D., Li, X., Liu, M., 2014. Thermodynamics, kinetics, and regeneration studies for adsorption of Cr(VI) from aqueous solutions using modified cellulose as adsorbent. Bioresources 9, 6998–7017.

Liu, P., Borrell, P.F., Bozic, M., Kokol, V., Oksman, K., Mathew, A.P., 2015a. Nanocelluloses and their phosphorylated derivatives for selective adsorption of Ag^{+}, Cu^{2+} and Fe^{3+} from industrial effluents. J. Hazard Mater. 294, 177–185.

Liu, L., Gao, Z.Y., Su, X.P., Chen, X., Jiang, L., Yao, J.M., 2015b. Adsorption removal of dyes from single and binary solutions using a cellulose-based bioadsorbent. ACS Sustain. Chem. Eng. 3, 432–442.

Liu, Y., Liu, A., Ibrahim, S.A., Yang, H., Huang, W., 2018. Isolation and characterization of microcrystalline cellulose from pomelo peel. Int. J. Biol. Macromol. 111, 717–721.

Liu, P., Zhu, C., Mathew, A.P., 2019. Mechanically robust high flux graphene oxide - nanocellulose membranes for dye removal from water. J. Hazard Mater. 371, 484–493.

Liu, Y., Liu, H., Shen, Z., 2021a. Nanocellulose based filtration membrane in industrial waste water treatment: a review. Materials 14, 5398.

Liu, Y., Qiao, L., Wang, A., Li, Y., Zhao, L., Du, K., 2021b. Tentacle-type poly(hydroxamic acid)-modified macroporous cellulose beads: synthesis, characterization, and application for heavy metal ions adsorption. J. Chromatogr. A 1645, 462098.

Liu, Y., Huang, Y., Huang, Q., Li, F., Liu, X., 2021c. Liquid-phase deposition functionalized wood sponges for oil/water separation. J. Mater. Sci. 56, 19075–19092.

Long, L.Y., Weng, Y.X., Wang, Y.Z., 2018. Cellulose aerogels: synthesis, applications, and prospects. Polymers 10, 623.

Long, S., Feng, Y., He, F., He, S., Hong, H., Yang, X., Zheng, L., Liu, J., Gan, L., Long, M., 2020. An ultralight, supercompressible, superhydrophobic and multifunctional carbon aerogel with a specially designed structure. Carbon 158, 137–145.

Lujan-Facundo, M.J., Iborra-Clar, M.I., Mendoza-Roca, J.A., Alcaina-Miranda, M.I., 2019. Pharmaceutical compounds removal by adsorption with commercial and reused carbon coming from a drinking water treatment plant. J. Cleaner Prod. 238, 117866.

Luo, M.K., Lin, H., Li, B., Dong, Y.B., He, Y.H., Wang, L., 2018. A novel modification of lignin on corncob-based biochar to enhance removal of cadmium from water. Bioresour. Technol. 259, 312–318.

Lv, Y., Li, Q., Hou, Y., Wang, B., Zhang, T., 2019. Facile preparation of an asymmetric wettability janus cellulose membrane for switchable emulsions' separation and antibacterial property. ACS Sustain. Chem. Eng. 7, 15002.

Ma, H., Hsiao, B.S., Chu, B., 2012a. Ultrafine cellulose nanofibers as efficient adsorbents for removal of UO_2^{2+} in water. ACS Macro Lett. 1, 213–216.

Ma, H., Burger, C., Hsiao, B.S., Chu, B., 2012b. Nanofibrous microfiltration membrane based on cellulose nanowhiskers. Biomacromolecules 13, 180–186.

Maatar, W., Boufi, S., 2015. Poly(methacylic acid-co-maleic acid) grafted nanofibrillated cellulose as a reusable novel heavy metal ions adsorbent. Carbohydr. Polym. 126, 199–207.

Mahfoudhi, N., Boufi, S., 2017. Nanocellulose as a novel nanostructured adsorbent for environmental remediation: a review. Cellulose 24, 1171–1197.

Mahmoodi, N.M., Oveisi, M., Taghizadeh, A., Taghizadeh, M., 2019. Novel magnetic amine functionalized carbon nanotube/metal-organic framework nanocomposites: from green ultrasound-assisted synthesis to detailed selective pollutant removal modelling from binary systems. J. Hazard Mater. 368, 746–759.

Manna, S., Roy, D., Saha, P., Gopakumar, D., Thomas, S., 2017. Rapid methylene blue adsorption using modified lignocellulosic materials. Process Saf. Environ. Prot. 107, 346–356.

Marghaki, N.S., Jonoush, Z.A., Rezaee, A., 2022. Chromium (VI) removal using microbial cellulose/nano-Fe_3O_4@polypyrrole: isotherm, kinetic and thermodynamic studies. Mater. Chem. Phys. 278, 125696.

Martins, N.C., Freire, C.S., Neto, C.P., Silvestre, A.J., Causio, J., Baldi, G., Sadocco, P., Trindade, T., 2013. Antibacterial paper based on composite coatings of nanofibrillated cellulose and ZnO. Coll. Surf. A Physicochem. Eng. Aspects 417, 111–119.

Mateo, S., Peinado, S., Morillas-Gutierrez, F., La Rubia, M.D., Moya, A.J., 2021. Nanocellulose from agricultural wastes: products and applications-A Review. Processes 9, 1594.

Mautner, A., 2020. Nanocellulose water treatment membranes and filters: a review. Polym. Int. 69, 741–751.

Mellah, A., Chegrouche, S., Barkat, M., 2006. The removal of uranium (VI) from aqueous solutions onto activated carbon: kinetic and thermodynamic investigations. J. Colloid Interface Sci. 296, 434–441.

Melo, B.C., Paulino, F.A.A., Cardoso, V.A., Pereira, A.G.B., Fajardo, A.R., Rodrigues, F.H.A., 2018. Cellulose nanowhiskers improve the methylene blue adsorption capacity of chitosan-g-poly(acrylic acid) hydrogel. Carbohydr. Polym. 181, 358–367.

Minh Trang, H., Tien Duc, P., Verheyen, D., Manh Khai, N., Thi Thuy, P., Zhu, J., Van der Bruggen, B., 2020. Fabrication of thin film nanocomposite nanofiltration membrane incorporated with cellulose nanocrystals for removal of Cu(II) and Pb(II). Chem. Eng. Sci. 228, 115998.

Mishra, R.K., Sabu, A., Tiwari, S.K., 2018. Materials chemistry and the futurist eco-friendly applications of nanocellulose: status and prospect. J. Saudi Chem. Soc. 22, 949–978.

Mo, L.T., Pang, H.W., Lu, Y.T., Li, Z., Kang, H.J., Wang, M.G., Zhang, S.F., Li, J.Z., 2021. Wood-inspired nanocellulose aerogel adsorbents with excellent selective pollutants capture, superfast adsorption, and easy regeneration. J. Hazard Mater. 415, 125612.

Mohammed, N., Grishkewich, N., Berry, R.M., Tam, K.C., 2015. Cellulose nanocrystal–alginate hydrogel beads as novel adsorbents for organic dyes in aqueous solutions. Cellulose 22, 3725–3738.

Mohammed, N., Grishkewich, N., Tam, K.C., 2018. Cellulose nanomaterials: promising sustainable nanomaterials for application in water/wastewater treatment processes. Environ. Sci. Nano 5, 623–658.

Moon, R.J., Martini, A., Nairn, J., Simonsen, J., Youngblood, J., 2011. Cellulose nanomaterials review: structure, properties and nanocomposites. Chem. Soc. Rev. 40, 3941–3994.

Mu, B., Liu, L., Li, W., Yang, Y., 2019. A water/cottonseed oil bath with controllable dye sorption for high dyeing quality and minimum discharges. J. Cleaner Prod. 236, 117566.

Mullin, J.V., Champ, M.A., 2003. Introduction/overview to in situ burning of oil spills. Elsevier, Spill Sci. Technol. Bullet. 323–330.

Mulyadi, A., Zhang, Z., Deng, Y., 2016. Fluorine-free oil absorbents made from cellulose nanofibril aerogels. ACS Appl. Mater. Interfaces 8 (4), 2732–2740.

Munirasu, S., Haija, M.A., Banat, F., 2016. Use of membrane technology for oil field and refinery produced water treatment-a review. Process Saf. Environ. Protect. 100, 183–202.

Muthu, S.S., Khadir, A., 2022. Membrane Based Methods for Dye Containing Wastewater. Springer.

Nasrollahzadeh, M., Sajjadi, M., Maham, M., Sajadi, S.M., Barzinjy, A.A., 2018. Biosynthesis of the palladium/sodium borosilicate nanocomposite using Euphorbia milli extract and evaluation of its catalytic activity in the reduction of chromium(VI), nitro compounds and organic dyes. Mater. Res. Bull. 102, 24–35.

Nguyen, V.T., Ha, L.Q., Nguyen, T.D.L., Ly, P.H., Dang Mao, N., Hoang, D., 2022. Nanocellulose and graphene oxide aerogels for adsorption and removal methylene blue from an aqueous environment. ACS Omega 7 (1), 1003–1013.

Nkalane, A., Oyewo, O.A., Leswifi, T., Onyango, M.S., 2019. Application of coagulant obtained through charge reversal of sawdust-derived cellulose nanocrystals in the enhancement of water turbidity removal. Mater. Res. Expr. 6, 105060.

Noor, M.H.M., Azli, M., Ngadi, N., Inuwa, I.M., Opotu, L.A., Mohameda, M., 2022. Optimization of sonication-assisted synthesis of magnetic Moringa oleifera as an efficient coagulant for palm oil wastewater treatment, Environm. Technol. Innovation 25, 102191.

Noreen, S., Khalid, U., Ibrahim, S.M., Javed, T., Ghani, A., Naz, S., Iqbal, M., 2020. ZnO, MgO and FeO adsorption efficiencies for direct sky Blue dye: equilibrium, kinetics and thermodynamics studies. J. Mater. Res. Technol. 9, 5881–5893.

Nqombolo, A., Munonde, T.S., Makhetha, T.A., Moutloali, R.M., Nomngongo, P.N., 2021. Cobalt/zinc based metal organic frameworks as an effective adsorbent for improved removal of As(V) and Cr(VI) in a wide pH range. J. Mater. Res. Technol. 12, 1845–1855.

O'Connell, D.W., Birkinshaw, C., O'Dwyer, T.F., 2008. Heavy metal adsorbents prepared from the modification of cellulose: a review. Bioresour. Technol. 99, 6709–6724.

Olivera, S., Muralidhara, H.B., Venkatesh, K., Guna, V.K., Gopalakrishna, K., Kumar, K.Y., 2016. Potential applications of cellulose and chitosan nanoparticles/composites in wastewater treatment: a review. Carbohydr. Polym. 153, 600–618.

Ozacar, M., Sengil, I.A., 2005. A kinetic study of metal complex dye sorption onto pine sawdust. Process Biochem. 40, 565–572.

Palliparambil, A.A., Bethireddy, S., Reddy, P.H., Rajulapati, S.B.J.I.J.O.S., 2016. Technology, removal of nickel by adsorption from aqueous solution by cellulose acetate-zero valent iron nanoparticle mixed matrix membrane. Indian J. Sci. Technol. 9, 1–6.

Pan, Z., Cao, S., Li, J., Du, Z., Cheng, F., 2019. Anti-fouling TiO_2 nanowires membrane for oil/water separation: synergetic effects of wettability and pore size. J. Membr. Sci. 572, 596–606.

Pang, J.B., Fu, F.L., Li, W.B., Zhu, L.J., Tang, B., 2019. Fe-Mn binary oxide decorated diatomite for rapid decolorization of methylene blue with H_2O_2. Appl. Surf. Sci. 478, 54–61.

Pang, Y., Yu, Z., Chen, H., Xiang, Q., Wang, Q., Xie, C., Liu, Y., 2022. Superhydrophobic polyurethane sponge based on sepiolite for efficient oil/water separation. J. Hazard Mater. 434, 128833.

Paximada, P., Dimitrakopoulou, E.A., Tsouko, E., Koutinas, A.A., Fasseas, C., Mandala, I.G., 2016. Structural modification of bacterial cellulose fibrils under ultrasonic irradiation. Carbohydr. Polym. 150, 5–12.

Pei, A.H., Butchosa, N., Berglund, L.A., Zhou, Q., 2013. Surface quaternized cellulose nanofibrils with high water absorbency and adsorption capacity for anionic dyes. Soft Matter 9, 2047–2055.

Peng, B.L., Yao, Z.L., Wang, X.C., Crombeen, M., Sweeney, D.G., Tam, K.C., 2020. Cellulose-based materials in wastewater treatment of petroleum industry. Green Energy Environ. 5, 37–49.

Phanthong, P., Reubroycharoen, P., Hao, X., Xu, G., Abudula, A., Guan, G., 2018. Nanocellulose: extraction and application. Carbon Res. Convers. 1, 32–43.

Pourfadakari, S., Jorfi, S., Ahmadi, M., Takdastan, A., 2017. Experimental data on adsorption of Cr(VI) from aqueous solution using nanosized cellulose fibers obtained from rice husk. Data Brief 15, 887–895.

Putro, J.N., Kurniawan, A., Ismadji, S., Ju, Y.-H., 2017. Nanocellulose based biosorbents for wastewater treatment: study of isotherm, kinetic, thermodynamic and reusability. Environ. Nanotechnol. Monit. Manag. 8, 134–149.

Qalyoubi, L., Al-Othman, A., Al-Asheh, S., 2021. Recent progress and challenges on adsorptive membranes for the removal of pollutants from wastewater. Part I: fundamentals and classification of membranes. Case Stud. Chem. Environ. Eng. 3, 100086.

Qasem, N.A.A., Mohammed, R.H., Lawal, D.U., 2021. Removal of heavy metal ions from wastewater: a comprehensive and critical review. NPJ Clean Water 4, 36.

Qiao, A., Cui, M., Huang, R., Ding, G., Qi, W., He, Z., Klemes, J.J., Su, R., 2021. Advances in nanocellulose-based materials as adsorbents of heavy metals and dyes. Carbohydr. Polym. 272, 118471.

Qin, F., Fang, Z., Zhou, J., Sun, C., Chen, K., Ding, Z., Li, G., Qiu, X., 2019. Efficient Removal of Cu^{2+} in water by carboxymethylated cellulose nanofibrils: performance and mechanism. Biomacromolecules 20, 4466−4475.

Qiu, K., Netravali, A.N., 2014. A review of fabrication and applications of bacterial cellulose based nanocomposites. Polym. Rev. 54, 598−626.

Qiu, S.J., Wang, X., Zhang, Q.R., Nie, G.Z., 2022. Development of MoS_2/cellulose aerogels nanocomposite with superior application capability for selective lead(II) capture. Sep. Purif. Technol. 284, 120275.

Quo, F., Hornsby, P., Sharma, H., Lyons, G., 2011. Preparation and characterisation of cellulose nanofibres. J. Mater. Sci. 46, 6029−6043.

Rafatullah, M., Sulaiman, O., Hashim, R., Ahmad, A., 2010. Adsorption of methylene blue on low-cost adsorbents: a review. J. Hazard Mater. 177, 70−80.

Ramesh, S., Kim, H.S., Kim, J.H., 2018. Cellulose-polyvinyl alcohol-nano-TiO_2 hybrid nanocomposite: thermal, optical, and antimicrobial properties against pathogenic bacteria. Polym.-Plast. Technol. Eng. 57, 669−681.

Renu, M.A., Singh, K., 2017. Heavy metal removal from wastewater using various adsorbents: a review. J. Water Reuse Desalinat. 7, 387−419.

Reshmy, R., Philip, E., Paul, S.A., Madhavan, A., Sindhu, R., Binod, P., Pandey, A., Sirohi, R., 2020. Nanocellulose-based products for sustainable applications-recent trends and possibilities. Rev. Environ. Sci. Bio-Technol. 19, 779−806.

Reshmy, R., Thomas, D., Philip, E., Paul, S.A., Madhavan, A., Sindhu, R., Binod, P., Pugazhendhi, A., Sirohi, R., Tarafdar, A., 2021. Potential of nanocellulose for wastewater treatment. Chemosphere 281, 130738.

Reshmy, R., Philip, E., Madhavan, A., Pugazhendhi, A., Sindhu, R., Sirohi, R., Awasthi, M.K., Pandey, A., Binod, P., 2022. Nanocellulose as green material for remediation of hazardous heavy metal contaminants. J. Hazard Mater. 424, 127516.

Rodrigues, F.H.A., Magalhaes, C.E.D.C., Medina, A.L., Fajardo, A.R., 2019. Hydrogel composites containing nanocellulose as adsorbents for aqueous removal of heavy metals: design, optimization, and application. Cellulose 26, 9119−9133.

Rojas, O.J., 2016. Cellulose chemistry and properties: fibers, nanocelluloses and advanced materials preface. In: Rojas, O.J. (Ed.), Cellulose Chemistry and Properties: Fibers, Nanocelluloses and Advanced Materials, pp. V−VI.

Rotaru, R., Savin, M., Tudorachi, N., Peptu, C., Samoila, P., Sacarescu, L., Harabagiu, V., 2018. Ferromagnetic iron oxide-cellulose nanocomposites prepared by ultrasonication. Polym. Chem. 9, 860−868.

Ruan, C.Q., Strømme, M., Lindh, J., 2018. Preparation of porous 2,3-dialdehyde cellulose beads crosslinked with chitosan and their application in adsorption of Congo red dye. Carbohydr. Polym. 181, 200−207.

Rueda-Marquez, J.J., Levchuk, I., Ibanez, P.F., Sillanpaa, M., 2020. A critical review on application of photocatalysis for toxicity reduction of real wastewaters. J. Cleaner Prod. 258, 120694.

Sadeek, S.A., Negm, N.A., Hefni, H.H.H., Wahab, M.M.A., 2015. Metal adsorption by agricultural biosorbents: adsorption isotherm, kinetic and biosorbents chemical structures. Int. J. Biol. Macromol. 81, 400−409.

Sai, H., Xing, L., Xiang, J., Cui, L., Jiao, J., Zhao, C., Li, Z., Li, F., Zhang, T., 2014. Flexible aerogels with interpenetrating network structure of bacterial cellulose-silica composite from sodium silicate precursor via freeze drying process. RSC Adv. 4, 30453–30461.

Sai, H., Fu, R., Xing, L., Xiang, J., Li, Z., Li, F., Zhang, T., 2015. Surface modification of bacterial cellulose aerogels' web-like skeleton for oil/water separation. ACS Appl. Mater. Interfaces 7, 7373–7381.

Saleem, F., Bhatti, H.N., Khan, A., Ben Farhat, L., Elqahtani, Z.M., Alwadai, N., Iqbal, M., 2022. Polypyrrole/magnetic/tea waste composites for PO_4^{3-} ions removal: adsorption-desorption, kinetics, and thermodynamics studies. Adsorpt. Sci. Technol. 2022, 4071162.

Sankaranarayanan, S., Lakshmi, D.S., Vivekanandhan, S., Ngamcharussrivichai, C., 2021. Biocarbons as emerging and sustainable hydrophobic/oleophilic sorbent materials for oil/water separation. Sustain. Mater. Technol. 28, e00268.

Scheuhammer, A., Braune, B., Chan, H.M., Frouin, H., Krey, A., Letcher, R., Loseto, L., Noel, M., Ostertag, S., Ross, P., Wayland, M., 2015. Recent progress on our understanding of the biological effects of mercury in fish and wildlife in the Canadian Arctic. Sci. Total Environ. 509, 91–103.

Schwarzenbach, R.P., Escher, B.I., Fenner, K., Hofstetter, T.B., Johnson, C.A., von Gunten, U., Wehrli, B., 2006. The challenge of micropollutants in aquatic systems. Science 313, 1072–1077.

Sehaqui, H., Zhou, Q., Ikkala, O., Berglund, L.A., 2011. Strong and tough cellulose nanopaper with high specific surface area and porosity. Biomacromolecules 12, 3638–3644.

Sekar, M., Sakthi, V., Rengaraj, S., 2004. Kinetics and equilibrium adsorption study of lead(II) onto activated carbon prepared from coconut shell. J. Colloid Interface Sci. 279, 307–313.

Shak, K.P.Y., Pang, Y.L., Mah, S.K., 2018. Nanocellulose: recent advances and its prospects in environmental remediation. Beilstein J. Nanotechnol. 9, 2479–2498.

Sheikhi, A., Safari, S., Yang, H., van de Ven, T.G.M., 2015. Copper removal using electrosterically stabilized nanocrystalline cellulose. ACS Appl. Mater. Interfaces 7, 11301–11308.

Shen, W., Chen, S.Y., Shi, S.K., Li, X., Zhang, X., Hu, W.L., Wang, H.P., 2009. Adsorption of Cu(II) and Pb(II) onto diethylenetriamine-bacterial cellulose. Carbohydr. Polym. 75, 110–114.

Shirazi, M.M.A., Bazgir, S., Meshkani, F., 2020. Electrospun nanofibrous membranes for water treatment. Adv. Membrane Technol. 57 (3), 467–504.

Silva, N.H.C.S., Figueira, P., Fabre, E., Pinto, R.J.B., Pereira, M.E., Silvestre, A.J.D., Marrucho, I.M., Vilela, C., Freire, C.S.R., 2020. Dual nanofibrillar-based bio-sorbent films composed of nanocellulose and lysozyme nanofibrils for mercury removal from spring waters. Carbohydr. Polym. 238, 116210.

Singh, K., Arora, J.K., Sinha, T.J.M., Srivastava, S., 2014. Functionalization of nanocrystalline cellulose for decontamination of Cr(III) and Cr(VI) from aqueous system: computational modeling approach. Clean Technol. Environ. Policy 16, 1179–1191.

Singh, K., Sinha, T.J.M., Srivastava, S., 2015. Functionalized nanocrystalline cellulose: smart biosorbent for decontamination of arsenic. Int. J. Miner. Process. 139, 51–63.

Singh, N.B., Nagpal, G., Agrawal, S., 2018. Rachna, water purification by using adsorbents: a review. Environ. Technol. Innov. 11, 187–240.

Sirvio, J.A., Hasa, T., Leiviska, T., Liimatainen, H., Hormi, O., 2016. Bisphosphonate nanocellulose in the removal of vanadium(V) from water. Cellulose 23, 689–697.

Sorriaux, M., Sorieul, M., Chen, Y., 2021. Bio-based and robust polydopamine coated nanocellulose/amyloid composite aerogel for fast and wide-spectrum water purification. Polymers 13 (19), 3442.

Srivastava, S., Kardam, A., Raj, K.R., 2012. Nanotech reinforcement onto cellulosic fibers: green remediation of toxic metals. Int. J. Green Nanotechnol. 4, 46–53.

Stephen, M., Catherine, N., Brenda, M., Andrew, K., Leslie, P., Corrine, G., 2011. Oxolane-2,5-dione modified electrospun cellulose nanofibers for heavy metals adsorption. J. Hazard Mater. 192, 922–927.

Taleb, K., Markovski, J., Veličković, Z., Rusmirović, J., Rančić, M., Pavlović, V., Marinković, A., 2019. Arsenic removal by magnetite-loaded amino modified nano/microcellulose adsorbents: effect of functionalization and media size. Arabian J. Chem. 12, 4675–4693.

Tan, K.B., Reza, A.K., Abdullah, A.Z., Amini Horri, B., Salamatinia, B., 2018. Development of self-assembled nanocrystalline cellulose as a promising practical adsorbent for methylene blue removal. Carbohydr. Polym. 199, 92–101.

Tang, F., Yu, H., Abdalkarim, S.Y.H., Sun, J., Fan, X., Li, Y., Zhou, Y., Tam, K.C., 2020. Green acid-free hydrolysis of wasted pomelo peel to produce carboxylated cellulose nanofibers with super absorption/flocculation ability for environmental remediation materials. Chem. Eng. J. 395, 125070.

Trache, D., Tarchoun, A.F., Derradji, M., Hamidon, T.S., Masruchin, N., Brosse, N., Hussin, M.H., 2020. Nanocellulose: from fundamentals to advanced applications. Front. Chem. 8, 392.

Tshikovhi, A., Mishra, S.B., Mishra, A.K., 2020. Nanocellulose-based composites for the removal of contaminants from wastewater. Int. J. Biol. Macromol. 152, 616–632.

Udomkun, P., Innawong, B., Jumrusjumroendee, N., 2018. Cellulose acetate and adsorbents supported on cellulose fiber extracted from waxy corn husks for improving shelf life of frying oil. Lwt-Food Sci. Technol. 97, 45–52.

Unuabonah, E.I., Taubert, A., 2014. Clay-polymer nanocomposites (CPNs): adsorbents of the future for water treatment. Appl. Clay Sci. 99, 83–92.

Vareda, J.P., Valente, A.J.M., Durães, L., 2019. Assessment of heavy metal pollution from anthropogenic activities and remediation strategies: a review. J. Environ. Manag. 246, 101–118.

Verma, S., Rao, B.T., Jayabalan, J., Rai, S., Phase, D., Srivastava, A., Kaul, R., 2019. Studies on growth of Au cube-ZnO core-shell nanoparticles for photocatalytic degradation of methylene blue and methyl orange dyes in aqueous media and in presence of different scavengers. J. Environ. Chem. Eng. 7, 103209.

Vilet, N.Z., Gué, E., Servent, A., Delalonde, M., Wisniewski, C., 2020. Filtration-compression step as downstream process for flavonoids extraction from citrus peels: performances and flavonoids dispersion state in the filtrate. Food Bioprod. Process. 120, 104–113.

Voicu, S.I., Thakur, V.K., 2021. Aminopropyltriethoxysilane as a linker for cellulose-based functional materials: new horizons and future challenges. Curr. Opin. Green Sustain. Chem. 30, 100480.

Voisin, H., Bergstrom, L., Liu, P., Mathew, A.P., 2017. Nanocellulose-based materials for water purification. Nanomaterials 7, 57.

Wan, C., Lu, Y., Cao, J., Sun, Q., Li, J., 2015. Preparation, characterization and oil adsorption properties of cellulose aerogels from four kinds of plant materials via a NAOH/PEG aqueous solution. Fibers Polym. 16, 302–307.

Wan, C., Jiao, Y., Wei, S., Zhang, L., Wu, Y., Li, J., 2019. Functional nanocomposites from sustainable regenerated cellulose aerogels: a review. Chem. Eng. J. 359, 459–475.

Wang, L.J., Li, J., 2013. Adsorption of CI Reactive Red 228 dye from aqueous solution by modified cellulose from flax shive: kinetics, equilibrium, and thermodynamics. Ind. Crop. Prod. 42, 153–158.

Wang, Y., Yadav, S., Heinlein, T., Konjik, V., Breitzke, H., Buntkowsky, G., Schneider, J.J., Zhang, K., 2014. Ultra-light nanocomposite aerogels of bacterial cellulose and reduced graphene oxide for specific absorption and separation of organic liquids. RSC Adv. 4, 21553–21558.

Wang, B., Liang, W., Guo, Z., Liu, W., 2015a. Biomimetic super-lyophobic and super-lyophilic materials applied for oil/water separation: a new strategy beyond nature. Chem. Soc. Rev. 44, 336–361.

Wang, S., Peng, X.W., Zhong, L.X., Tan, J.W., Jing, S.S., Cao, X.F., Chen, W., Liu, C.F., Sun, R.C., 2015b. An ultralight, elastic, cost-effective, and highly recyclable superabsorbent from microfibrillated cellulose fibers for oil spillage cleanup. J. Mater. Chem. A 3, 8772–8781.

Wang, D.C., Yu, H.Y., Fan, X.M., Gu, J.P., Ye, S.N., Yao, J.M., Ni, Q.Q., 2018. High aspect ratio carboxylated cellulose nanofibers cross-linked to robust aerogels for superabsorption-flocculants: paving way from nanoscale to macroscale. ACS Appl. Mater. Interfaces 10, 20755–20766.

Wang, J., Liu, M., Duan, C., Sun, J., Xu, Y., 2019a. Preparation and characterization of cellulose-based adsorbent and its application in heavy metal ions removal. Carbohydr. Polym. 206, 837–843.

Wang, X., Jiang, S., Cui, S., Tang, Y., Pei, Z., Duan, H., 2019b. Magnetic-controlled aerogels from carboxylated cellulose and $MnFe_2O_4$ as a novel adsorbent for removal of Cu(II). Cellulose 26, 5051–5063.

Wang, K., Liu, X., Tan, Y., Zhang, W., Zhang, S., Li, J., 2019c. Two-dimensional membrane and three-dimensional bulk aerogel materials via top-down wood nanotechnology for multi-behavioral and reusable oil/water separation. Chem. Eng. J. 371, 769–780.

Wang, J., Zhang, D., Chu, F., 2021. Wood-derived functional polymeric materials. Adv. Mater. 33, 2001135.

Wang, P.-L., Ma, C., Yuan, Q., Mai, T., Ma, M.G., 2022. Novel $Ti_3C_2T_x$ MXene wrapped wood sponges for fast cleanup of crude oil spills by outstanding Joule heating and photothermal effect. J. Colloid Interface Sci. 606, 971–982.

Wei, J., Yang, Z., Sun, Y., Wang, C., Fan, J., Kang, G., Zhang, R., Dong, X., Li, Y., 2019a. Nanocellulose-based magnetic hybrid aerogel for adsorption of heavy metal ions from water. J. Mater. Sci. 54, 6709–6718.

Wei, Y.Q., Sun, M.M., Fang, H.Y., 2019b. Dienzyme-assisted salting-out extraction of flavonoids from the seeds of Cuscuta chinensis Lam. Ind. Crop. Prod. 127, 232–236.

WHO, 2019. Drinking-water. Available at: https://www.who.int/en/news-room/factsheets/detail/drinking-water. (Accessed 16 January 2021).

Wu, M.B., Zhang, C., Pi, J.K., Liu, C., Yang, J., Xu, Z.K., 2019. Cellulose nanocrystals as anti-oil nanomaterials for separating crude oil from aqueous emulsions and mixtures. J. Mater. Chem. A 7, 7033–7041.

Wu, J., Tian, J., Qian, Z., Huang, J., Sun, D., 2023. Highly robust separation for aqueous oils enabled by balsa wood-based cellulose aerogel with intrinsic superior hydrophilicity. Sep. Purif. Technol. 315, 123688.

Xiang, B., Sun, Q., Zhong, Q., Mu, P., Li, J., 2022. Current research situation and future prospect of superwetting smart oil/water separation materials. J. Mater. Chem. A 10, 20190–20217.

Xiao, J., Lv, W., Song, Y., Zheng, Q., 2018. Graphene/nanofiber aerogels: performance regulation towards multiple applications in dye adsorption and oil/water separation. Chem. Eng. J. 338, 202–210.

Xiong, Y., Wang, C., Wang, H.W., Yao, Q.F., Fan, B.T., Chen, Y.P., Sun, Q.F., Jin, C.D., Xu, X.J., 2017. A 3D titanate aerogel with cellulose as the adsorption-aggregator for highly efficient water purification. J. Mater. Chem. A 5, 5813–5819.

Xu, Y., Wu, T., Cui, Z., Kang, L., Cai, Y., Li, J., Tian, D., 2022. In situ growth of COFs within wood microchannels for wastewater treatment and oil-water separation. Sep. Purif. Technol. 303, 122275.

Yamaki, S.B., Barros, D.S., Garcia, C.M., Socoloski, P., Oliveira, O.N., Atvars, T.D.Z., 2005. Spectroscopic studies of the intermolecular interactions of Congo red and tinopal CBS with modified cellulose fibers. Langmuir 21, 5414–5420.

Yan, H., Zhang, W.X., Kan, X.W., Dong, L., Jiang, Z.W., Li, H.J., Yang, H., Cheng, R.S., 2011. Sorption of methylene blue by carboxymethyl cellulose and reuse process in a secondary sorption. Coll. Surf. A-Physicochem. Eng. Asp. 380, 143–151.

Yang, R., Aubrecht, K.B., Ma, H.Y., Wang, R., Grubbs, R.B., Hsiao, B.S., Chu, B., 2014. Thiol-modified cellulose nanofibrous composite membranes for chromium (VI) and lead (II) adsorption. Polymer 55, 1167–1176.

Yang, Y., Guo, Z., Liu, W., 2022. Special superwetting materials from bioinspired to intelligent surface for on-demand oil/water separation: a comprehensive review. Small 18, 2204624.

Yin, X., He, Y., Wang, Y., Yu, H., Chen, J., Gao, Y., 2020. Bio-inspired antifouling cellulose nanofiber multifunctional filtration membrane for highly efficient emulsion separation and application in water purification. Korean J. Chem. Eng. 37, 1751–1760.

Yu, X., Tong, S., Ge, M., Wu, L., Zuo, J., Cao, C., Song, W., 2013. Adsorption of heavy metal ions from aqueous solution by carboxylated cellulose nanocrystals. J. Environ. Sci. 25, 933–943.

Yu, H.Y., Zhang, D.Z., Lu, F.F., Yao, J., 2016. New Approach for single-step extraction of carboxylated cellulose nanocrystals for their use as adsorbents and flocculants. ACS Sustain. Chem. Eng. 4, 2632–2643.

Yuan, B., Li, L., Murugadoss, V., Vupputuri, S., Wang, J., Alikhani, N., Guo, Z., 2020. Nano-cellulose-based composite materials for wastewater treatment and waste-oil remediation. ES Food Agroforestry 1, 41–52.

Zeng, X., Wang, Y., He, X., Liu, C., Wang, X., Wang, X., 2021. Enhanced removal of Cr(VI) by reductive sorption with surface-modified Ti_3C_2Tx MXene nanocomposites. J. Environ. Chem. Eng. 9, 106203.

Zhan, C., Sharma, P.R., He, H., Sharma, S.K., McCauley-Pearl, A., Wang, R., Hsiao, B.S., 2020. Rice husk based nanocellulose scaffolds for highly efficient removal of heavy metal ions from contaminated water. Environ. Sci.-Water Res. Technol. 6, 3080–3090.

Zhang, H.Y., Yan, X.J., Jiang, Y., Cong, J., 2011. Development and characteristic of bacterial cellulose for antimicrobial wound dressing. Adv. Mater. Res. 152, 978–987.

Zhang, X.F., Zhao, J.Q., Cheng, L., Lu, C.H., Wang, Y.R., He, X., Zhang, W., 2014. Acrylic acid grafted and acrylic acid/sodium humate grafted bamboo cellulose nanofibers for Cu^{2+} adsorption. Rsc Adv 4, 55195–55201.

Zhang, C., Liang, H.Q., Xu, Z.K., Wang, Z., 2019. Harnessing solar-driven photothermal effect toward the water-energy nexus. Adv. Sci. 6, 1900883.

Zhang, J., Zhong, T., Xiang, Y., Zhang, X., Feng, X., 2021. Microfibrillated cellulose reinforced poly(vinyl imidazole) cryogels for continuous removal of heavy metals. J. Appl. Polym. Sci. 138, e51456.

Zhao, Y., Xue, Z., Wang, X., Wang, L., Wang, A., 2012. Adsorption of Congo red onto ligno-cellulose/montmorillonite nanocomposite. J. Wuhan Univ. Technol.-Mater. Sci. Ed. 27, 931–938.

Zhao, M., Tao, Y., Wang, J., He, Y., 2020. Facile preparation of superhydrophobic porous wood for continuous oil-water separation. J. Water Process Eng. 36, 101279.

Zheng, W., Huang, J., Li, S., Ge, M., Teng, L., Chen, Z., Lai, Y., 2021. Advanced materials with special wettability toward intelligent oily wastewater remediation. ACS Appl. Mater. Interfaces 13, 67–87.

Zhou, S., Hao, G., Zhou, X., Jiang, W., Wang, T., Zhang, N., Yu, L., 2016. One-pot synthesis of robust superhydrophobic, functionalized graphene/polyurethane sponge for effective continuous oil–water separation. Chem. Eng. J. 302, 155–162.

Zhou, C., Ni, J., Zhang, D., Sun, C., 2019. Cellulosic adsorbent functionalized with macrocyclic pyridone pentamer for selectively removing metal cations from aqueous solutions. Carbohydr. Polym. 217, 1–5.

Zhu, G., Xu, H., Dufresne, A., Lin, N., 2018. High-adsorption, self-extinguishing, thermal, and acoustic-resistance aerogels based on organic and inorganic waste valorization from cellulose nanocrystals and red mud. ACS Sustain. Chem. Eng. 6, 7168–7180.

Zhu, Z., Fu, S., Lucia, L.A., 2019. A fiber-aligned thermal-managed wood-based superhydrophobic aerogel for efficient oil recovery. ACS Sustain. Chem. Eng. 7, 16428–16439.

Zhu, M., Liu, Y., Chen, M., Sadrzadeh, M., Xu, Z., Gan, D., Huang, Z., Ma, L., Yang, B., Zhou, Y., 2021. Robust superhydrophilic and underwater superoleophobic membrane optimized by Cu doping modified metal-organic frameworks for oil-water separation and water purification. J. Membr. Sci. 640, 119755.

Zong, P., Cao, D., Cheng, Y., Wang, S., Zhang, J., Guo, Z., Hayat, T., Alharbi, N.S., He, C., 2019. Carboxymethyl cellulose supported magnetic graphene oxide composites by plasma induced technique and their highly efficient removal of uranium ions. Cellulose 26, 4039–4060.

Zou, Q., Wang, W., Zhang, T., Liu, Y., 2021. Simultaneous removal of Cr(VI), Cd, and Pb from aqueous solution by iron sulfide nanoparticles: influencing factors and interactions of metals. Chin. J. Chem. Eng. 40, 245–255.

Chapter 5

Cellulose composites for electronic devices

Chang Ma,[1] Wentao Cao[2] and Mingguo Ma[3,*]

[1]*College of Mechanical Engineering, Tianjin University of Commerce, Tianjin, PR China;* [2]*Department of Prosthodontics Shanghai, Stomatological Hospital & School of Stomatology, Fudan University, Shanghai, PR China;* [3]*MOE Engineering Research Center of Forestry Biomass Materials and Bioenergy, Research Center of Biomass Clean Utilization, Beijing Key Laboratory of Lignocellulosic Chemistry, College of Materials Science and Technology, Beijing Forestry University, Beijing, PR China*

[*]*Corresponding author: E-mail: mg_ma@bjfu.edu.cn*

5.1 Introduction

Wearable materials are one of the rapidly developing directions in recent years, which can be widely used in fields, such as human physiological activity monitoring, health management, disease prevention, artificial skin, and human−computer interaction (Keum et al., 2020; Khan and Andreescu, 2020; Peng et al., 2020; Yoon et al., 2020). With the development of Internet technology, rapid progress has been conducted on the preparation, design, and applications of wearable devices (Ling et al., 2020). For example, wearable sensors, including pressure sensors, strain sensors, and gas sensors, can track and monitor health status (measuring respiratory moisture, heart rate, body temperature, blood pressure, and other physiological information), joint activity, as well as small signals, such as pulse, throat sound, and micro-expression in real-time and noninvasive manner (Rao et al., 2020; Tai et al., 2020; Heo et al., 2020). There are also wearable energy storage devices and electromagnetic shielding materials for integrated wearable systems, which are used as power and protection components for wearable devices (Simon and Gogotsi, 2020). To achieve skin fit, wearable devices should have flexibility, lightweight, durability, skin friendliness, and a certain degree of mechanical robustness (Wang et al., 2019).

Surface Science and Advanced Functional Applications of Cellulose Composites
https://doi.org/10.1016/B978-0-443-13675-7.00005-1

Compared with previously widely reported carbon-based materials, metal-based materials, and conductive polymer-based materials, MXene has unique advantages (Lukatskaya et al., 2013), such as excellent conductivity, large specific surface area, unique layered structure, good aqueous solution dispersion, and abundant terminal groups (Khazaei et al., 2013; Qiuming Peng et al., 2014). This makes MXene a promising wearable material (Fu et al., 2019; Gao et al., 2020; Huang et al., 2020; Yang et al., 2018; Sun N et al., 2020).

In this chapter, we introduced the synthesis and property of advanced functional cellulose-based nanocomposites for electronic devices. Three typical examples, such as multiple response hydrophobic air-laid paper (AP)/MXene smart nanocomposite, carbon nanotube (CNT)–enhanced AP/MXene smart nanocomposite, and cellulose nanofiber/MXene liquid electrode friction nanogenerator, were selected to present cellulose-based nanocomposites for electronic devices. We try to contribute great content to this book "***Surface Science and Advanced Functional Applications of Cellulose Composites***."

5.2 Synthesis and property of advanced functional cellulose-based nanocomposites for electronic devices

5.2.1 Multiple response hydrophobic air-laid paper/MXene smart nanocomposite

With the rapid development of electronic technology, electronic devices have become more intelligent, compact, and portable. This trend has led to the prosperous development of wearable products, and many new types of wearable devices have emerged, such as electronic skins, sensors, smartwatches, and smart textile products (Jin et al., 2017; Lee et al., 2020; Xu et al., 2020). Therefore, wearable functional materials with good conductivity and flexibility have attracted the attention (Shahzad et al., 2016). Various innovative wearable materials have specific functions and can be applied in various fields, such as energy collection and storage (Liubing Dong et al., 2016), electromagnetic shielding (Geng et al., 2019), thermal regulation (Luo et al., 2020), and sensors (Wang H et al., 2020). However, achieving high integration, low power consumption, and versatility without affecting inherent performance still remains a challenge. Therefore, developing a multipurpose electronic device that can respond to various external stimuli in complex environments is of great significance for expanding the application range of electronic products.

In response to the aforementioned issues, we used the "impregnation drying" method to obtain a wearable hydrophobic AP/MXene intelligent composite paper. This composite smart paper has the ability to respond to light, electricity, and machinery, which improves the durability and hydrophobicity of the composite paper while retaining its original breathability. It can also have good working stability. Therefore, multifunctional smart paper with good photothermal and electrothermal effects, and sensing performance is expected to be applied as wearable electronic devices and portable devices in the fields of thermal management and human motion signal monitoring, which has promising applications in human–computer interaction and other fields.

Untreated AP was soaked in MXene dispersion. Polydimethylsiloxane (PDMS) was used as a hydrophobic surface modifier, which was soaked and dried at 60°C for 100 min to make it water resistant, thus preparing Si-MAP. The sample is named based on the number of times immersed in MXene dispersion. For example, MAP-3 represents AP soaked in MXene dispersion for three drying cycles. Finally, Si-MAP-3 was prepared from MAP-3 by coating with siloxane PDMS.

The MXene air-laid paper (MAP) is made through an "immersion drying" method. During this process, the initial AP is immersed in a certain concentration of MXene nanosheet suspension, and then placed in a vacuum drying oven to dry at room temperature. Finally, a thin layer of silicone resin is applied to the MAP to make the fabric hydrophobic, thereby preventing oxidation of the MXene layer and improving the long-term durability and stability of the composite material, thus obtaining Si-MAP. The synthesized intelligent Si-MAP composite paper exhibits excellent waterproof properties, satisfactory conductivity, excellent optical/electrical dual-power heating performance, and fast pressure response, making it potentially valuable in wearable thermal management and human motion monitoring. As an active material, MXene nanosheet dispersion is prepared by etching the precursor MAX (Ti_3AlC_2) phase powder. Hydrochloric acid (HCl) and lithium fluoride (LiF) are used as etchants to selectively remove the Al layer and obtain accordion-like multilayer MXene. Secondly, the MXene sheets were peeled and separated through high-power ultrasound and further centrifugation treatment. MXene dispersion shows the typical Tyndall effect, which shows that the stripped single-layer MXene dispersion has satisfactory uniformity and stability. The MXene nanosheets were tested through various characterization methods, such as X-ray diffraction (XRD), X-ray photoelectron spectroscopy (XPS), and transmission electron microscopy (TEM). The diffraction peak of (002) shifts from 9.7° in the MAX phase to 8.8° in multilayer MXene and then to 7.2° in single-layer MXene nanosheets, while other characteristic peaks weaken or even disappear. This indicates that due to the removal of the Al layer and the detachment of MXene nanosheets, the corresponding interlayer spacing continues to expand. The XPS spectrum shows that the Al2p peak disappears in the MXene spectrum after etching, indicating that the Al layer has been removed. In addition, high-resolution spectral information of Ti and F elements indicated the presence of −OH, −F, and −O groups in the etching reaction. These active terminal functional groups of MXene can endow MXene nanosheets with hydrophilic properties, thereby enhancing the interfacial adhesion with the original dust-free paper substrate. In addition, the TEM images further confirm that the as-obtained MXene has a single-layer sheet-like structure.

To prepare MXene intelligent composite paper, AP is used as a flexible substrate, which is made of cellulose and polyester fibers and has porous, lightweight, and breathable characteristics (Ma et al., 2020). Unlike pure polyester fiber fabrics and traditional cellulose paper, such as A4 printing paper and filter paper, AP combines the advantages of superhydrophilicity, high tensile strength, and excellent folding resistance (over 90,000 cycles). Therefore, it can absorb a large amount of conductive MXene nanosheet dispersion without any pretreatment, and due to van der Waals forces and hydrogen bonding between cellulose and MXene, its fibers can be tightly wrapped by MXene nanosheets to complete the composite.

Obviously, the surface of the original AP fiber is smooth, while the surface of MAP fiber and Si-MAP fiber is wrinkled due to the encapsulation of MXene nanosheets, and the roughness of the wrinkling depends on the number of "immersion drying" cycles. It is worth noting that, after composite and hydrophobic treatment with MXene nanosheets, the structure of the entire dust-free paper and most of the pores between the fibers are well maintained. This conformal composite also ensures the breathability of the composite paper, and the slight decrease in the breathability of Si-MAPs may be due to the disappearance of some voids between the fiber bundles after silane coating. In addition to endowing the composite paper with hydrophobicity, the silane layer on the fibers can also protect the internal MXene layer, increasing the durability of the composite paper. The EDS spectra of Ti and Si show the distribution of organic silicon layer and MXene coating on the surface of Si-MAP, intuitively indicating that both silane and MXene coatings have excellent uniformity. In addition, as a wearable device, Si-MAP exhibits the same excellent flexibility as the original AP during bending and twisting processes. Si-MAP can also be cut into different shapes for sewing or other purposes. With a simple pair of scissors, complex shapes, such as lightning, skirts, and water droplets, can be easily cut out of Si-MAP composite paper. The aforementioned characteristics make the prepared Si-MAPs have certain potential in the field of wearable electronic devices.

Si-MAP exhibits satisfactory hydrophobicity, with a water contact angle (CA) of up to 135.6°. The uncoated original dust-free paper is hydrophilic and dyed blue after soaking in methylene blue staining water for a few seconds. In contrast, Si-MAPs are not dyed, which further proves that Si-MAPs have outstanding waterproof ability. When dyed alkaline solution (pink, pH 12), pure water (yellow, pH 7), and acidic solution (blue, pH 2) drop onto the surface of Si-MAPs, respectively, the droplets stably stand on the composite paper and form spheroids, indicating that Si-MAPs have certain corrosion resistance and chemical durability. In addition, as a candidate material for wearable devices, Si MAP should also have the ability to resist common liquid stains in daily life. Six common stains were selected for the test, including clean water, dye water, milk, coffee, ink, and soy sauce (see Fig. 5.1). It is clear that Si-MAP with good hydrophobicity can resist the aforementioned common stains, and after simple wiping, Si-MAPs can restore a clean initial state. Furthermore, considering the inevitable impact of wear and tear on material performance in daily use, stability is an essential component of wearable materials, and having sufficient stability can ensure that they can adapt to complex usage environments in practical applications. Therefore, this study also used ultrasonic testing, rolling testing, artificial sweat testing, cleaning agent testing, wear testing, and tape testing to simulate wear during actual use. In ultrasonic testing, each sample is immersed in water and subjected to ultrasonic vibration of different durations at a power of 60 W. In the rolling test, wear test, and tape test, Si-MAPs samples were subjected to different cycles of treatment; in the artificial sweat test and detergent test, samples are rinsed with artificial sweat and commercial detergent. It shows the change in the water contact angle values of the samples after the durability test mentioned before: after each round of experiments, the CA value remains above 125°, reflecting the excellent durability and stability of Si-MAPs, and further demonstrating the potential for practical application of Si-MAPs composite paper in the field of wearable devices. Based on the aforementioned test results and SEM morphology characterization, the good hydrophobicity of Si-MAPs is mainly due to the following two reasons: (1) the hydrophobic coating of polydimethylsiloxane (PDMS) helps to reduce surface energy while protecting the MXene layer; (2) MXene coating is beneficial for increasing the roughness of the fiber surface, reducing the solid–liquid contact area, and thus enhancing hydrophobicity.

There have been reports on two-dimensional MXene nanosheets in the field of photothermal conversion, such as tumor hyperthermia, photothermal driving

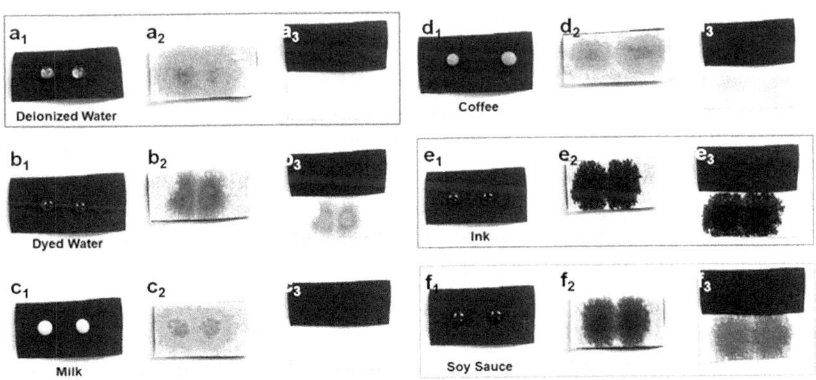

FIGURE 5.1 Digital images for different kinds of stains on the Si-MAPs (1), AP (2), and after a simple wipe (3), respectively. *Reproduced from Ma et al. (2020) with permission from American Chemical Society.*

materials, and water treatment (Fan et al., 2019; Zhao et al., 2020; Lin H et al., 2017). Further exploration of the light response behavior of Si-MAP for wearable thermal management and wearable heating devices has more practical significance for composite smart paper. To this end, the original AP, Si-MAP-3, Si-MAP-5, and Si-MAP-10 were exposed to 808 nm NIR light sources with different power densities (0.25, 0.50, 0.75, 1.00, and 1.25 W cm^{-2}) for 180 s, and the surface temperature of the sample was measured using an infrared camera. Si-MAPs exhibit significant and rapid photoinduced heating behavior and can reach the maximum equilibrium surface temperature within 100 s. On the contrary, under the same light source, the original AP did not show significant temperature changes. Therefore, we attribute this result to the high photothermal conversion performance of MXene.

To study the factors affecting the photothermal performance of composite paper, we designed and conducted a series of experiments. For three samples with different "immersion drying" times, the maximum equilibrium temperature values gradually increase with the increase of NIR laser power density. Taking Si-MAP-10 as an example, under low power density (0.25 W cm^{-2}) illumination, the maximum surface equilibrium temperature can reach 34.6°C, while a high power density of 1.25 W cm^{-2} can generate a surface equilibrium temperature of approximately 104.9°C. On the other hand, when the NIR laser power density remains constant (such as 1 W cm^{-2}), the equilibrium temperature and heating rate both increase with the increase of sample impregnation times. These results indicate that the equilibrium temperature can be adjusted by adjusting the incident light intensity and the number of impregnations used to prepare Si-MAP samples. To further investigate the photothermal stability of Si-MAPs, we subjected Si-MAP-10 to five cycles of on/off irradiation under an external light source intensity of 1 W cm^{-2}. In each cycle, the tested sample is exposed to the light source for 60 s, and then the light source is turned off, with a cooling period of 60 s. During this process, an infrared camera is used to record the temperature of the sample. The surface temperature of Si-MAP can achieve a rapid and stable increase, and when the light source is turned off, the surface temperature of Si-MAP also decreases rapidly. In addition, the fluctuation of peak temperature during this process is very small, which not only demonstrates the sensitive light responsiveness of MXene composite smart paper, but also further highlights its stable photothermal performance and excellent photothermal conversion behavior. These excellent characteristics make it possible for Si-MAP to be used as a wearable or even attached photoheater to the skin. To observe the surface temperature changes of Si-MAP-10 more intuitively, we used infrared thermal imaging photos to reveal the specific heating process. As the illumination time prolonged, the temperature increased from 28.5 to 33.4°C within 0.5 s, reaching 53.1°C at 1.5 s, 74.9°C at 5.0 s, and 91.4°C at 100.0 s. The test results reveal the enormous potential of MXene composite smart paper with practical application stability for wearable photothermal clothing.

Electric heaters are receiving increasing attention from researchers due to their widespread applications in daily life and industrial production. In previous literature, traditional electric heaters were mainly composed of expensive metal nanowires, conductive carbon materials, such as CNTs or graphene, or polymers (Kim et al., 2013; Zhou et al., 2016; Wang et al., 2017). Due to the excellent conductivity of MXene, MXene-based composite materials can also be developed for use in electric heaters. According to Joule's law, conductivity is one of the important factors affecting the Joule heating performance of materials. Therefore, we first tested the conductivity of Si-MAPs. The original dust-free paper is insulated and then immersed in MXene nanosheet dispersion to make it conductive, and the conductivity of Si-MAPs shows an upward trend with the increase of the number of "immersion drying" cycles. Si-MAP-10 has a conductivity of 865 S m^{-1} measured by a four-point probe resistance testing system. In addition, Si-MAPs still maintain their original conductivity during cyclic bending and torsion deformation, which proves that composite paper has satisfactory flexibility and stability in practical applications. In summary, the Si-MAPs series has the potential applications to be used as wearable electric response heaters.

To demonstrate the effects of impregnation times and external voltage on the electric heating behavior of Si-MAP, the electric heating performance of MXene intelligent composite paper was tested. Measuring instruments and methods are similar to photoheating: square (2.5 × 2.5 cm) Si-MAPs are connected to a DC power supply through copper wire and conductive silver paste to provide external voltage. The temperature changes on the surface of Si-MAP are monitored and recorded by an infrared camera. Considering human health and safety, the input voltage must be below 36 V (standard safety voltage). More importantly, the purpose of wearability requires that the working voltage should allow the Si-MAPs heater to be powered by micro-supercapacitors or button cell for easy wearing. Therefore, this study selected ultralow input voltages (1, 2, 3, and 4 V). It is not difficult to find that under the same external working voltage, the equilibrium temperature of Si-MAP-10 is always the highest among the three samples, which also conforms to the law of Joule. No significant temperature increase was observed on the surface of the sample under an applied voltage of 1 V. However, when the input voltage increases from 2 to 3 V and 4 V, the equilibrium temperature of Si-MAP-10 can increase from 32.8 to 67.0°C and 118.7°C, respectively. Therefore, it is easy to adjust the maximum temperature by changing the input voltage. The temperature curve of Si-MAP-10 under gradually increasing external voltage applied at intervals of 50 s shows that the sample rapidly generates heat energy, and the temperature increases linearly, confirming the excellent electrothermal response ability of the sample. To further evaluate the cycling stability of Si-MAP, temperature changes were tested during five periodic power on/off cycles at an external voltage of 4 V. In each cycle, the sample was powered on for 60 s and then powered off for cooling. The electric heating behavior of Si-MAP-10 did not show a significant decline throughout the entire five cycles, indicating its good electrothermal stability. To observe the changes in temperature more intuitively, Si-MAP achieves a wide temperature range (16.4−118.7°C) under low external voltage. Compared with similar MXene-based electric heaters reported earlier, the Si-MAP we prepared can reach higher temperatures at lower input voltages, indicating that Si-MAP has better electrothermal conversion efficiency. Si-MAP can combine the advantages of fast electric thermal response, excellent electric heating ability, and low external working voltage requirements (less than 4 V), combined with breathability and portability, making MXene intelligent composite paper have sufficient potential in the field of wearable thermal management devices.

In addition to the aforementioned applications in wearable thermal management devices, flexible Si-MAPs also have the potential as pressure sensing materials for human motion monitoring, thanks to their excellent conductivity, special fiber network structure, and good breathability. Fig. 5.2a shows a

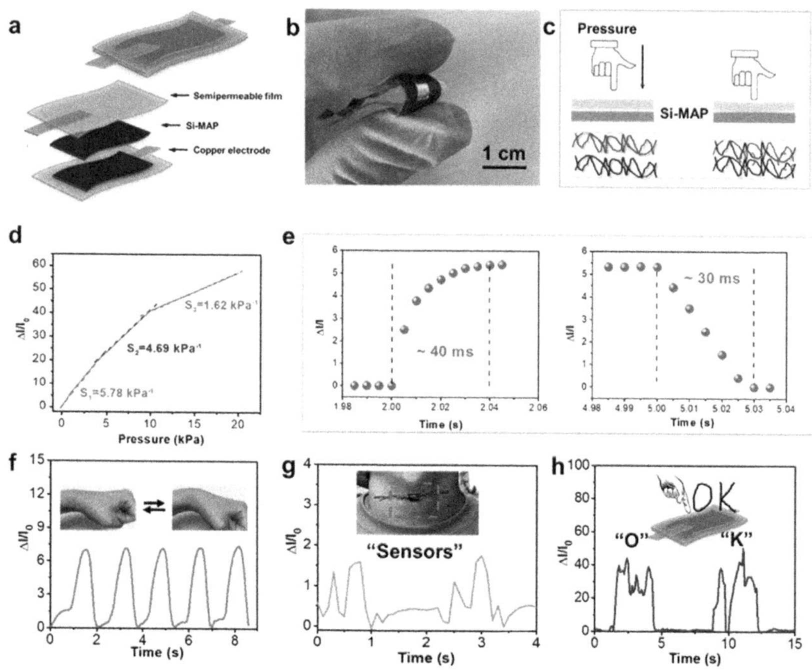

FIGURE 5.2 Electromechanical properties of the Si-MAPs pressure human motion monitoring. (a) Schematic illustration of the flexible two-layer Si-MAPs pressure human motion monitoring; (b) digital photograph of the flexible Si-MAPs pressure human motion monitoring; (c) schematic diagrams of the structural change of two Si-MAPs layers under pressure; (d) sensitivity of the Si-MAPs pressure sensor; (e) response time and recovery time; relative current changes of (f) wrist bending, (g) speaking "sensors" and (h) writing "OK" on the sensor with finger. *Reproduced from Ma et al. (2020) with permission from American Chemical Society.*

schematic diagram of a pressure sensor based on Si-MAPs, where two layers of Si-MAP-10 are stacked together to form a pressure-sensitive structure. The top and bottom layers of this two-layer structure are bonded to the copper foil electrode through silver paste to improve the stability of the connection. Finally, the entire device is encapsulated in two medical semipermeable polyurethane films, ensuring that the assembled Si-MAPs pressure sensor is flexible, waterproof, breathable, and wearable. It should be emphasized that a simple packaging process can help improve mechanical strength and reduce damage in actual daily use. Fig. 5.2b is a photo of a flexible Si-MAP-based pressure sensor assembled according to the aforementioned steps.

To further analyze the sensing mechanism of the pressure-sensitive structure composed of two layers of Si-MAPs, we established an abstract structural model. Fig. 5.3 shows the equivalent circuit diagram of Si-MAPs pressure sensors. The total resistance (Rt) of the sensor component consists of three parts: Re, Rp, and Rc, where Re is the resistance of the copper film electrode, Rp is the resistance of the two-layer Si-MAPs structure, and Rc is the contact interface between the resistance copper electrode and the Si-MAPs piezoresistive structure.

Therefore, the pressure sensitivity of Si-MAPs sensors is mainly due to the variation of Rp with pressure. During pressure loading/unloading, Fig. 5.2c shows a cross-sectional schematic diagram of the two-layer piezoresistive structure of Si-MAPs. In the initial state, the gap between the two layers of Si-MAP is relatively large and the contact area is small. When pressure is applied to the Si-MAPs composite paper sensor, the MXene layers on the dust-free paper fibers will approach each other, generating more conductive points (red dashed circles), leading to a sharp decrease in Rp. After releasing the pressure load, Si-MAPs returned to their original state, causing Rp to return to its initial value. For a fixed external voltage (1 V), the change in resistance will generate a corresponding signal in the current according to Ohm's law. Fig. 5.2d shows the high sensitivity of Si-MAPs pressure sensors over a large response range of 0–20 kPa. It is obvious that this curve can be divided into

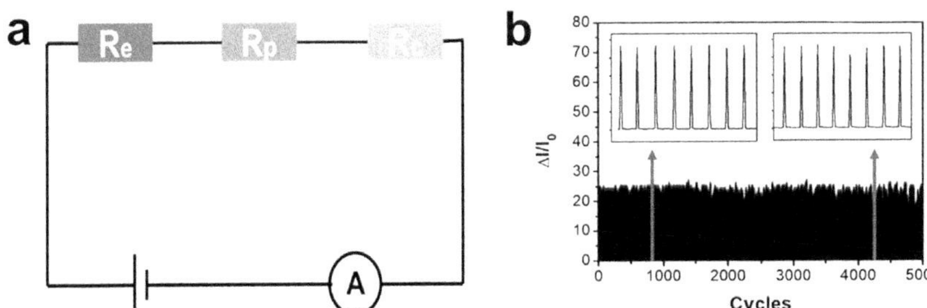

FIGURE 5.3 (a) The equivalent circuit diagram of the Si-MAPs-based pressure sensor; (b) durability characteristic tests of Si-MAPs sensor for 5000 cycles. *Reproduced from Ma et al. (2020) with permission from American Chemical Society.*

three linear regions, each with a different slope. For the low pressure range, the sensitivity of the pressure sensor based on Si-MAPs is 5.78 kPa^{-1}. The medium voltage range is 4.69 kPa^{-1}. Within the high-pressure range, the sensitivity is 1.62 kPa^{-1}. The response time and recovery time of Si-MAP based pressure sensors are ~40 and ~30 ms, respectively (Fig. 5.2e). The durability test results under loading/unloading cycles indicated that there is no significant difference in the results after 5000 repeated pressure/release cycles. Therefore, Si-MAPs pressure sensors have fast response ability, high stability, and durability, indicating that flexible pressure sensors based on smart paper have a longer working life.

Si-MAP pressure sensors have the advantages of strong flexibility, high sensitivity, fast response speed, and long working life and can be widely used in various fields. Here, we will focus on its potential applications in wearable devices. To demonstrate the practical application potential of pressure sensors based on Si-MAPs in human motion monitoring, a series of real-time human activity detection tests were conducted. For example, as shown in Fig. 5.2f, by connecting the sensor to the wrist, human joint movements can be recognized and recorded, and the signal response is stable and repeatable. In addition, the sensor can also capture small movements of the human body, such as irregular vibrations of the vocal cords during vocalization. Fig. 5.2g shows that three distinguishable peaks represent the multisyllabic word "sensor." In addition to the basic human motion signals, it is worth mentioning that sensors based on Si MAPs can also serve as a sensing platform for fingertip recognition features. By writing on Si-MAP (Fig. 5.2h), complex and unique waveforms can be generated, which will greatly expand the application field of this sensor in wearable products such as smartwatches. Compared with previous work, Si-MAP has multiple functions, which is beneficial for its application in different situations. In addition to the light/electric dual response heater, it can also be assembled into sensors. In addition, this intelligent textile also has excellent stability in harsh environments. Therefore, it can be applied to various scenarios, such as meeting the needs of patients who need health monitoring and warmth preservation, or workers or soldiers who need to record or transmit some information and maintain good working conditions. In such complex and special environments and needs, MXene intelligent composite paper is undoubtedly an ideal choice.

In this section, we utilized an adjustable "impregnation drying" method to design and prepare a hydrophobic MXene intelligent composite paper with multiple response capabilities, which was then coated with PDMS to exhibit significant electrical/photothermal response and mechanical sensitivity. Under relatively low external light source irradiation power (1.25 W cm^{-2}) and external DC power supply working voltage (4V), Si-MAP can reach equilibrium temperatures of 104.9 and 118.7°C, respectively. Thanks to the fiber structure of Si-MAP, pressure sensors based on Si-MAP exhibit good response performance, with a wide pressure response range (0–25 kPa) and high sensitivity (for low pressure range, the sensitivity of Si-MAPs based pressure sensors is 5.78 kPa^{-1}; for medium pressure range, 4.69 kPa^{-1}; and for high pressure range, 1.62 kPa^{-1}), fast response time (30–40 ms), and excellent durability (>5000 cycles). Due to the use of PDMS coating, Si-MAPs intelligent composite paper can meet the waterproof requirements of daily life (contact angle is about 135.6°). And after the modification of hydrophobic coating, the breathability and flexibility of composite paper are well preserved, which greatly expands the application range of multiresponse composite paper. Si-MAPs intelligent composite paper can be applied in many fields, such as wearable light/electric dual drive heaters, human motion monitoring, or smart watches, and will provide new inspiration and ideas for the next generation of wearable electronic products.

5.2.2 Carbon nanotube–enhanced air-laid paper/MXene smart nanocomposite

MXene, as a two-dimensional material, may have smaller working windows and insufficient sensitivity due to the interaction between sheets (hydrogen bonding or van der Waals forces) that hinder effective slip between layers. Previous studies have shown that introducing one-dimensional (1D) nanomaterials into MXene sheets can significantly improve the sensitivity of sensors, as the high aspect ratio and bridging interconnection of 1D nanomaterials make it easier to form effective percolation and conductive networks (Duan et al., 2020; Ma et al., 2021). CNTs, also known as Bucky tubes, are a 1D carbon-based nanomaterial with excellent conductivity, mechanical properties, and heat transfer properties. These characteristics make CNTs show great potential in the field of electronic devices (Baughman et al., 2002). Some researchers have also utilized doping CNTs to enhance the performance of MXene-based sensors. For example, Cai et al. (2018) utilized 1D CNTs to link MXene layers, which endows sensor devices with high sensitivity and wide response windows. Not only that, CNTs also have excellent broadband spectral absorption performance and are often used as light absorbing materials (Xiong et al., 2018).

Therefore, this section introduces the synergistic effect between CNTs and MXene, further improving the photothermal, electrothermal, and sensing properties of composite smart paper. The specific analysis of the impact of the introduction of CNTs on the microstructure and various properties of composite paper is conducted.

The untreated AP was soaked in MXene dispersion and then soaked it in CNTs dispersion. These two "impregnation drying" processes form a group, and the sample is named based on the number of times used the "impregnation drying" group. For example, MCAP-3 represents AP soaked and dried three times in MXene dispersion and CNTs aqueous dispersion. Using the same method and steps, the composite paper prepared by soaking only in MXene dispersion is named MAP, while the composite paper prepared by soaking only in CNTs dispersion is named CAP.

The etching of MXene nanosheets is a prerequisite and key for the preparation of intelligent composite paper. MXene is a 2D sheet-like nanomaterial obtained by selectively removing the Al layer of the precursor ternary compound MAX phase using LiF and HCl as etchants and further ultrasonic layering. A picture of MXene nanoflake dispersion shows the typical Tyndall effect, indicating that the MXene nanoflake dispersion we prepared has colloidal stability and good dispersibility. TEM photo of MXene shows the single-layer sheet-like structure. XRD showed that MAX had numerous characteristic peaks, such as (002), (004), (101), and (104), before etching. However, after the etching and dispersion process, the (002) characteristic peak moved from $2\theta = 9.5$ degrees to $2\theta = 6.1$ degree to the left, and other characteristic peaks disappeared. This means that the spacing between layers after the removal of the Al layer increased, and MXene etching was successful. The XPS spectrum further shows that compared with MAX, MXene introduces the F element, which is caused by the −F functional group of MXene generated during the etching process. These active terminal functional groups of MXene can endow MXene nanosheets with hydrophilic properties, thereby enhancing the interfacial adhesion with the original dust-free paper substrate and contributing to the stability of the composite smart paper.

On the basis of the previous chapter, this chapter introduces CNTs, aiming to further improve the optical and electrical heating performance and sensing response performance of composite smart papers. Therefore, we first observed and compared the microstructure of composite smart paper. Fig. 5.4 shows SEM images and photos of composite smart paper. As shown in Fig. 5.4a, the surface of the AP fibers is very smooth before compounding. The composite paper fibers composed solely of MXene are covered with wrinkled layers formed by stacking 2D MXene nanosheets (Fig. 5.4b). The surface of the composite paper fibers composed solely of CNT is a typical 1D CNTs layer (Fig. 5.4c), while the fiber surface of MXene/CNT composite smart paper is interwoven with two-dimensional MXene nanosheets and one-dimensional CNTs (Fig. 5.4d−f), and as the number of impregnations increases, the coverage area has significantly increased. The enlarged structure in the figure can be seen (Fig. 5.4g), and the one-dimensional CNT link on the fiber surface of MXene/CNT composite smart paper acts as a 2D MXene layer, playing a bridging role between MXene nanosheets. Due to the excellent conductivity of CNT itself, it increases the conductive pathway between MXene nanosheets and reduces the stacking of pure MXene nanosheets. This is more helpful in improving response ability compared with composite paper that simply combines MXene and CNT, which will be analyzed and demonstrated in subsequent performance tests. The EDS diagram of MXene/CNT intelligent composite paper (MCAP-9) after nine impregnations. It can be clearly seen that the presence of CNT and MXene does not block the gaps in the dust-free

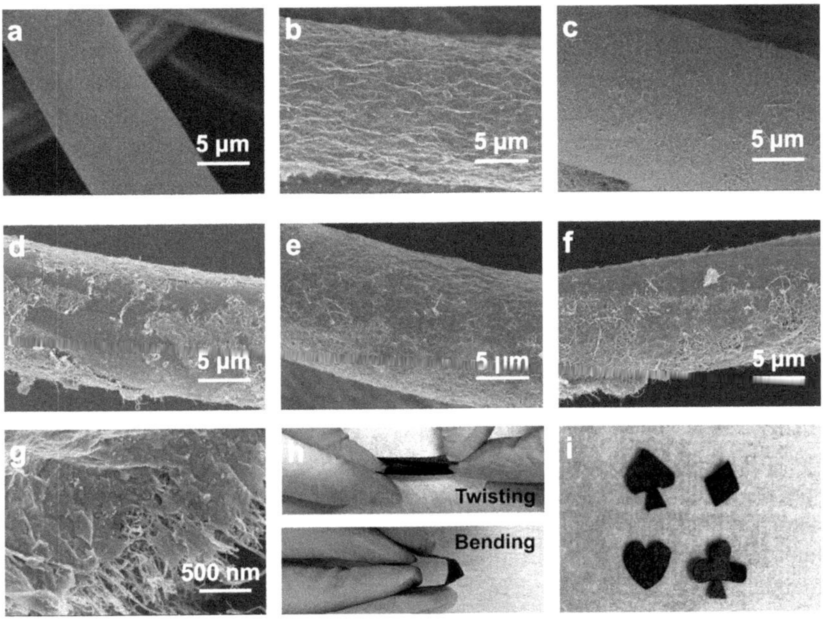

FIGURE 5.4 SEM images of (a) pristine AP fabrics; (b) MAP; (c) CAP; (d) MCAP-3; (e) MCAP-6; (f) MCAP-9; (g) MCAP-9 at high magnification. (h) Si-MAP with excellent flexibility during bending and twisting. (i) Si-MAP in various shapes.

paper, which also contributes to the breathability and practicality of the composite paper. The EDS diagram shows that C and Ti elements are uniformly dispersed on the fibers of the dust-free paper, indicating that CNTs and MXene are evenly distributed on the fiber surface. In addition, the composite MXene/CNT composite smart paper still inherits the good flexibility of the dust-free paper substrate, which can be twisted and bent freely, which contributes to the application and development of composite smart paper in the field of wearable materials. Moreover, MXene/CNT composite smart paper can be easily cut to different shapes without obvious limitations, which is conducive to personalized design of composite paper and the construction of wearable functional integrated devices.

According to previous reports, both MXene and CNTs have good photothermal conversion capabilities: under illumination conditions, the electron phonon relaxation process of the MXene surface plasma can convert the energy of the external light source into heat, while CNTs have high light absorption performance, which is of great significance for the multifunctional exploration of intelligent composite paper. Therefore, we tested the photoinduced heating ability of MXene/CNT composite smart paper. Using an 808 nm near-infrared (NIR) laser as the light source, with a diameter of approximately 13 mm, the MXene/CNT is irradiated. Composite papers with different

MXene/CNT ink impregnation times exhibit significant rapid photothermal conversion behavior under light irradiation. Among them, at an incident power of 1.25 W cm^{-2}, the maximum equilibrium temperature of MCAP-3 can reach 110.2°C, MCAP-6 can reach 158.7°C, and MCAP-9 can reach a higher equilibrium temperature of 187.6°C. This indicates that with the increase of ink impregnation times, the photothermal conversion effect of MXene/CNT intelligent composite paper is improved. By controlling the incident light power, the magnitude of temperature changes can also be adjusted. For example, when the incident light power is 0.25 W cm^{-2}, the equilibrium temperature of MCAP-9 is 104.7°C. When the incident light power is increased to 0.5 W cm^{-2}, the equilibrium temperature of MCAP-9 increases to 120.1°C. As the incident light power further increases, the equilibrium temperature of MCAP-9 increases from 142.7°C, 159.2 to 187.6 °C from 0.75 W cm^{-2}, 1 W cm^{-2} to 1.25 W cm^{-2}. Different MXene/CNT intelligent composite paper samples can exhibit good heating curves and good insulation state. When the light source is turned off, the composite paper can quickly cool. The cyclic heating and cooling of composite paper (MCAP-9) under multiple cycles of light irradiation (0.25 W cm^{-2}) show that it can still maintain a similar heating curve to the initial one after five cycles, and there is no significant attenuation change at a constant temperature of around 105°C. This indicates that MXene/CNT smart composite paper has stable performance in photoheating. To further illustrate the synergistic effect of MXene and CNTs, we prepared composite paper (MAP-9) using MXene alone and composite paper (CAP-9) using CNTs alone using the same method, and compared the performance of MXene/CNT intelligent composite paper (MCAP-9) with the same method. The results are shown in Fig. 5.4f. Under the same incident light source power density of 1.25 W cm^{-2}, MAP-9 showed an equilibrium temperature of 98.6°C, CAP-9 showed an equilibrium temperature of 122.8°C, and MCAP-9 had a higher equilibrium temperature than both, reaching 187.6°C. This is because the addition of CNTs further improved the material's light absorption performance. The thermal imaging images captured by an infrared camera further intuitively demonstrate the ability of the material to undergo rapid photoinduced heating. The temperature gradually increases from the beginning of external light irradiation and can reach 34.6°C within 0.1 s, 70.6°C within 1 s, and 155.2°C within 5 s, which is 85% of the equilibrium temperature. The three-dimensional thermal diagram indicated that only the portion illuminated by the light spot and a small amount of surrounding areas have the ability to respond to light heating and diffuse in a circular shape, which also provides the possibility for precise heating of wearable thermal managers for the human body.

In recent years, electric heating devices have received widespread attention in fields such as human thermal management and wearable devices. The principle of electric heating is Joule's law, so the electric heating performance is also known as Joule heating performance. In this study, in addition to photoheating, we also studied and tested the electroheating ability of MXene/CNT smart composite paper. It shows the temperature changes over time caused by the heat generated by different MXene/CNT smart composite paper samples under different external voltages. As the number of impregnations increases, the saturation temperature of MXene/CNT smart composite paper samples also increases: under an external voltage of 4 V, MCAP-3 can reach an equilibrium temperature of 82.4°C, MCAP-6 can reach an equilibrium temperature of 122.5°C, and MCAP-9 can reach an equilibrium temperature exceeding 170°C. When using samples with the same immersion frequency, the temperature is related to the voltage of the external power supply. For example, the equilibrium temperature of sample MCAP-9 can reach 32.6°C when connected to an external 1 V power supply, 49.3°C when connected to an external 2V power supply, 86.4°C when connected to 3 V, and 170.8°C when connected to 4V. This is because according to Joule's law, the heat generated by energizing a material is directly proportional to the magnitude of the external voltage and inversely proportional to the magnitude of the material's resistance. Therefore, the temperature of the material after being energized can be adjusted by setting different external power supply voltages as needed. It is evident that MCAP-9 has a higher equilibrium temperature and stronger electric heating ability than MAP-9 and CAP-9, which also demonstrates the synergistic effect of MXene and CNT in electric heating. The electrical heating performance of the MXene/CNT intelligent composite paper sample did not show significant changes over multiple cycles, highlighting its stability in operation.

In addition to having dual optical and electrical responsiveness, the assembled MXene/CNT intelligent composite paper can also be used as a pressure sensor and can monitor human motion signals. To test its sensing performance, we assembled MCAP-9. Fig. 5.5a shows photos of a flexible sensor based on MCAP-9. The assembly process involves placing two 1 cm × 1 cm MCAP-9 pieces opposite each other to form a pressure-sensitive element, which is then connected using silver paste, aluminum foil, etc. Finally, it is packaged using 3M company's medical semipermeable film. When no external pressure is applied, the gap between the two layers of MCAP-9 in the pressure-sensitive element is relatively large and the contact area is small. When pressure is applied to MCAP-9, the conductive fibers of MCAP-9 come closer to each other, generating more conductive points, leading to a decrease in overall resistance. After releasing the pressure load, the pressure-sensitive components composed of MCAP-9 return to their original state, causing the resistance to recover to its previous initial value, resulting in a corresponding relationship between pressure changes and resistance changes. Therefore, in this study, we analyzed the relative change in current Δ I/I0 of the pressure sensor under different pressure loads at an external voltage of 1 V, where I0 is the initial current value of the pressure sensor and Δ I is the current change value of the pressure sensor under

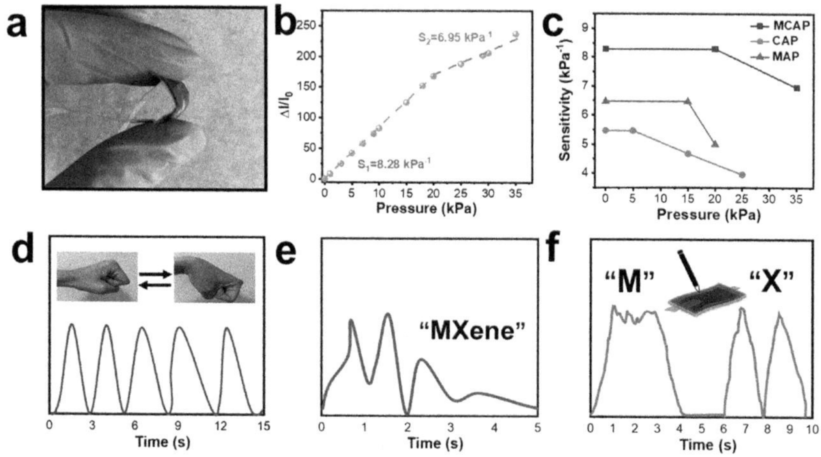

FIGURE 5.5 Electromechanical properties of the MCAP-9 pressure human motion monitoring. (a) Digital photograph of the flexible MCAP-9 pressure human motion monitoring; (b) Sensitivity of the MCAP pressure sensor; (c) sensitivity comparison of the MCAP, CAP, and MAP pressure sensor; (d) relative current changes of wrist bending; (e) relative current changes of speaking "MXene"; (f) relative current changes of writing "MX" on the sensor with pencil.

pressure. Fig. 5.5b shows the current changes of the composite paper pressure sensor under different pressures, which is called sensitivity (S).

Sensitivity is one of the most important features of pressure sensors, and when designing wearable sensors, sensitivity should be emphasized because high sensitivity helps to accurately capture subtle signals of human motion. Sensitivity can be adjusted through the morphology control strategy of the conductive network, and the sensitivity of the MXene pressure sensor can be improved through structural optimization and reasonable design. On the other hand, a wide response range (working window) is also an important parameter of the sensor, and a large working window can ensure sufficient detection range to meet the needs of full-dimensional human motion detection. By designing a special structure for sensor response-sensitive components, the working window of the sensor can also be extended.

In this study, it can be clearly seen that the relative current change value of the MCAP pressure sensor can be divided into two linear parts, that is, it has different sensitivity within two ranges. The pressure sensor assembled by MCAP has a wide pressure response range of 0–35 kPa, and within the pressure range of 0–20 kPa, the sensitivity can reach 8.28 kPa^{-1}. In the pressure range of 20–35 kPa^{-1}, the sensitivity can reach 6.95 kPa^{-1}. To explore the role of CNT and MXene, we used the same preparation strategy and assembly methods to assemble CAP-9 and MAP-9 into pressure sensors and tested their sensitivity. As shown in Fig. 5.5c, the pressure sensor assembled by CAP-9 has a pressure response range of 0–25 kPa^{-1}, while the pressure sensor assembled by MAP-9 has a pressure response range of 0–20 kPa^{-1}, both of which are narrower than MCAP pressure sensors and have significantly lower sensitivity than MCAP pressure sensors. We speculate that the wide pressure response range and high sensitivity of MCAP are mainly attributed to the synergistic effect of 2D MXene nanosheets and one-dimensional CNT nanotubes. The previous SEM results mentioned that the insertion of CNTs into layered MXene layers plays a bridging role, which can improve the response ability of pressure sensors to a certain extent. Moreover, the insertion of CNTs also avoids the stacking of 2D MXene nanosheets, increases the conductive path between layers, and further improves sensitivity.

The pressure sensor based on intelligent composite paper has excellent flexibility and sensitivity, and to test its practicality, we use it as a wearable device that can attach to the skin for real-time monitoring of human motion signals. As shown in Fig. 5.5d, during the repeated bending of the wrist, the sensor displays a clear and regular electrical signal. In addition, attaching the strain sensor to the throat can also monitor the movement of the throat and small deformations of related muscles. When the subject emits "MXene," the output signal of the sensor is shown in Fig. 5.5e. The pressure sensor can also be applied in fields such as handwriting recognition (Fig. 5.5f), which also expands the application range of the pressure sensor. Finally, Fig. 5.6 shows the application of the MCAP flexible pressure sensor in monitoring the pulse signal of the human wrist. As shown in the photo, the MCAP sensor is fixed to the human wrist and records its relative current changes after being connected to an electrochemical workstation. Due to the sufficiently low detection limit and high sensitivity of MCAP, the three typical characteristic peaks of human pulse waveform, namely percussion wave (P-wave), tidal wave (T-wave), and heavy pulse wave (D-wave), are clearly distinguishable. This further indicates that flexible MCAP pressure sensors have broad application prospects in the field of wearable health monitoring equipment for the human body.

This section introduces a synergistic effect between 1D nanomaterial CNTs and 2D nanomaterial MXene, further improving the photothermal,

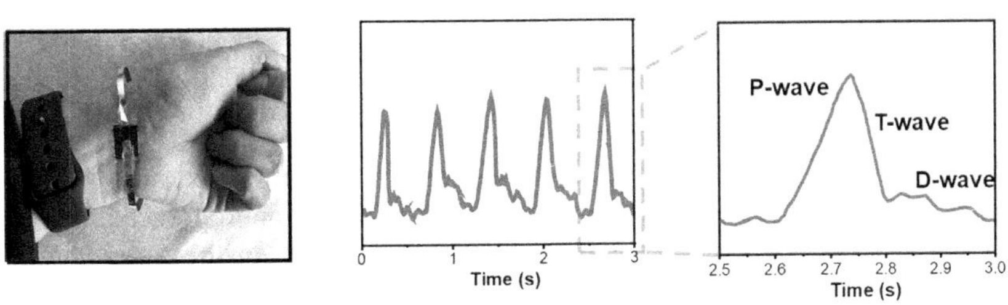

FIGURE 5.6 Relative current changes of pulse.

electrothermal, and sensing properties of composite smart paper. The impact of the introduction of CNTs on the microstructure and various properties of composite paper is analyzed in detail. The introduction of CNTs further enhances the photothermal, electrothermal, and sensing properties of composite paper. Compared with composite paper control samples prepared using only MXene or CNTs using the same method, the MCAP constructed in this section achieved higher equilibrium temperatures for photoheating and electroheating under the same external light source irradiation and voltage conditions. The bridging effect of 1D CNTs insertion between MXene layers endows MXene and CNTs with a synergistic effect, allowing the pressure sensor assembled by MCAP to exhibit a wide pressure response range and better sensitivity. It can also accurately and real-time monitor various human motion signals such as wrist bending, throat muscle tremors, vocalizations, handwriting, and human pulse beats.

5.2.3 Cellulose nanofiber/MXene liquid electrode friction nanogenerator

With the rapid development of wearable/portable electronic devices, the demand for high-performance power sources is also increasing (Shabangoli et al., 2018; Xia et al., 2018). Although researchers have made some progress in utilizing green energy sources such as solar, thermal, and wind energy (Ding et al., 2016; Liu and Chen, 2020; Suarez et al., 2016), these green energy sources are highly dependent on the external environment and difficult to achieve sustained energy collection and conversion. The mechanical energy generated by the movement of organisms is less affected by environmental factors and is considered a renewable and sustainable source of green energy. Based on the coupling effect of friction electrification and electrostatic induction, friction nanogenerator (TENGs) can effectively transform the mechanical energy generated by biological movement into electrical energy (Fan et al., 2019). TENGs have made significant progress in applications such as human–machine interfaces, drug delivery, implantable electronics, and health monitoring due to their wide range of material sources, light weight, ease of manufacturing, and high energy conversion efficiency (Chandrashekar et al., 2015; Yang et al., 2019; Wang ZL et al., 2015). At present, TENGs mainly include four working modes: vertical contact separation, planar sliding, single electrode, and independent layer (Yi et al., 2019). Among them, single-electrode TENGs have the characteristics of simple structure and easy portability, and have a broader application prospect in wearable electronics (Mule et al., 2019; Shankaregowda et al., 2019). Choosing appropriate electrode materials is one of the most important prerequisites for ensuring the normal operation of single electrode TENGs. Liquids with high electron mobility and fluidity, such as liquid metal (Tang W et al., 2015), physiological saline (Zou et al., 2019), and graphite oxide dispersion (Wu et al., 2019b), are considered as effective electrode materials, which can be combined with flexible elastomers to build tensile single-electrode TENGs. However, the electrical output performance of liquid electrode–based TENGs is generally low, and the synthesis process is relatively complex, which to some extent limits the further development of liquid electrode–based TENGs.

To solve the aforementioned problems, this section will prepare a stretchable, shape-adaptive friction nanogenerator (CM-TENG) with the TOCNFs/MXene mixture as the liquid electrode. We analyzed the power generation mechanism of CM-TENG and studied the effects of factors such as MXene content in liquid electrodes, surface roughness of silica gel, and different contact frequencies on the electrical output performance of CM-TENG. Moreover, we have designed an energy collection and conversion system that converts the collected mechanical energy into electrical energy, which can charge electronic devices such as electronic watches, temperature sensors, and luminous hair bands. Specifically, the CM-TENG can serve as a self-driving mechanical sensor that can monitor various human joint motion behaviors and has important development prospects in fields such as medical health monitoring.

The tensile TOCNFs/MXene liquid electrode–based friction nanogenerator (CM-TENG) was prepared by a typical mold casting method. Firstly, liquid silicone was poured into a premade PTFE mold, precured at room temperature for 2 h, and then cured in an 80°C oven for 1 h. The cured silicone film was removed from the mold. At the same time, using sandpaper with different roughness as a template, a silicone film with rough surface was prepared according to the aforementioned method. The smooth surface of silicone with an oxygen plasma cleaning machine (IoN 40, PVA TePla) was treated for 1 min to activate its surface. Then, we bond the edges of the two layers of silicone together to form a sealing space and connect the internal and external environments with conductive copper wires. Different proportions of TOCNFs/MXene mixtures were injected into a silicone-sealed chamber as liquid electrodes to obtain flexible CM-TENG.

After mixing and stirring MXene nanosheet dispersion and TOCNFs dispersion at different ratios at room temperature for 12 h, a stable and dispersed TOCNFs/MXene mixture can be obtained, indicating that MXene and TOCNFs have good compatibility in the aqueous phase (Fig. 5.7a). One of the main reasons why the TOCNFs/MXene mixture can maintain stable dispersion is the steric hindrance effect between the carboxyl groups on the surface of TOCNFs. In addition, the Coulombic repulsion between MXene nanosheets (−24.9 mV) and TOCNFs (−63.4 mV) is also an important factor in maintaining stable dispersion of the mixture (Fig. 5.7b). After storing the TOCNFs/MXene mixture in a 4°C refrigerator for ~6 months, it can still maintain a relatively stable Zeta potential (~50 mV) (Fig. 5.8a). On the contrary, the absolute zeta potential of MXene dispersion significantly weakened after ~6 months, indicating a decrease in colloidal stability.

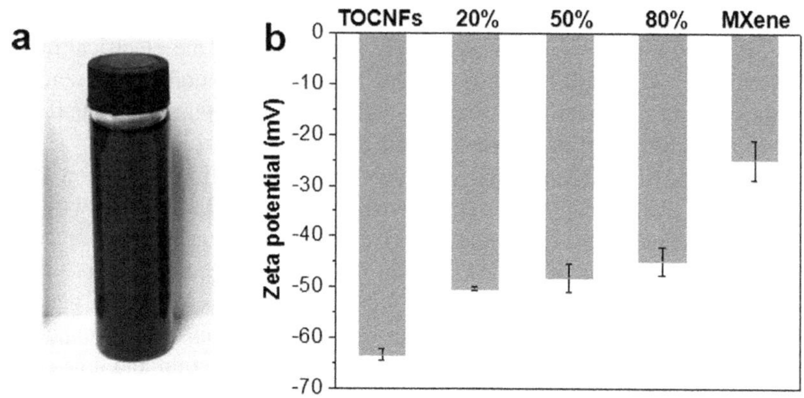

FIGURE 5.7 (a) Optical image of CNFs/MXene mixture. (b) Zeta potential diagram of TOCNFs/MXene mixture with various MXene contents. *Reproduced from Cao et al. (2020) with permission from Wiley-VCH.*

Fig. 5.8b and c compares the state of MXene dispersion and TOCNFs/MXene mixture after approximately 6 months of storage in a 4°C refrigerator. It can be seen that after several months of storage, MXene dispersion exhibits visible precipitation clusters. In contrast, the TOCNFs/MXene mixture showed almost no color change or aggregation. In addition, the rich functional groups on the surface of MXene nanosheets, such as—OH, —F, and = O, can form hydrogen bonds with TOCNFs. Therefore, TOCNFs can not only serve as dispersants to ensure stable dispersion of the mixture, but also as connectors to promote hydrogen bonding between MXene nanosheets. Finally, we injected the TOCNFs/MXene mixture into a silicone vacuum using a syringe to obtain a stretchable and shape-adaptive CM-TENG. The CM-TENG has excellent flexibility and can be easily folded, stretched, and twisted. In addition, when the palm lightly taps CM-TENG, the generated electricity can simultaneously light up 230 connected LED lights.

The power generation mechanism of CM-TENG is mainly based on the coupling effect of friction power generation and electrostatic induction. Considering that the plasma-activated silicone rubber film has abundant hydrophilic OH groups, MXene nanosheets with surface terminal functional groups can be adsorbed on the activated surface of silicone rubber under hydrogen bonding. At the same time, the presence of TOCNFs and water molecules in the liquid electrode can also promote the connection between MXene nanosheets and the surface of activated silicone rubber. In addition to the conductive TOCNFs/MXene dispersion connected to the floor, the skin is also grounded as a part of the human body, thus constructing a single electrode TENG. When human skin comes into contact with CM-TENG, due to the influence of contact electrification, electrons transfer from the skin to the surface of silicone rubber and the adsorbed MXene nanosheets until saturation

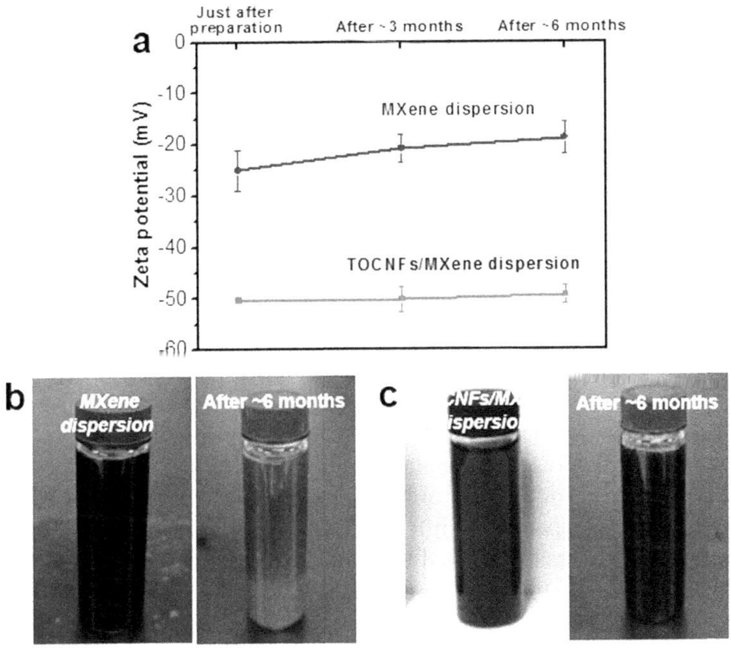

FIGURE 5.8 (a) Zeta potential values of MXene dispersion and TOCNFs/MXene mixture after being placed for different time. Optical images of (b) MXene dispersion and (c) TOCNFs/MXene mixture after ~6 months. *Reproduced from Cao et al. (2020) with permission from Wiley-VCH.*

(state i). The combination of silicone rubber film and adsorbed MXene nanosheets is considered as a frictional charged layer. Among them, MXene nanosheets show higher electronegativity, which can enhance the attraction of the silicone layer to electrons, thus attracting more mobile electrons on the skin surface and generating more negative charges. When the skin is away from silicone rubber, positive charges are generated in the TOCNFs/MXene dispersion, causing instantaneous electrons to flow from the electrode to the ground through an external load, forming an electrical signal (state ii). When the skin and silicone rubber are completely separated, the induced positive charge in the TOCNFs/MXene dispersion will be completely balanced by the negative charge on the surface of the silicone rubber, thus achieving electrostatic equilibrium (state iii). When the skin approaches silicone rubber again, the positive charge sensed in the liquid electrode is neutralized by free electrons flowing from the ground, forming a reverse electrical output signal (state iv). Therefore, repeated contact separation movements between human skin and CM-TENG will generate alternating current. To quantitatively verify the working mechanism, we used COMSOL software to simulate the potential distribution of CM-TENG in contact separation mode. When the human skin approaches or moves away from silicone rubber, the potential difference of the friction layer also changes accordingly. Under the action of changing potential differences, free electrons will be driven and transferred to achieve electrostatic equilibrium. During this process, alternating current can be generated in the external circuit.

To evaluate the electrical output performance of the nanogenerator, we use a commercial PTFE film as a substitute for human skin. Driven by a linear motor, we use it with CM-TENG (with an area of 8×4 cm^2) for cyclic contact separation. Due to the rough micro/nanostructure of the friction layer surface being an important factor affecting the electrical output performance of TENG, we prepared silica gel thin films with rough surfaces using sandpaper as a template. The surface SEM images of silicone rubber films were obtained using various mesh sizes of sandpaper as templates. Compared with the untreated silicone rubber film with a smooth surface, the surface of the silicone film prepared using sandpaper as a template will generate a large number of microporous structures, resulting in a significant increase in the surface area of the silicone film. We also investigated the relationship between CM-TENG prepared with sandpaper of different mesh sizes and its electrical output performance. The results showed that the higher the mesh size of sandpaper, the better the electrical output performance of CM-TENG prepared using this template. This is mainly because the silicone film prepared by sandpaper with a larger mesh size has a rougher surface and more pores, which can increase the contact area between the friction layers. The electrical output performance of CM-TENG prepared using 320 grit sandpaper as a template is nearly twice that of untreated CM-TENG. In this chapter, the highest values of open circuit voltage (Voc), short circuit current (Isc), and transfer charge quantity (Qsc) of CM-TENG can reach 300 V, 5.5 μA, and 120 nC, respectively.

In addition, we also systematically studied the electrical output characteristics of CM-TENG with different d-$Ti_3C_2T_x$ MXene nanosheet contents. We prepared a mixed liquid level liquid electrode with MXene nanosheets and TOCNFs with a solid content ratio of 20:80, 50:50, and 80:20, and the resulting CM-TENG was represented as CM-TENG-20%, CM-TENG-50%, and CM-TENG-80%, respectively. In addition, we also prepared friction nanogenerator (C-TENG) with pure TOCNFs dispersion as liquid electrode. The electrical output performance of CM-TENG is much higher than that of C-TENG. This is mainly due to the superhigh electronegativity of MXene nanosheets, which can provide additional charge storage layers for the silicone layer to increase electrical output. Moreover, as the weight ratio of MXene nanosheets to CNFs further increases to 80:20, the Voc value of CM-TENG can reach ~300 V, approximately five times that of C-TENG.

The frequency of contact separation between friction layers is also an important factor affecting the output performance of CM-TENG. When the contact frequency between the friction layers increases from 0.5 to 3.0 Hz, the Voc and Qsc values of CM-TENG remain basically unchanged, at ~300 V and ~120 nC, respectively. Unlike this, as the working frequency increases, the

Isc value of CM-TENG increases from ~5.5 to ~7.8 μA. This is mainly due to the higher working frequency changing the contact time between the two friction layers, resulting in higher charge flow and ultimately higher current output. In addition, we also analyzed the response time of CM-TENG to external stimuli. The response time of CM-TENG at a constant operating frequency of 2 Hz is approximately 26 ms. The ultrashort response time ensures that CM-TENG can detect signals in real-time under external forces. To verify that a single-electrode CM-TENG can be driven by human skin, we slapped CM-TENG to monitor its electrical output performance. CM-TENG can generate a stable output voltage of ~350 V at a working frequency of approximately 2.5 Hz. In addition, we also investigated the effect of environmental humidity on the electrical output performance of CM-TENG. The results indicate that the peak output voltage of CM-TENG decreases with increasing environmental relative humidity. This is mainly because the relative dielectric constant of air decreases with increasing relative humidity, resulting in a corresponding decrease in the output voltage of CM-TENG. Because the comfortable relative humidity range for the human body is 45%–65%, CM-TENG can still generate stable electrical signal output even at a relative humidity of ~70%. The aforementioned results indicate that CM-TENG can still maintain normal operation in different humidity environments. In addition, we also compared the electrical output performance of the prepared stretchable CM-TENG with other liquid electrode–based TENGs. The TENGs based on TOCNFs/MXene liquid electrodes prepared in this section exhibit higher Voc values compared with other liquid electrodes, including liquid metal (Tang et al., 2015; Yang et al., 2018), NaCl solution (Zou et al., 2019; Wang et al., 2017; Liu et al., 2018), and others (Shi et al., 2019; Yuan et al., 2019; Wu et al., 2019), based on TENGs. The excellent electrical output performance of CM-TENG is mainly attributed to the high electronegativity of MXene nanosheets and their inherent high conductivity.

The deformation ability and durability are important factors that affect the working life of devices. Therefore, in this chapter, we also studied the electrical output capability of CM-TENG under various deformation conditions. CM-TENG has excellent flexibility, can be easily stretched to 120% deformation, and can fully recover to its initial state. The electrical output performance of CM-TENG under different tensile strain states indicated that the Voc, Isc, and Qsc values of CM-TENG show a trend of first increasing and then decreasing. The trend of change in output performance is mainly caused by the synergistic effect of the contact area and thickness of the friction layer during the working process of CM-TENG. When the horizontal strain of CM-TENG increases, the thickness of silica gel will become thinner according to the Poisson effect. Reducing the thickness of the elastomer can shorten the distance between the surface charge of the silicone gel and the TOCNFs/MXene electrode, thereby increasing the output current of CM-TENG. Meanwhile, when the device is stretched, the contact area between the friction layers first increases and then decreases, and the electrical output performance of CM-TENG also changes accordingly. Therefore, as the tensile strain level of CM-TENG increases from 0% to 100%, the surface area and thickness of the silicone layer increase, resulting in a gradual improvement in the electrical output performance of CM-TENG. When CM-TENG is stretched to 120% deformation, as the surface area of the silicone layer decreases, the electrical output performance of CM-TENG also decreases. In addition, we also investigated the changes in electrical output performance of CM-TENG under twisted and folded states. After various deformations, the electrical output performance of CM-TENG can still be restored to its initial value, indicating its good mechanical durability. Moreover, when we tap the CM-TENG in the stretched and twisted state with our hands, the generated electricity can still light up about 230 LEDs, indicating that CM-TENG has the possibility of application under extreme conditions. We also compared CM-TENG with several other MXene-based TENGs (Dong et al., 2018; Wu et al., 2019; Jiang et al., 2019a,b) in terms of stretchability, Voc peak, and output power density. It can be seen that other work involves preparing MXene as a solid film for electrode or friction layers, while this work uses highly conductive MXene dispersion as a liquid electrode to prepare single-electrode TENG. In addition, MXene nanosheets with high electronegativity can be adsorbed on the surface of silica gel to improve the electron capture ability of silica gel. Therefore, the CM-TENG constructed in this chapter can effectively combine the advantages of silicone and MXene, while achieving excellent mechanical tensile properties and high electrical output performance.

In addition, we also analyzed the stability of CM-TENG under continuous operation. The output voltage of CM-TENG can always be stable at around 300 V during ~6000 contact separation cycles at a frequency of 2 Hz. Moreover, after 1000 50% strain tensile cycles, the output voltage of CM-TENG did not show significant changes. Therefore, CM-TENG with TOCNFs/MXene dispersion as electrode exhibits excellent mechanical stability and good reliability in continuous working environments. At the same time, single-electrode TENG also needs to have the ability to perceive different objects to be better applied in practical life. We selected several commonly used materials, such as aluminum (Al), regular A4 paper (Paper), latex (Latex), Kapton film, and polyethylene terephthalate (PET), to test the power generation performance of CM-TENG and compared the output voltage of different materials. The contact between silicone and different materials will result in different electrical output performance of CM-TENG. This is mainly because the power generation performance of TENG is affected by many factors, including the electronegativity, viscosity, surface roughness, etc. of the material itself, which will affect the amount of charge induced by contact friction. The research results of this chapter indicate that CM-TENG can generate the highest output voltage when in contact with PET film, while the electrical performance when in contact with aluminum foil is poor.

Based on the excellent performance of CM-TENG mentioned before, we found that the device exhibits great potential in energy collection. In this chapter, we use an external resistor as the load to evaluate the output power performance of CM-TENG. As the external resistance increases from 105 to 109 Ω, the output voltage of CM-TENG shows a positive correlation trend. At the same time, we observed a decreasing trend in the output current of CM-TENG as the external resistance value increases. Under an external resistive load of 50 M Ω, CM-TENG can achieve a maximum instantaneous power density value of 504.9 mW m^{-2}. This electrical output capability can drive some small electronic products, providing suitable innovative strategies for the sustainable energy supply demand of wearable electronic products.

In addition, to investigate the effectiveness of single-electrode TENG as a sustainable power source for wearable devices, we constructed a self-charging energy conversion system based on CM-TENG, including commercial capacitors, flexible electronic devices, and bridge rectifiers. CM-TENG is used to obtain mechanical energy, and then the generated alternating current is output as direct current using a rectifier bridge to charge the capacitor. We studied the voltage across 22 μF commercial capacitors of CM-TENG at different operating frequencies of 0.5−3.0 Hz to evaluate its charging capacity. It can be observed that as the working frequency increases, the charging speed of the capacitor by CM-TENG gradually accelerates. For a 1 μF capacitor, the voltage at both ends of the capacitor can reach 3 V within 50 s. For a 10 μF capacitor, the voltage at both ends can reach 1.5 V in 100 s. In addition, the self-charging system based on CM-TENG also has excellent continuous working ability. We slapped CM-TENG with our hands at a frequency of ~2 Hz to study its ability to charge capacitors. From the figure, it can be seen that by tapping CM-TENG with your hand for approximately 55 s, the voltage at both ends of the 10 μF capacitor can reach 3 V. Subsequently, we further explored the ability of CM-TENG's self-charging system to directly drive commercial electronic products. CM-TENG can convert the mechanical energy of hand tapping into electrical energy, providing power for electronic watches, temperature sensors, or luminous hair bands. For example, tapping the CM-TENG with your hand for about 60 s can generate electricity that can drive the electronic watch to operate normally for about 60 s. Therefore, the aforementioned research indicates that energy conversion systems based on CM-TENG have great potential in providing sustainable and recyclable energy for wearable electronic products.

The CM-TENG prepared in this chapter has excellent stretchability, deformability, and durability, making it a self-powered mechanical sensor for real-time monitoring of various human activities. Real-time and continuous monitoring of human movement can provide effective biofeedback for human physiological evaluation. Firstly, we attach CM-TENG to the finger joint to monitor finger bending motion. As shown in the figure, by recording the changes in the output voltage signal of CM-TENG, we can achieve precise and real-time monitoring of finger bending motion. The peak output voltage of CM-TENG varies with the variation of finger bending angle. Compared with a smaller bending angle (30°), when the finger bending angle (90°) is larger, it can cause the CM-TENG sensor to produce higher output voltage values. This is mainly because a larger bending angle will increase the contact area between the silicone and the skin, resulting in an increase in the output voltage value of CM-TENG. In addition, as the finger bending frequency increases from 0.5 to 1.5 Hz, the maximum output voltage value of CM-TENG remains almost unchanged, while the peak value of the voltage signal increases from 1 to 3 within 2 s.

In addition, we can also identify fingers tapping CM-TENG with different pressures by identifying different voltage output signals (Fig. 5.9a). The CM-TENG sensor can be attached to the wrist joint to monitor the bending movement of the tester's wrist. As shown in Fig. 5.9b, when the wrist is bent, the output voltage of CM-TENG can reach 25 V. After the wrist is bent and restored, the output voltage value drops to zero. In addition to joint movements with significant deformation, the CM-TENG sensor can also capture subtle physiological signals in the human body. As shown in Fig. 5.9c, we attached CM-TENG to the human neck to monitor small deformations of the skin and

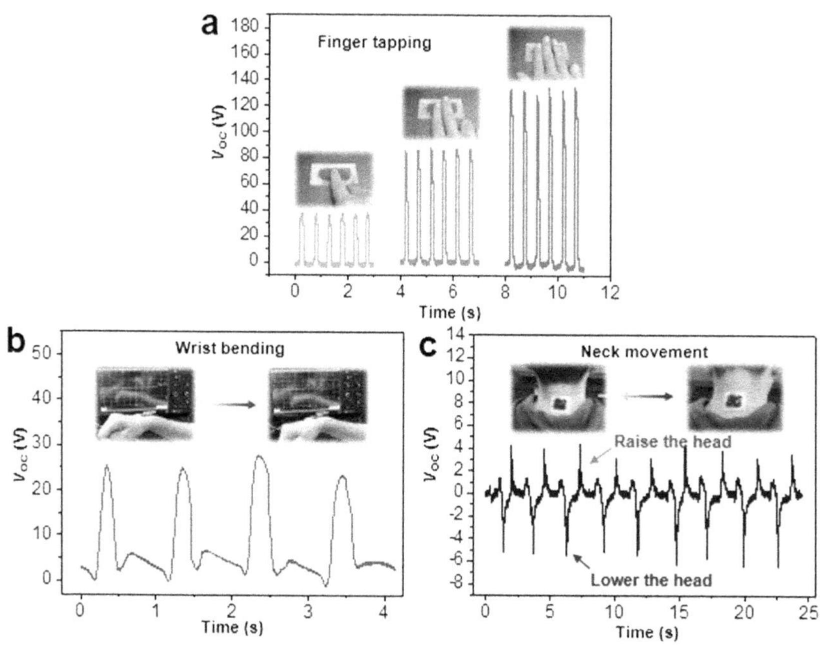

FIGURE 5.9 Optical images and corresponding output voltage signal of CM-TENG under (a) finger tapping, (b) wrist bending, and (c) neck movement. *Reproduced from Cao et al. (2020) with permission from Wiley-VCH.*

muscles. It can be seen that by recording and analyzing the output voltage signal of the sensor, the movement of a person's head up or down can be accurately monitored. At the same time, we also used the CM-TENG sensor to test human arm bending and knee bending movements. When the angle of joint bending is not changed but only the speed of bending is changed, the CM-TENG sensor can achieve precise monitoring of the above movements, and its output voltage signal peak remains almost unchanged. In addition, we have integrated CM-TENG with multi-channel wireless signal transmission modules to construct a system for collecting human motion signals and wireless transmission. The results indicate that using Bluetooth technology can obtain real-time and accurate wrist bending motion signals. Therefore, self-powered CM-TENG sensors can monitor the bending angle of human joints by analyzing the recorded peak voltage output and calculate the motion speed by calculating the number of peaks.

In this section, the cellulose nanofiber/MXene (TOCNFs/MXene) mixture is used as the conductive liquid electrode to prepare a stretchable, shape-adaptive friction nanogenerator (CM-TENG), which has potential applications in soft robot, human kinematics, biomechanics, and other fields. TOCNFs enhance the stability of MXene aqueous dispersion, resulting in excellent flowability of the TOCNFs/MXene mixture, while endowing CM-TENG with durability and stability for long-term operation. With the TOCNFs/MXene mixture with high conductivity and high electronegativity as the liquid electrode, the CM-TENG constructed shows good flexibility (stretching, twisting, curling, etc.) and excellent electrical output performance (V_{oc} = 300 V). A CM-TENG-based energy conversion system has been designed, which can collect human mechanical energy and convert it into electrical energy, which can be stored in capacitors to power electronic products. CM-TENG can be attached to human skin and clothing as a self-powered electrical sensor, monitoring the frequency and amplitude of various physiological movements remotely, accurately, and in real time.

5.3 Chapter summary

In summary, this chapter introduced the synthesis and property of advanced functional cellulose-based nanocomposites for electronic devices via three typical examples, such as multiple response hydrophobic AP/MXene smart nanocomposite, CNT-enhanced AP/MXene smart nanocomposite, and cellulose nanofiber/MXene liquid electrode friction nanogenerator. It should be pointed out, however, these are still some problems requiring improvement. The preparation of cellulose-based nanocomposites may involve multiple steps and complex chemical reactions, which increases the difficulty and cost of their preparation. For large-scale production, this complexity may become a limiting factor. Although nanocellulose has excellent mechanical properties and flexibility, when combined with other nanomaterials, it may face the problem of unstable performance. For example, the interface compatibility and long-term stability between different materials need to be further optimized. Although there has been some research progress in nanocellulose-based conductive composites, ensuring a balance between high conductivity and mechanical properties remains a challenge. In addition, compared with traditional conductive materials, the conductivity of these composites may still need to be improved. Integrating cellulose-based nanocomposites into existing electronic manufacturing processes may face some technical challenges. For example, compared with traditional inorganic semiconductor materials, these materials may have differences in processing temperature, pressure, and compatibility. Electronic devices require long-term stable operation, so the reliability and durability of cellulose-based nanocomposites are crucial. At present, further research and verification are needed on the performance of these materials in long-term use and harsh environments. To achieve the widespread application of cellulose-based nanocomposites in electronic devices, it is also necessary to solve the problems of standardized and scaled production. This includes establishing unified quality standards, optimizing production processes, and ensuring the stability of the supply chain. In summary, although cellulose-based nanocomposites have broad applications in the field of electronic devices, achieving their commercial applications still requires solving a series of technical and production problems.

Acknowledgments

The financial support from the National Key R&D Program of China (2023YFC3905701) and National Natural Science Foundation of China (52202108) is gratefully acknowledged.

References

Baughman, R.H., Zakhidov, A.A., de Heer, W.A., 2002. Carbon nanotubes-the route toward applications. Science 297 (5582), 787–792.

Cai, Y., Shen, J., Ge, G., Zhang, Y., Jin, W., Huang, W., Shao, J., Yang, J., Dong, X., 2018. Stretchable Ti 3 C 2 T x MXene/carbon nanotube composite based strain sensor with ultrahigh sensitivity and tunable sensing range. ACS Nano 12 (1), 56–62. https://doi.org/10.1021/acsnano.7b06251.

Cao, W.T., Ouyang, H., Xin, W., Chao, S., Ma, C., Li, Z., Chen, F., Ma, M.G., 2020. A stretchable highoutput triboelectric nanogenerator improved by MXene liquid electrode with high electronegativity. Adv. Funct. Mater. 30 (50). https://doi.org/10.1002/adfm.202004181.

Chandrashekar, B.N., Deng, B., Smitha, A.S., Chen, Y., Tan, C., Zhang, H., Peng, H., Liu, Z., 2015. Roll-to-Roll green transfer of CVD graphene onto plastic for a transparent and flexible triboelectric nanogenerator. Adv. Mater. 27 (35), 5210–5216. https://doi.org/10.1002/adma.201502560.

Ding, R., Liu, H., Zhang, X., Xiao, J., Kishor, R., Sun, H., Zhu, B., Chen, G., Gao, F., Feng, X., Chen, J., Chen, X., Sun, X., Zheng, Y., 2016. Flexible piezoelectric nanocomposite generators based on formamidinium lead halide perovskite nanoparticles. Adv. Funct. Mater. 26 (42), 7708–7716. https://doi.org/10.1002/adfm.201602634.

Dong, L., Xu, C., Li, Y., Pan, Z., Liang, G., Zhou, E., Kang, F., Yang, Q.-H., 2016. Breathable and wearable energy storage based on highly flexible paper electrodes. Adv. Mater. 28 (42), 9313–9319. https://doi.org/10.1002/adma.201602541.

Dong, Y., Mallineni, S.S.K., Maleski, K., Behlow, H., Mochalin, V.N., Rao, A.M., Gogotsi, Y., Podila, R., 2018. Metallic MXenes: a new family of materials for flexible triboelectric nanogenerators. Nano Energy 44, 103–110. https://doi.org/10.1016/j.nanoen.2017.11.044.

Duan, L., D'Hooge, D.R., Cardon, L., 2020. Recent progress on flexible and stretchable piezoresistive strain sensors: from design to application. Prog. Mater. Sci. 114, 100617.

Fan, X., Ding, Y., Liu, Y., Liang, J., Chen, Y., 2019. Plasmonic Ti 3 C 2 T x MXene enables highly efficient photothermal conversion for healable and transparent wearable device. ACS Nano 13 (7), 8124–8134. https://doi.org/10.1021/acsnano.9b03161.

Fu, Z., Wang, N., Legut, D., Si, C., Zhang, Q., Du, S., Germann, T.C., Francisco, J.S., Zhang, R., 2019. Rational design of flexible two-dimensional MXenes with multiple functionalities. Chem. Rev. 119 (23), 11980–12031. https://doi.org/10.1021/acs.chemrev.9b00348.

Gao, L., Li, C., Huang, W., Mei, S., Lin, H., Ou, Q., Zhang, Y., Guo, J., Zhang, F., Xu, S., Zhang, H., 2020. MXene/polymer membranes: synthesis, properties, and emerging applications. Chem. Mater. 32 (5), 1703–1747. https://doi.org/10.1021/acs.chemmater.9b04408.

Geng, L., Zhu, P., Wei, Y., Guo, R., Xiang, C., Cui, C., Li, Y., 2019. A facile approach for coating Ti3C2Tx on cotton fabric for electromagnetic wave shielding. Cellulose 26 (4), 2833–2847. https://doi.org/10.1007/s10570-019-02284-5.

Heo, J.S., Hossain, M.F., Kim, I., 2020. Challenges in design and fabrication of flexible/stretchable carbon-and textile-based wearable sensors for health monitoring: a critical review. Sensors 20 (14), 3927.

Huang, W., Hu, L., Tang, Y., Xie, Z., Zhang, H., 2020. Recent advances in functional 2D MXene-based nanostructures for next-generation devices. Adv. Funct. Mater. 30 (49). https://doi.org/10.1002/adfm.202005223.

Jiang, C., Li, X., Yao, Y., Lan, L., Shao, Y., Zhao, F., Ying, Y., Ping, J., 2019a. A multifunctional and highly flexible triboelectric nanogenerator based on mxene-enabled porous film integrated with laser-induced graphene electrode. Nano Energy 66, 104121.

Jiang, C., Wu, C., Li, X., Yao, Y., Lan, L., Zhao, F., Ye, Z., Ying, Y., Ping, J., 2019b. All-electrospun flexible triboelectric nanogenerator based on metallic MXene nanosheets. Nano Energy 59, 268–276.

Jin, H., Matsuhisa, N., Lee, S., Abbas, M., Yokota, T., Someya, T., 2017. Enhancing the performance of stretchable conductors for E-textiles by controlled ink permeation. Adv. Mater. 29 (21). https://doi.org/10.1002/adma.201605848.

Keum, K., Kim, J.W., Hong, S.Y., Son, J.G., Lee, S.S., Ha, J.S., 2020. Flexible/stretchable supercapacitors with novel functionality for wearable electronics. Adv. Mater. 32 (51). https://doi.org/10.1002/adma.202002180.

Khan, R., Andreescu, S., 2020. MXenes-based bioanalytical sensors: design, characterization, and applications. Sensors 20 (18). https://doi.org/10.3390/s20185434.

Khazaei, M., Arai, M., Sasaki, T., Chung, C.Y., Venkataramanan, N.S., Estili, M., Sakka, Y., Kawazoe, Y., 2013. Novel electronic and magnetic properties of two-dimensional transition metal carbides and nitrides. Adv. Funct. Mater. 23 (17), 2185–2192. https://doi.org/10.1002/adfm.201202502.

Kim, T., Kim, Y.W., Lee, H.S., Kim, H., Yang, W.S., Suh, K.S., 2013. Uniformly interconnected silver-nanowire networks for transparent film heaters. Adv. Funct. Mater. 23 (10), 1250–1255. https://doi.org/10.1002/adfm.201202013.

Lee, J., Llerena Zambrano, B., Woo, J., Yoon, K., Lee, T., 2020. Recent advances in 1D stretchable electrodes and devices for textile and wearable electronics: materials, fabrications, and applications. Adv. Mater. 32 (5). https://doi.org/10.1002/adma.201902532.

Lin, H., Wang, X., Yu, L., Chen, Y., Shi, J., 2017. Two-dimensional ultrathin MXene ceramic nanosheets for photothermal conversion. Nano Lett. 17 (1), 384–391.

Ling, Y., An, T., Yap, L.W., Zhu, B., Gong, S., Cheng, W., 2020. Disruptive, soft, wearable sensors. Adv. Mater. 32 (18). https://doi.org/10.1002/adma.201904664.

Liu, Y., Chen, Y., 2020. Integrated perovskite/bulk-heterojunction organic solar cells. Adv. Mater. 32 (3). https://doi.org/10.1002/adma.201805843.

Liu, R., Kuang, X., Deng, J., Wang, Y.C., Wang, A.C., Ding, W., 2018. Shape memory polymers for body motion energy harvesting and self-powered mechanosensing. Adv. Mater. 30 (8), 1705195.

Lukatskaya, M.R., Mashtalir, O., Ren, C.E., Dall'Agnese, Y., Rozier, P., Taberna, P.L., Naguib, M., Simon, P., Barsoum, M.W., Gogotsi, Y., 2013. Cation intercalation and high volumetric capacitance of two-dimensional titanium carbide. Science 341 (6153), 1502–1505. https://doi.org/10.1126/science.1241488.

Luo, J., Huo, L., Wang, L., Huang, X., Li, J., Guo, Z., Gao, Q., Hu, M., Xue, H., Gao, J., 2020. Superhydrophobic and multi-responsive fabric composite with excellent electro-photo-thermal effect and electromagnetic interference shielding performance. Chem. Eng. J. 391. https://doi.org/10.1016/j.cej.2019.123537.

Ma, C., Yuan, Q., Du, H.S., Ma, M.G., Si, C.L., Wan, P.B., 2020. Multiresponsive MXene ($Ti_3C_2T_x$)-decorated textiles for wearable thermal management and human motion monitoring. ACS Appl. Mater. Interfaces 12 (30), 34226–34234.

Ma, C., Ma, M.G., Si, C., Ji, X.X., Wan, P., 2021. Flexible MXene-based composites for wearable devices. Adv. Funct. Mater. 31 (22), 2009524.

Mule, A.R., Dudem, B., Patnam, H., Graham, S.A., Yu, J.S., 2019. Wearable single-electrode-mode triboelectric nanogenerator via conductive polymer-coated textiles for self-power electronics. ACS Sustain. Chem. Eng. 7 (19), 16450–16458. https://doi.org/10.1021/acssuschemeng.9b03629.

Peng, Q., Guo, J., Zhang, Q., Xiang, J., Liu, B., Zhou, A., Liu, R., Tian, Y., 2014. Unique lead adsorption behavior of activated hydroxyl group in two-dimensional titanium carbide. J. Am. Chem. Soc. 136 (11), 4113–4116. https://doi.org/10.1021/ja500506k.

Peng, B., Zhao, F., Ping, J., Ying, Y., 2020. Recent advances in nanomaterial-enabled wearable sensors: material synthesis, sensor design, and personal health monitoring. Small 16 (44). https://doi.org/10.1002/smll.202002681.

Rao, Z., Ershad, F., Almasri, A., Gonzalez, L., Wu, X., Yu, C., 2020. Soft electronics for the skin: from health monitors to human–machine interfaces. Adv. Mater. Technol. 5 (9). https://doi.org/10.1002/admt.202000233.

Shabangoli, Y., Rahmanifar, M.S., El-Kady, M.F., Noori, A., Mousavi, M.F., Kaner, R.B., 2018. Thionine functionalized 3D graphene aerogel: combining simplicity and efficiency in fabrication of a metal-free redox supercapacitor. Adv. Energy Mater. 8 (34). https://doi.org/10.1002/aenm.201802869.

Shahzad, F., Alhabeb, M., Hatter, C.B., Anasori, B., Hong, S.M., Koo, C.M., Gogotsi, Y., 2016. Electromagnetic interference shielding with 2D transition metal carbides (MXenes). Science 353 (6304), 1137–1140. https://doi.org/10.1126/science.aag2421.

Shankaregowda, S.A., Ahmed, R.F.S.M., Nanjegowda, C.B., Wang, J., Guan, S., Puttaswamy, M., 2019. Single-electrode triboelectric nanogenerator based on economical graphite coated paper for harvesting waste environmental energy. Nano Energy 66, 104141.

Shi, J., Chen, X., Li, G., Sun, N., Jiang, H., Bao, D., Xie, L., Peng, M., Liu, Y., Wen, Z., Sun, X., 2019. A liquid PEDOT:PSS electrode-based stretchable triboelectric nanogenerator for a portable self-charging power source. Nanoscale 11 (15), 7513−7519. https://doi.org/10.1039/c9nr01271k.

Simon, P., Gogotsi, Y., 2020. Perspectives for electrochemical capacitors and related devices. Nat. Mater. 19 (11), 1151−1163. https://doi.org/10.1038/s41563-020-0747-z.

Suarez, F., Nozariasbmarz, A., Vashaee, D., Öztürk, M.C., 2016. Designing thermoelectric generators for self-powered wearable electronics. Energy Environ. Sci. 9 (6), 2099−2113. https://doi.org/10.1039/c6ee00456c.

Sun, N., Guan, Z., Zhu, Q., Anasori, B., Gogotsi, Y., Xu, B., 2020. Enhanced ionic accessibility of flexible MXene electrodes produced by natural sedimentation. Nano-Micro Lett. 12 (1), 89.

Tai, H., Duan, Z., Wang, Y., Wang, S., Jiang, Y., 2020. Paper-based sensors for gas, humidity, and strain detections: a review. ACS Appl. Mater. Interfaces 12 (28), 31037−31053. https://doi.org/10.1021/acsami.0c06435.

Tang, W., Jiang, T., Fan, F.R., Yu, A.F., Zhang, C., Cao, X., Wang, Z.L., 2015. Liquid-metal electrode for high-performance triboelectric nanogenerator at an instantaneous energy conversion efficiency of 70.6. Adv. Funct. Mater. 25 (24), 3718−3725.

Wang, Z.L., Chen, J., Lin, L., 2015. Progress in triboelectric nanogenerators as a new energy technology and self-powered sensors. Energy Environ. Sci. 8 (8), 2250−2282.

Wang, R., Xu, Z., Zhuang, J., Liu, Z., Peng, L., Li, Z., 2017. Highly stretchable graphene fibers with ultrafast electrothermal response for low-voltage wearable heaters. Adv. Electr. Mater. 3 (2), 1600425.

Wang, X., Yin, Y., Yi, F., Dai, K., Niu, S., Han, Y., Zhang, Y., You, Z., 2017. Bioinspired stretchable triboelectric nanogenerator as energy-harvesting skin for self-powered electronics. Nano Energy 39, 429−436.

Wang, C., Xia, K., Wang, H., Liang, X., Yin, Z., Zhang, Y., 2019. Advanced carbon for flexible and wearable electronics. Adv. Mater. 31 (9), 1801072.

Wang, H., Li, S., Wang, Y., Wang, H., Shen, X., Zhang, M., 2020. Bioinspired fluffy fabric with in situ grown carbon nanotubes for ultrasensitive wearable airflow sensor. Adv. Mater. 32 (11), 1908214.

Wu, Y., Luo, Y., Qu, J., Daoud, W.A., Qi, T., 2019. Liquid single-electrode triboelectric nanogenerator based on graphene oxide dispersion for wearable electronics. Nano Energy 64. https://doi.org/10.1016/j.nanoen.2019.103948.

Xia, Q., Sun, S., Xu, J., Zan, F., Yue, J., Zhang, Q., Gu, L., Xia, H., 2018. Self-standing 3D cathodes for all-solid-state thin film lithium batteries with improved interface kinetics. Small 14 (52). https://doi.org/10.1002/smll.201804149.

Xiong, Z.C., Zhu, Y.J., Qin, D.D., Chen, F.F., Yang, R.L., 2018. Flexible fire-resistant photothermal paper comprising ultralong hydroxyapatite nanowires and carbon nanotubes for solar energy-driven water purification. Small 14 (50). https://doi.org/10.1002/smll.201803387.

Xu, T., Zhang, Z., Qu, L., 2020. Graphene-based fibers: recent advances in preparation and application. Adv. Mater. 32 (5), 1901979.

Yang, Q., Wang, Y., Li, X., Li, H., Wang, Z., Tang, Z., 2018. Recent progress of MXene-based nanomaterials in flexible energy storage and electronic devices. Energy Environ. Mater. 1 (4), 183−195.

Yang, Y., Sun, N., Wen, Z., Cheng, P., Zheng, H., Shao, H., 2018. Liquid-metal based superstretchable and structure-designable triboelectric nanogenerator for wearable electronics. ACS Nano 12 (2), 2027.

Yang, X., Xu, L., Lin, P., Zhong, W., Bai, Y., Luo, J., Chen, J., Wang, Z.L., 2019. Macroscopic self-assembly network of encapsulated high-performance triboelectric nanogenerators for water wave energy harvesting. Nano Energy 60, 404–412. https://doi.org/10.1016/j.nanoen.2019.03.054.

Yi, F., Zhang, Z., Kang, Z., Liao, Q., Zhang, Y., 2019. Recent advances in triboelectric nanogenerator-based health monitoring. Adv. Funct. Mater. 29 (41). https://doi.org/10.1002/adfm.201808849.

Yoon, J., Cho, H.Y., Shin, M., Choi, H.K., Lee, T., Choi, J.W., 2020. Flexible electrochemical biosensors for healthcare monitoring. J. Mater. Chem. B 8 (33), 7303–7318. https://doi.org/10.1039/d0tb01325k.

Yuan, Z., Du, X., Niu, H., Li, N., Shen, G., Li, C., Wang, Z.L., 2019. Motion recognition by a liquid filled tubular triboelectric nanogenerator. Nanoscale 11 (2), 495–503. https://doi.org/10.1039/c8nr08406h.

Zhao, X., Zha, X.J., Tang, L.S., Pu, J.H., Ke, K., Bao, R.Y., Liu, Z.y., Yang, M.B., Yang, W., 2020. Self-assembled core-shell polydopamine@MXene with synergistic solar absorption capability for highly efficient solar-to-vapor generation. Nano Res. 13 (1), 255–264. https://doi.org/10.1007/s12274-019-2608-0.

Zhou, J., Mulle, M., Zhang, Y., Xu, X., Li, E.Q., Han, F., Thoroddsen, S.T., Lubineau, G., 2016. High-ampacity conductive polymer microfibers as fast response wearable heaters and electromechanical actuators. J. Mater. Chem. C 4 (6), 1238–1249. https://doi.org/10.1039/c5tc03380b.

Zou, Y., Tan, P., Shi, B., Ouyang, H., Jiang, D., Liu, Z., Li, H., Yu, M., Wang, C., Qu, X., Zhao, L., Fan, Y., Wang, Z.L., Li, Z., 2019. A bionic stretchable nanogenerator for underwater sensing and energy harvesting. Nat. Commun. 10 (1). https://doi.org/10.1038/s41467-019-10433-4.

Chapter 6

Advanced cellulose-based electrodes and electrolytes for supercapacitor energy storage devices

Dandan Li, Tian Mai and Mingguo Ma*

MOE Engineering Research Center of Forestry Biomass Materials and Bioenergy, Research Center of Biomass Clean Utilization, Beijing Key Laboratory of Lignocellulosic Chemistry, College of Materials Science and Technology, Beijing Forestry University, Beijing, PR China

**Corresponding author. E-mail: mg_ma@bjfu.edu.cn*

6.1 Introduction

The rapid development of modern society has resulted in resource scarcity, serious environmental pollution, and other issues. Therefore, it is imminent to develop green, safe, and high-efficiency energy storage technology (Kim et al., 2022). Supercapacitors have gained attention for their combination of high-power density and long-term cycle life (Wang et al., 2019). The selection of suitable electrode materials and well-designed structures have a great influence on the electrochemical performance (high energy density and Coulombic efficiency, long stability, etc.) of flexible supercapacitors (Rostami et al., 2022; Zhou et al., 2021a).

The cellulose consists of linear macromolecular polysaccharides composed of glucose molecules via β-1,4 glycosidic bonds, which are high in carbon content and chemically reactive and are ideally suited for conversion into carbon precursor materials for storage devices (Xu et al., 2021; Zhang et al., 2018). When utilized as supercapacitor electrodes, the intertwined carbon fibers produce a large number of holes and a large surface area, which facilitates rapid ion transport during charging/discharging to increase the capacitive performance of the supercapacitor device (Chen et al., 2014).

Surface Science and Advanced Functional Applications of Cellulose Composites
https://doi.org/10.1016/B978-0-443-13675-7.00006-3

Meanwhile, with the progress of smart wearable devices, there is an urgent requirement to develop supercapacitors with bendable, foldable, and stretchable energy storage devices (Sun et al., 2021). The construction of flexible supercapacitors with excellent performance requires the flexibility of different components such as electrodes and electrolytes (Xu et al., 2021; Fu et al., 2023). Cellulose is a natural green material originating from plants and microorganisms, which possesses excellent properties including high surface chemical activity, unique network structure, excellent mechanical strength, and high specific surface area (Wang et al., 2017b). Although cellulose has poor electrical conductivity, it can be used as a flexible skeleton compounded with conductive electrode materials to obtain flexible supercapacitor electrodes with high specific capacitance and high electrochemical stability (Han et al., 2019b; Kim et al., 2022; Manjakkal et al., 2020).

As a channel for ion transport, the electrolyte directly influences the energy density and long-term stability of supercapacitors. In various flexible electrolytes, as a novel electrolyte material, hydrogel electrolyte has many outstanding features, which involve better ionic conductivity, excellent flexibility, and viscosity (Gao et al., 2022; Yang et al., 2020). In addition, benefiting from the densely distributed hydroxyl groups on the cellulose molecular chain, the introduction of cellulose materials into hydrogel electrolytes enables multifunctionality, which contributes to the development of electrochemical supercapacitor energy storage devices in the area of wearables (Fu et al., 2021, 2023).

Herein, this paper reviews the recent research achievements in the exploitation of cellulose materials for supercapacitors. The types of cellulose-based electrode are classified into carbonized cellulose-based materials, cellulose-based film electrodes, cellulose-based hydrogel electrodes, and cellulose-based aerogel electrodes. Meanwhile, the hydrogel electrolytes have different properties (mechanically enhanced, wide temperature range, self-healing, and adhesion) in various application scenarios as illustrated. And finally, we provide an insight into future works.

6.2 Cellulose-based electrode types

Electrode materials are the key to improving the electrochemical performance of supercapacitors. Cellulose has great advantages as flexible electrodes due to flexibility, hydrophilicity, and rich surface chemistry. However, cellulose inherently is not electrically conductive. Therefore, the most straightforward method is to carbonize cellulose into carbon materials. Carbonization modification of cellulose enables the preparation of carbon composites (Chen et al., 2013), heteroatom-doped carbon (Chen et al., 2014), and porous carbon (Wang et al., 2015; Wu et al., 2015), which can be used as electrically conductive carbon materials for the preparation of electrodes. The other way to render cellulose electrically conductive is to compound it with conductive materials through electrostatic adsorption, covalent bonding, and cross-linking. Cellulose is combined with commonly used electrode materials (carbon materials, conductive polymers, metal oxides, and conductive materials) to prepare composite electrodes. Because of the high aspect ratio, excellent mechanical properties, flexibility, and hydrophilicity, cellulose is easily incorporated with conductive carbon nanofillers (such as graphene and CNTs), which enables the composite electrode to have an ultralow density and high specific surface area (Zheng et al., 2015; Hsu et al., 2019). Moreover, cellulose-based electrode materials with water swelling properties enable supercapacitors to achieve excellent electrolyte absorption performance. Meanwhile, cellulose with hydroxyl and carboxyl groups can be chemically bonded to conductive polymers, such as PANI and PPy (Han et al., 2019b; Cheng et al., 2020). This leads to flexible supercapacitors to obtain the high capacitance of these pseudocapacitor materials. Cellulose-based electrode types include thin films, hydrogels, and aerogels, which have flexible properties. Therefore, the design diversification of cellulose-based electrode structures provides more possibilities for flexible supercapacitors in different application fields.

6.2.1 Carbonized cellulose-based electrode materials

Cellulose is the most abundant renewable organic material available in nature with large carbon content, which is the most suitable raw material source for carbon-based energy storage (Xu et al., 2019). Under the protection of inert gases, such as N_2 and Ar_2, cellulose could still maintain the 3D network shape and high specific surface area characteristics after high-temperature carbonization. Yu et al. (2014) used bacterial cellulose (BC) as raw material and obtained porous network carbon nanofibers with a specific surface area as high as 506 m^2/g by carbonization at 800°C high temperature calcination. Using this newly available carbon nanofiber as the electrode material for supercapacitors could overcome the challenges of poor ion storage capacity and slow ion transport rates in supercapacitors.

With the natural porous structure of cellulose, it is used as an electrode material not only facilitates the transport of ions and electrons, but also makes it easy to realize the doping of heteroatoms to enhance the capacitance value of the electrode. By doping heteroatoms (N, P, B, S, and I) on the surface of carbon materials, the electron-donor properties and surface chemical properties of carbon fiber materials could be effectively regulated, which could lead to better energy storage performance. The high specific surface area of cellulose-based carbon nanofibers provides excellent and extensive binding sites for the introduction of heteroatom, which is a suitable template material for elemental doping (Chen et al., 2013). For example, Wu et al. (2015) converted cellulose nanocrystals (CNCs) into highly porous nitrogen (N)-doped carbon nanorods as electrode materials for supercapacitors. The melamine-formaldehyde (MF) precursor was mixed with CNCs for in situ condensation on the CNCs surface. Subsequently, an abundant number of micro-, meso-, and macroporous N-doped carbon nanorods (N-MFCNC900) were obtained by pyrolysis at a carbonization temperature of 900°C. The optimum capacitance of 3285 F/g was obtained for the N- MFCNC900 electrode at 0.01 V/s in a 1 M H_2SO_4 electrolyte. In addition, Chen et al. (2014) investigated the different heteroatom (N, P, and B)-doped cellulose carbon nanofiber electrodes. P-doped, N-P-co-doped, and B-P-codoped cellulose-based carbon material electrodes were synthesized by immersing BC membranes in H_3PO_4, $NH_4H_2PO_4$, and H_3PO_4/H_3BO_4 under nitrogen by carbonization reaction at 800°C, respectively. Moreover, the supercapacitor with N-P-codoped carbon nanofiber electrode material had good electrochemical performance with a maximum power density of 186.03 kW/kg.

Also, to obtain carbon electrodes with good flexibility, the strategy of carbonized cellulose-based fabrics was commonly employed. Chen et al. (2021c) used a flexible cellulose textile as the 3D porous framework to accommodate one-dimensional (1D) active materials (polypyrrole nanotubes and carbon nanotubes (CNTs)). The active nitrogen-doped carbon interface (i-Cell) was obtained by carbonization of dopamine after in situ polymerization on cellulose textile. Subsequently, 1D polypyrrole (PPy) and CNTs active materials were deposited on i-Cell using the electrophoretic deposition method. The aim was to improve the active material mass loading to commercial levels. This strategy resulted in flexible electrodes with high electrical conductivity, large porosities, and great electrolyte affinity properties, which facilitated fast electron transport and ion migration, leading to improved electrochemical performance (Chen et al., 2021c). As a result, the flexible anode electrode of CNTs deposited on i-Cells (CNT-i-Cell) had a large specific surface area (139.7 $m^{2/}g$) and an ultrahigh capacitance value of 9501 mF/cm^2 at high loading (30.1 mg/cm^2). Moreover, the flexible cathode electrode of PPy deposited on i-Cells (PPy-i-Cell) also achieved excellent capacitance performance of 6212 mF/cm^2 at a high loading capacity (25 mg/cm^2). Assembled flexible wearable asymmetric supercapacitor had a maximum energy density of 1.42 mWh/cm^2, which had a broad application perspective in large-capacity flexible energy storage devices.

6.2.2 Cellulose-based film electrode

Flexible paper-based electrodes have received attention in the energy storage field for their abundant regulatable multiporous network structure and excellent pliability (Wang et al., 2015). As a renewable biomass material with a large aspect ratio as well as excellent mechanical properties, cellulose is beneficial for the design and construction of self-supported porous paper-based electrode materials (Kim et al., 2022). In one respect, cellulose materials provide outstanding flexibility and self-supporting properties for paper-based electrodes. Alternatively, the loose structure in the cellulose paper-based electrodes offered multiple porous channels during the rapid diffusion of electrolyte ions, which enhanced the electrical performance of the supercapacitor. Compared with the traditional method of preparing polymer binder-based electrode materials, the flexible cellulose paper-based electrodes were prepared without the need for additional binders. It allowed greater involvement of electroactive substances in the electrochemical reaction process, improving the energy storage capability of the electrode. Until now, several most common techniques are described in the following, including vacuum filtration, coating method, and electrostatic spinning.

6.2.2.1 Vacuum filtration

Vacuum filtration is a typical process for preparing cellulose paper-based electrode materials (Li et al., 2014; Hsu et al., 2019). For example, Hsu et al. (2019) introduced nanocellulose as a flexible loading substrate to enhance mechanical properties, and added reduced graphene oxide (rGO) to improve the electrical conductivity during the in situ polymerization of polyaniline (PANI). Subsequently, a two-dimensional (2D) nanocellulose-based PANI/rGO electrode film was prepared by vacuum filtration to have good pliability and electrical conductivity (Fig. 6.1a). After assembling the flexible

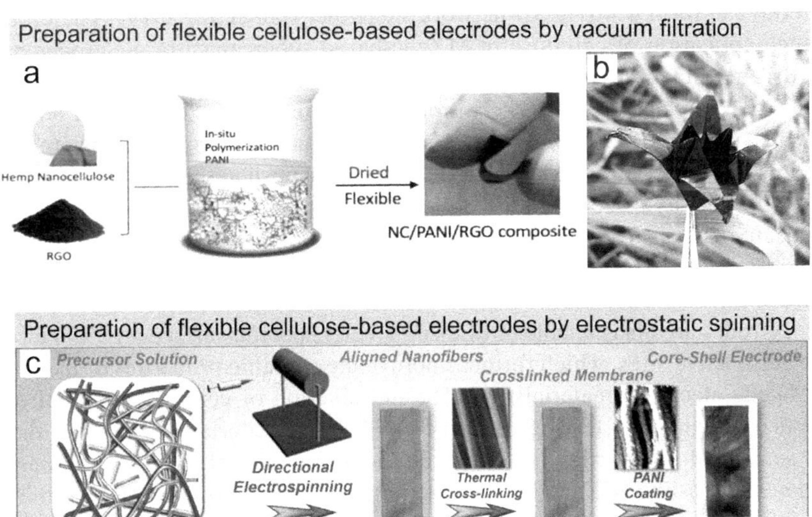

FIGURE 6.1 (a) Schematic diagram of the formed NC/PANI/RGO composite film by vacuum filtration method. (b) Photograph of CNF@c-MOF nanopaper showed good flexibility. (c) Schematic diagram of the electrode film prepared by electrostatic spinning technique. *(a) Reprinted with permission from (HSu et al., 2019). Copyright 2019, American Chemical Society. (b) Reprinted with permission from (Zhou et al., 2019). Copyright 2019, American Chemical Society. (c) Reprinted with permission from (Han et al., 2019b). Copyright 2019, American Chemical Society.*

electrode into a sandwich-type supercapacitor, it obtained the electrochemical performance of the specific capacitance of 79.71 F/g at the current density of 0.2 A/g.

In recent years, besides the common carbon-based, metal oxide, and conductive polymer electrode materials, novel conductive materials (including conductive metal organic frameworks (MOFs) and MXenes) composited with cellulose have received increasing attention as flexible electrodes for supercapacitors (Zhou et al., 2019, 2021a; Rostami et al., 2022; Jiao et al., 2019). As a 2D transition metal material, MXene benefits from a metal core layer that facilitates fast electron transport, and the transition metal oxide—like surface enables the contribution of redox active sites for pseudocapacitive energy storage (Zheng et al., 2023). However, the strong hydrogen bonding force and interlayer van der Waals force of MXene easily produce interlayer stacking and flocculation phenomenon, which seriously reduces the electron transport ability and results in decreased electrochemical performance (Jiang et al., 2020). Moreover, the high brittleness of pure MXene film leads to poor mechanical properties and short life of the MXene-based energy storage device, thus seriously limiting the application of MXene in supercapacitors (Jiang et al., 2020; Zhu et al., 2023b). With enriched pores, greater specific surface area, and more oxygen-containing reactive groups, cellulose composite with MXene material can form a strong 3D network skeleton (Tian et al., 2019). Therefore, the application of cellulose in MXene-based supercapacitor electrodes has attracted much attention from researchers (Chen et al., 2022; Tang et al., 2022; Lyu et al., 2023). The introduction of cellulose has several advantages. (1) The incorporation of cellulose can effectively alleviate the interlayer stacking phenomenon of MXene nanosheets caused by van der Waals and hydrogen bonding forces, and expand the MXene interlayer spacing. (2) The porous MXene/cellulose composite electrode promotes electrolyte penetration and provides an effective and stable channel for rapid diffusion/transfer of electrolyte ions. (3) MXene/cellulose composite electrodes are also capable of offering greater specific surface area and more electroactive sites, which further improve the galvanic properties of the pure MXene electrode materials. (4) The introduction of cellulose material enhances further the tearing resistance of MXene materials to improve their mechanical properties as electrodes. Wang et al. (2019) prepared honeycomb-like MXene/BC composite films by combining 1D BC fibers with 2D $Ti_3C_2T_x$ flakes as flexible supercapacitor electrodes. The BC fibers enhanced the interaction of MXene nanosheets and constructed a 3D porous network structure by interweaving, which provided a carrier for the adsorption and rapid transport of electrolyte ions. At high MXene loadings (5 mg/cm^2), the MXene/BC film electrode had high capacitive performance (416 F/g) and excellent mechanical properties. In addition, a flexible asymmetric pseudocapacitor was produced using MXene/BC and PANI/BC thin films as electrodes in H_2SO_4 electrolyte. At a power density of 2.12 mW/cm^2, the supercapacitor reached a maximal energy density of 252 μWh/cm^2. Similarly, Jiao et al. (2019) regulated the interlayer spacing by using BC nanofiber as a spacer to hinder the repacking of MXene sheets. Moreover, by a straightforward laser-cutting kirigami patterning technique, it further manufactured all-solid-state microsupercapacitor arrays (MSCAs) with bendable, twistable, and stretchable properties. The BC fiber structure effectively increased the c-lattice parameter of the MXene/BC composite films, which led to the exposure of more active centers on the MXene surface. The contact opportunity between the active electrode material and the electrolyte ions was further increased, thus improving the electrochemical performance of MSCAs. Furthermore, the area capacitance (111.5 mF/cm^2) of MSCAs remained stable at elongations up to 100%. At present, the research of cellulose-based MXene composites in the area of energy storage has just started. In the future, cellulose nanomaterials need to be fully utilized to solve the bottleneck problem of MXene by taking advantage of their performance characteristics.

MOFs have the properties of high crystallinity, large accessible surface area, and high porosity. However, the inherent low conductivity of MOFs hinders charge transfer within the framework, thus greatly limiting the electrochemical performance (Sheberla et al., 2016). Additionally, the crystallization of MOFs usually leads to mechanical brittleness, which renders it difficult to obtain freestanding MOF films and hampers the application in flexible supercapacitor devices (Du et al., 2018). To solve these challenges, the researchers have expanded the studies of MOF materials for flexible supercapacitor electrodes by using cellulose material as a flexible and environmentally friendly substrate (Zhang et al., 2021a). For example, Zhou et al. (2019) employed cellulose nanofibers (CNFs) as substrates for the growth of conductive metal–organic frameworks (c-MOFs) nanolayers to form CNF@c-MOF hybrid nanofibers by interfacial interaction. Self-supported CNF@c-MOF nanopapers were assembled by the filtration technique and had high electrical conductivity (100 S/cm), graded micropores, and excellent mechanical properties (Fig. 6.1b). The CNF@c-MOF nanopaper electrode had a specific capacitance of 103 F/g, and the supercapacitor assembled with CNF@c-MOF nanopaper electrode had a capacitance degradation rate of 1% after 10,000 charge/discharge cycles. In addition, the CV curve of the flexible device remained nearly unchanged when bent at 90, 120, and 180 degrees under the working conditions, indicating the CNF@c-MOF electrode had good mechanical stability and excellent electrochemical properties.

6.2.2.2 Electrostatic spinning

Currently, the cellulose materials processed by electrostatic spinning combined with carbonization technology are mainly utilized as electrode films in supercapacitors (Simotwo et al., 2017; Han et al., 2019b; Cai et al., 2016). As shown in Fig. 6.1c, the precursor solution was compounded from CNC, CNT, polyvinyl alcohol (PVA), and polyacrylic acid (PAA). The cross-linked nanofiber film (CNT−CNC/PVA−PAA) was produced by directional electrostatic spinning (Han et al., 2019b). In particular, PAA and PVA constituted the cross-linked matrix of the CNT−CNC/PVA−PAA film; the CNC and CNT endowed the CNT−CNC/PVA−PAA film with excellent mechanical and electrochemical properties, respectively. Subsequently, thermal treatment of the CNT−CNC/PVA−PAA film was carried out to obtain the core layer material, and the polyaniline (PANI) was employed as the shell layer material to enhance the electrical conductivity. Finally, electrodes nanofibrous film (PANI@CNT−CNC/PVA−PAA) with the core−shell structure was used to assemble high-performance flexible symmetric solid-state supercapacitors. Electrochemical tests have shown that supercapacitors under bending conditions retain up to 90% of their capacitance after 2000 cycles (Han et al., 2019b). Electrostatically spinning and carbonization of cellulose-based electrode materials with controllable size and good flexibility had great potential for future application in flexible energy storage devices.

6.2.2.3 *Coating and impregnation methods*

In addition, the coating and impregnation methods are also used to prepare 2D flexible electrode film by loading the active materials. Cellulose-based textiles (e.g., cotton fabrics and nylon fabrics) with hydroxy, amide, and other functional groups are used as flexible substrates (Khadem et al., 2022). For example, Manjakkal et al. (2020) prepared electrodes by coating poly (3,4-ethylenedioxythiophene): poly(styrenesulfonic acid) (PEDOT: PSS) ink on cellulose/polyester cloth. Meanwhile, the cellulose/polyester cloth was used as separator and sweat as electrolyte. Moreover, the energy density was 1.36 Wh/kg of the assembled textile-based supercapacitor in artificial sweat. Finally, the potential of textile-based self-powered supercapacitors was displayed. When using authentic human sweat (under fully wet conditions), it has been observed with energy and power densities of 0.25 Wh/kg and 30.62 W/kg, respectively. Therefore, textile-based supercapacitors with sweat electrolytes were a viable solution that could be used in self-powered wearable systems that require low power sensors. In Fig. 6.2, Fan et al. prepared high-performance flexible electrodes and supercapacitors using MXene (active material) and 3D needle-punched waste denim felts (flexible substrate) by impregnation and carbonization techniques (Fan et al., 2023). Waste denim cotton was used as raw material and opened them into waste denim fibers. Subsequently, the 3D needle-punched waste denim felts with high specific surface area were prepared using a needling process. MXene flakes were impregnated into the flexible substrate and carbonized into flexible

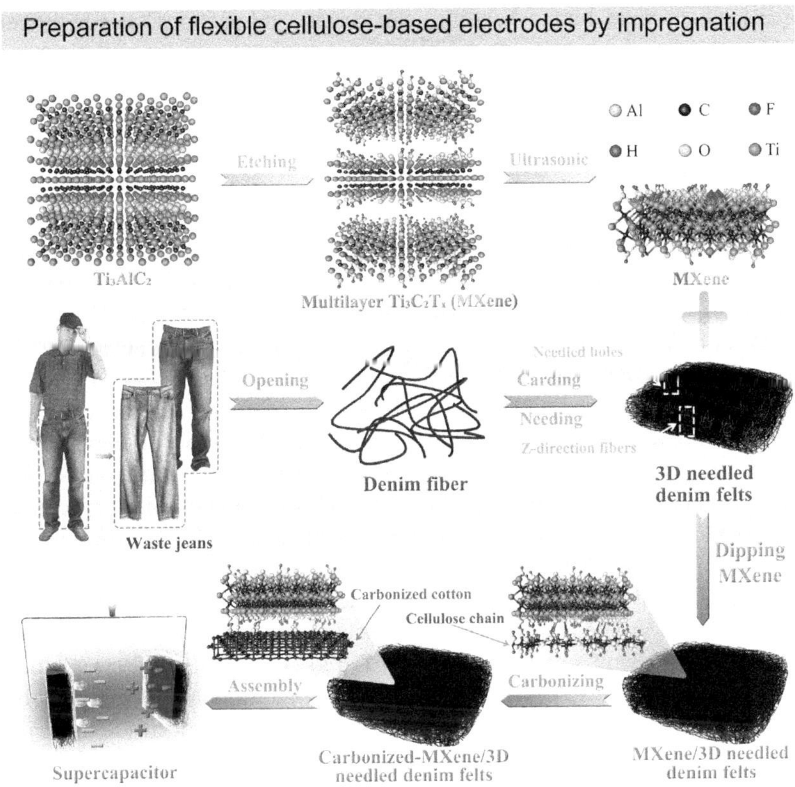

FIGURE 6.2 Schematic diagram of MXene/3D needled denim felts to prepare electrodes and supercapacitors by impregnation method. *Reprinted with permission from (Fan et al., 2023). Copyright 2023, Springer.*

supercapacitor electrodes. The obtained supercapacitor achieved remarkable maximum specific capacitance with superior long-term cycling stability.

6.2.3 Cellulose-based hydrogel electrode

The hydrophilic hydroxyl groups of cellulose can form nanoscale 3D network structures through hydrogen bonding, leading to the formation of hydrogels with unique chemical structures and mechanical properties (Fourmann et al., 2021). Doping of conductive materials into cellulose matrices allows the preparation of conductive cellulose-based hydrogel electrodes, which have been widely investigated in the field of supercapacitors based on excellent conductivity and sustainability (Hsu et al., 2020; Wu et al., 2020; Cheng et al., 2020; Han et al., 2019a). Generally, conductive cellulose-based hydrogels are formed by the interaction of different covalent and/or noncovalent bonds (Nele et al., 2020). Herein, we summarized the three most common interaction forces, which were borate bonding, electrostatic interactions, and hydrogen bonding.

6.2.3.1 Borate bonding

As the most common dynamic covalent bond, the reversible reaction of borate bonding endows hydrogels with self-healing properties and thus has received increasing attention in the preparation of hydrogel electrodes. For example, Han et al. (2019a) reported multifunctional electroconductive hydrogels (ECHs). Initially, polyaniline@cellulose nanofibers (PANI@CNF) conductive complexes were fabricated by in situ polymerization of polymerizing aniline (ANI) monomers with CNF as a template in an acidic environment. Subsequently, the PANI@CNF compounds were uniformly incorporated into the PVA hydrogel system to form a physically cross-linked network with hydrogen bonding (Fig. 6.3a and b). Then, the integration of borax cross-linked PVA gels constitutes dynamic borate bonding to form the second network. The nicely distributed PANI@CNF compounds built efficiently constructed an efficient pathway in the PVA channel, offering well electroconductivity value (5.19 S/m) for the ECHs. Finally, ECHs were prepared with collaborative properties, such as inherent ductility, well mechanical strength, great viscoelasticity, and rapid self-healing ability (Fig. 6.3c). The ECHs had very well self-healing ability (15 s) (Fig. 6.3d). This was attributed to the dynamic borate and hydrogen bonding that gave ECHs the property to self-heal. In addition, the ECHs were used as electrode materials

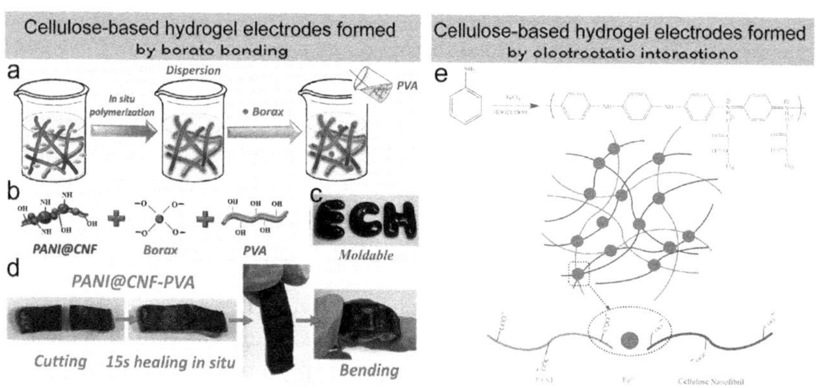

FIGURE 6.3 (a) Schematic diagram of the synthesis process of PANI@CNF-PVA composite hydrogels electrodes. (b) Schematic diagram of multiple complexations with dynamic cross-linking among PANI@CNF, borax, and PVA. (c) Photograph of the PANI@CNF-PVA hydrogels. (d) Photograph of self-healing property for PANI@CNF-PVA hydrogels. (e) Schematic diagram of hydrogel formation by cross-linking of PANI and TOCN via interactions between Fe^{3+} and carboxylate anions. *(d) Reprinted with permission from (Han et al., 2019a). Copyright 2019, Elsevier. (e) Reprinted with permission from (Liu et al., 2019). Copyright 2019, American Chemical Society.*

with a specific capacitance of 226.1 F/g at 0.2 A/g (Han et al., 2019a). Therefore, the integration of unique properties led to the preparation of ECHs with numerous promising applications in software, self-healing, and implants for electronics.

6.2.3.2 Electrostatic interactions

Hydrogel networks are also stabilized by electrostatic interactions. It is worth noting that electrostatic interactions are not isolated. Instead, synergize with other noncovalent interactions (e.g., hydrogen bonding, van der Waals interactions) to maintain the cellulose-based hydrogels network (Zhang et al., 2017; Zhou et al., 2021b). As shown in Fig. 6.3e, Fe^{3+} was used as an initiator to prepare PANI bearing carboxylate anion; meanwhile, Fe^{3+} also acted as a cross-linking agent to form electrostatic interactions with PANI and TEMPO-oxidized cellulose nanofibril (TOCNF) (Liu et al., 2019). This led to the creation of mechanically stable and electrically conductive cellulose-based conductive hydrogel networks. When the prepared hydrogel was used as a supercapacitor electrode, the test showed that it displayed a specific capacitance value of 3060 mF/cm^2 at 0.5 mA/cm^2. Since cellulose-based materials contain abundant carboxyl groups, the formation of cellulose-based hydrogels through electrostatic interactions and coordination with multivalent metal ions (Fe^{3+}, Zn^{2+}, Cu^{2+}) in the future will have potential applications.

6.2.3.3 Hydrogen bonding

Hydrogen bonding is one of the most common noncovalent interaction forces in the preparation of cellulose-based hydrogels (Zhang et al., 2020; Lu et al., 2022; Cheng et al., 2020). For example, the hydrophobic segments of PPy conductive polymer had the shortcoming of easily flocculating in the hydrogel matrix, which could influence the performance of hydrogel electrodes. To solve the problem, Cheng et al. (2020) employed carboxylate anions of carboxymethyl cellulose sodium (CMC) as a good stabilizer to cross-link the conductive PPy chains, forming a hydrophilic dendritic network structure. In the PAAm system, PAAm/CMC/PPy hydrogels were obtained by hydrogen bonding. The hydrogels had strong interfacial adhesion, which mainly derived from metal complex and H-bonds. In terms of electrochemical performance, the assembled supercapacitor had a specific capacitance of 126.38 F/g at 0.2 A/g (Cheng et al., 2020). Besides, Lu et al. (2022) fabricated a flexible, stretchable cellulose/phytic acid/polyaniline (CPP) hydrogel electrode with sandwich structure through in situ growth of polyaniline onto cellulose/phytic hydrogel. As-prepared CPP hydrogel electrodes and hydrogel electrolytes were combined to form an integrated symmetric supercapacitor. In practical applications, the charged symmetric supercapacitor was able to power multiple electronic devices. Therefore, these all-hydrogel supercapacitors paved the way for the development of flexible electronic skin.

6.2.4 Cellulose-based aerogel electrode

Compared with other aerogel materials, cellulose-based aerogel electrodes were prepared with increased plasticity due to the good mechanical properties of cellulose (Zhang et al., 2019; Zheng et al., 2015). The pore structure of the aerogel electrode can be adjusted by chemically modifying cellulose to enhance the electrochemical behavior of the cellulose-based aerogel electrode (Rostami et al., 2022).

For example, the successful preparation process of $Ti_3C_2T_X$/SWCNT/CNF composite aerogel was shown in Fig. 6.4a. The carboxyl groups on the TEMPO-oxidized CNFs chains form hydrogen bonding forces with SWCNT and MXene, which effectively form a homogenous dispersion (Fig. 6.4b). Subsequently, under the action of hydrogen bonding, the $Ti_3C_2T_x$/SWCNT/CNF aqueous dispersion was converted into a hybrid hydrogel (Fig. 6.4c). During the subsequent freezing process, the uniform dispersion between $Ti_3C_2T_X$, SWCNT, and CNF contributed to the formation of an evenly $Ti_3C_2T_X$/SWCNT/CNF aerogel structure (Fig. 6.4d). Moreover, in addition to increasing the transmitted stress efficiency, the rich honeycomb pore structure of $Ti_3C_2T_X$/SWCNT/CNF carbon aerogel promoted the rapid migration of electrolyte ions (Fig. 6.4e). As the electrode of the supercapacitor, the capacitance value of the solid-state symmetric supercapacitor was 746.68 mF/cm^2 at 0.1 mA/cm^2. Furthermore, the supercapacitor maintained about 91.23% of the original capacitance after 10,000 consecutive cycles (Lyu et al., 2023). Similarly, Liu et al. (2022a) prepared ordered lamellar honeycomb structure CNF/CNT/RGO composite carbon aerogel electrodes by freezing process through hydrogen bonding. The capacitance value of the

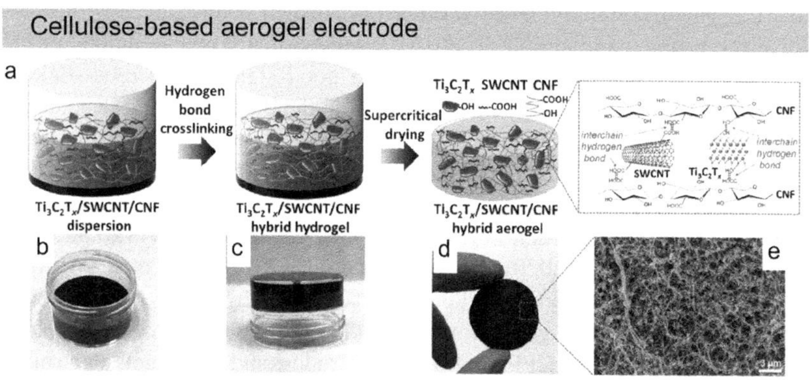

FIGURE 6.4 (a) Schematic diagram of the two steps to prepare $Ti_3C_2T_X$/SWCNT/CNF aerogels. (b) Photographs of the aqueous dispersion of $Ti_3C_2T_X$/SWCNT/CNF. (c) Photographs of the $Ti_3C_2T_X$/SWCNT/CNF aerogels. (d) Photographs of the $Ti_3C_2T_X$/SWCNT/CNF aerogels. (e) Microstructure of $Ti_3C_2T_x$/SWCNT/CNF hybrid aerogel. *Reprinted with permission from (Lyu et al., 2023). Copyright 2023, Elsevier.*

solid-state symmetric supercapacitor was 109.4 mF/cm^2 at 0.4 mA/cm^2. Furthermore, the solid-state symmetric supercapacitor maintained about 88% of the original capacitance after 5000 consecutive cycles at 50% compressive strain, which demonstrated the outstanding cycling stability of the device under harsh compressive conditions (Liu et al., 2022a).

Therefore, the design of compressible supercapacitor electrodes must obtain excellent compressibility as well as exhibit well electrochemical properties. Cellulose-based aerogel has the virtues of high mechanical strength and adaptability of pore structure, which will be promising for the research of compressible electrodes.

6.3 Performances of cellulose-based hydrogel electrolyte

As part of the supercapacitor, the electrolyte has a significant role to play in the electrochemical performance of the supercapacitor, including the potential window, cycling stability, and ionic conductivity (Amaral et al., 2022; Park et al., 2022). Conventional liquid electrolytes have high ionic conductivity and stable electrochemical properties (Huang et al., 2023). However, there are serious limitations in practical applications, such as liquid leakage, volume expansion, electrode corrosion, self-discharge, poor low-temperature performance, and difficulty in designing different shapes (Zhao et al., 2020). Hydrogels have attracted extensive attention as potential electrolyte candidates for assembling flexible all-solid-state supercapacitors due to their unique network structure, excellent mechanical properties, and ultrahigh water content (Zhang et al., 2021b; Ni et al., 2022). The secure polymer network in the hydrogel electrolyte facilitates ion transport (Peng et al., 2019). The soft and moist hydrogel interface ensures close contact with the electrodes, which gives the supercapacitor flexibility (Peng et al., 2016). Meanwhile, hydrogel electrolyte promotes the development of miniaturization, integration, and wearability of electrochemical energy storage devices (Zhong et al., 2015). Nevertheless, hydrogel electrolytes still face serious challenges in practical applications of solid-state supercapacitors, such as low ionic conductivity, unsatisfactory performance under extreme operating conditions ($<-20°C$, $>50°C$), weak interfacial adhesion (poor electrode/electrolyte interfacial contact due to mechanical deformations), and poor mechanical tolerance (weak mechanical strength) (Moon et al., 2015; Huang et al., 2017; Yang et al., 2021a). Recently, cellulose-based nanofillers with abundant hydroxyl groups have been shown to be employed as reinforcing agents to optimize the mechanical properties of hydrogel electrolytes (Chen et al., 2020; Juan et al., 2022; Guan et al., 2021). Cellulose-based nanofibers are easily combined with PAM and PVA long chains in the form of hydrogen bonding, and the abundant hydrogen bonding endows the hydrogels with good water retention, spontaneous healing, and tensile properties (Li et al., 2021; Chen et al., 2019). In addition, the cellulose-based hydrogel electrolytes with 3D network structure and nanopore size enable the electrolyte ion transport to be more ordered, the charge transfer resistance to be substantially reduced, and the electrolyte ion conductivity to be further improved (Zhang et al., 2023). Finally, the cellulose-based hydrogel electrolyte possesses cheap, degradable, and environmentally friendly properties, which has great research value and wide application prospects in the study of wearable supercapacitor energy storage devices (Zhu et al., 2022).

6.3.1 Mechanically enhanced cellulose-based hydrogel electrolytes

Most of the conventional hydrogel electrolytes represented by polyacrylamide (PAM) and polyacrylate (PANa) are physically cross-linked using hydrogen bonding as the primary method (Zhang et al., 2023; Ma et al., 2019). However, the physically cross-linked polymer network has low mechanical strength. In the presence of complex deformation, the hydrogel electrolyte is susceptible to cracking, which affects the cycling stability of supercapacitors. To solve the problem of insufficient mechanical properties of such hydrogel electrolytes with a single network, a feasible approach is to introduce a second network for the enhancement of mechanical properties (Gao et al., 2022).

Herein, the researchers found that BC nanofiber has a high aspect ratio, natural porosity networks, well mechanical properties, and abundant -OH groups (Nakayama et al., 2004; Wang et al., 2017a). Therefore, BC nanofiber is easy to self-assemble into a tangled mesh structure through hydrogen bonding, which can provide energy dissipation to the hydrogel network, thus endowing the hydrogel electrolyte with strong toughness. For example, Li et al. (2021) prepared BC-reinforced PAM hydrogel electrolytes. BC aerogels enriched with oxygen-rich species on the cellulose chains were chosen as the backbone network structure. Acrylamide (AM) monomers were uniformly adsorbed in BC gels through strong hydrogen bonding and electrostatic interactions, and polymerized in situ to form PAM gel networks. Due to the addition of BC nanofibers, the prepared BC/PAM hydrogel electrolyte had a maximum tensile stress of 330 kPa, which was more than four times of pure PAM hydrogel. In addition, the BC/PAM hydrogel readily stretched to a strain range of 1300% without significant cracking or fracture. The prepared BC/PAM hydrogels were capable of being deformed randomly, as demonstrated by the large elongation of the hydrogels in the knotted shape. Assembled all-solid-state supercapacitors using BC/PAM gel electrolytes exhibited superior electrochemical performance (energy density up to 50.1 $\mu Wh/cm^2$) and the capacitance retention of 97.5% after 2000 cycles of bending, showing great potential for wearable applications (Li et al., 2021).

Furthermore, the BC fibers are well suited for use in alkaline supercapacitors thanks to well-tolerated alkalinity. For example, Wang et al. (2023) used a free radical polymerization method to fill BC aerogels with PANa/KOH hydrogels (Fig. 6.5a). Pure PANa/KOH hydrogel undergoes volume expansion after water absorption, leading to a rapid decrease in mechanical properties. Correspondingly, the introduced BC aerogel network was capable of hydrogen bonding with the PANa molecular chains, thus effectively limiting the volume expansion of the PANa hydrogel (Fig. 6.5b). Therefore, the prepared alkaline BC-based hydrogel electrolyte obtained from the preparation had good mechanical performance (with a stress of 1.1 MPa), and excellent antiswelling properties. Besides, the polar groups contained in the BC fiber network provided a wider range of jumping sites for electrolyte ion transfer, which increased the ionic electrical conductivity. The alkaline BC-based hydrogel electrolyte had an ionic conductivity of 1.91 S/m at 25°C, and interestingly, the ionic conductivity reached 1.35 S/m at −25°C (Figs. 6.5c). This could be explained by the chemical functional groups contained in PANa and BC effectively attracting counterions, thus locking up free water and increasing the freezing point of water, which locked up free water and increased the freezing point of water. Therefore, the prepared alkaline BC-based hydrogel had good antifreezing properties. Because of these excellent properties, the asymmetric alkaline supercapacitors constructed using alkaline BC-based hydrogel electrolytes exhibited an ultrahigh energy density of 25.63 W h/kg and excellent stability of over 92% at 10,000 cycles. Also, the assembled supercapacitors were well adapted to a variety of extreme conditions, including weight bearing, bending, and hammering (Wang et al., 2023). It provided reliable solutions for powering flexible wearable devices.

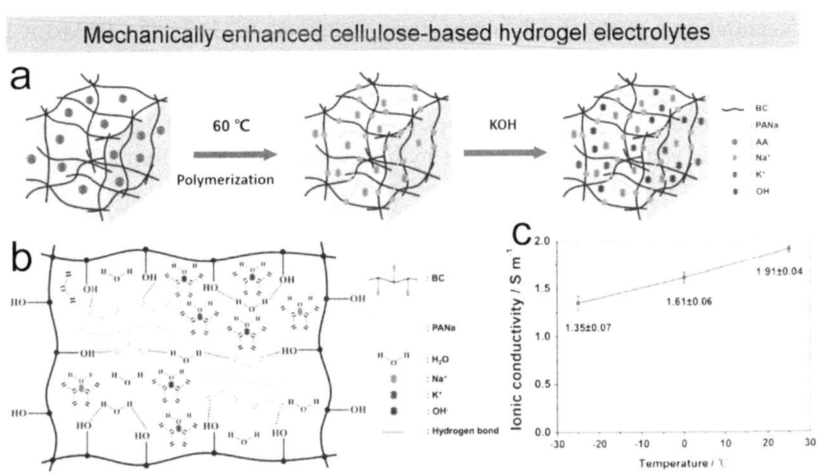

FIGURE 6.5 (a) Schematic diagram of the process of the PANa/BC/KOH hydrogel electrolyte obtained via free radical polymerization. (b) Schematic illustration of PANa/BC/KOH hydrogel electrolyte. (c) Ionic conductivity of PANa/BC/KOH hydrogels at 25, 0, and −25°C. *Reprinted with permission from (Wang et al., 2023). Copyright 2023, Elsevier.*

6.3.2 Wide temperature range of cellulose-based hydrogel electrolytes

Hydrogel electrolytes are susceptible to external environmental effects due to the large amount of water filling the network structure of the hydrogel (Chen et al., 2021b; Yang et al., 2022). On the one hand, the soft hydrogel electrolyte is susceptible to freezing at low temperatures, leading to a decrease in conductivity and flexibility, and ultimately failure of the device (Zhu et al., 2023a). On the other hand, the hydrogel electrolyte undergoes water loss by evaporation at high temperatures, and the volume of the hydrogel shrinks accordingly, and the internal 3D network collapses (Liu et al., 2022b). Similarly, the flexibility and electrical conductivity are greatly reduced, making the energy storage device fail. Therefore, the development of cellulose-based hydrogel electrolytes with a wide temperature range became the key to solving the problem.

Currently, the common method is to introduce high concentrations of salts into the dispersed solution system of the hydrogel to reduce the hydrogen bonding within the water molecules, and to lower the freezing point (Yang et al., 2020). As shown in Fig. 6.6a, the flexible and antifreezing hydrogel electrolyte (PAM/LiCl/WSCA) was prepared. The water-soluble cellulose acetate (WSCA) had good water solubility and abundant hydroxyl groups, which were capable of forming abundant hydrogen bonds with PAM chains and were used to enhance the mechanical properties of hydrogel electrolytes. In addition, the high concentration of LiCl concentration enabled the hydrogel to exhibit excellent antifreezing capabilities. Fig. 6.6b exhibited the improved freezing resistance of PAM/LiCl/WSCA hydrogels with the increase of LiCl concentration. As the LiCl content was increased to 50%, the PAM/LiCl/WSCA hydrogel consistently showed a soft and flexible state even when the temperature dropped to $-80°C$. It was attributed to the fact that the presence of large amounts of Li^+ and Cl^- ions effectively strengthened the interactions between the water molecules in the hydrogel, preventing the formation of ice

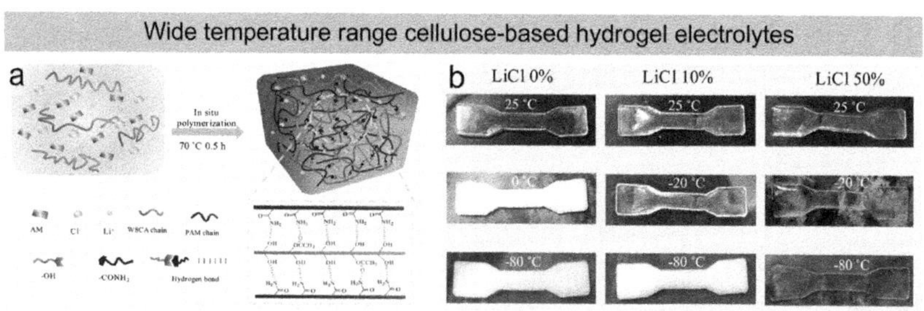

FIGURE 6.6 (a) Schematic diagram of the preparation process of PAM/LiCl/WSCA hydrogel. (b) Morphological photographs of hydrogels at low temperatures under different ratios of LiCl. *Reprinted with permission from (Zhang et al., 2022a). Copyright 2022, Elsevier.*

and the evaporation of water. In addition, supercapacitors assembled with PAM/LiCl/WSCA hydrogel as electrolytes showed excellent capacitance retention stability after 500 folding cycles. Furthermore, the supercapacitor retained 64.64% of the capacitance at −40°C (Zhang et al., 2022a).

Except for highly concentrated salt solutions, the freezing point is lowered by introducing organic freeze-resistant agents (e.g., ethylene glycol (EG), glycerol, and DMSO) into the dispersed solution system of the hydrogel to reduce the hydrogen bonding within the water molecules (Jiang et al., 2021; Wang et al., 2021; Yan et al., 2023). For instance, cotton was used as the raw material for the polymetric framework, which was converted to dissolve cellulose in the presence of abundant Zn^{2+} ions through alkaline pretreatment and following milling. And tetraethyl orthosilicate (TEOS) was used as the cross-linker, which was readily hydrolyzed to form Si $(OH)_4$. The TEOS was cross-linked with the -OH groups of cellulose by covalent siloxane bonding (-Si-O-Si-) to generate a 3D porous frame network with well thermostability. With the addition of the antifreezing agent glycerol, the as-prepared hydrogels were rendered frost-resistant. The ionic conductivities of the prepared hydrogel electrolytes were 38.5, 24.1, and 19.4 mS/cm at temperatures of 60, −20, and −40°C, respectively. A flexible Zn-ion energy storage device was assembled with hydrogel electrolyte, providing high durability under various harsh environments. In simulated low temperatures, the Zn-ion energy storage device sealed in solid ice charges a cell phone. In addition, a Zn-ion energy storage device submerged in boiled water (97.1°C) powered the electronics. These demonstrations showed the excellent reliability of the batteries at low and high temperatures (Chen et al., 2021a).

In general, the wide temperature range of cellulose-based hydrogel electrolytes is valuable for the studies of supercapacitors and battery energy storage devices. It provided technical support and solutions for the practical application of energy storage devices. It also promoted the development and application of electrochemical energy storage technology.

6.3.3 Self-healing of cellulose-based hydrogel electrolytes

The cellulose-based hydrogel electrolyte has the function of self-healing, which can repair the broken parts and restore the original electrolyte structure and performance (Chen et al., 2019). Self-healing of cellulose-based hydrogel electrolytes has the following benefits (Juan et al., 2022; Chen et al., 2023): (1) Hydrogels can self-heal at the point of breakage, improving the stability and service life of the hydrogel electrolyte. (2) It reduces the risk of electrolyte leakage and enhances the safety of the device. (3) The self-healing hydrogel electrolyte has good mechanical and ionic transport properties. At present, the self-healing cellulose-based hydrogel electrolytes healing mechanism can be categorized into dynamic reversible covalent bonding and dynamic noncovalent bonding mechanisms (Heidarian et al., 2021).

Dynamic covalent bonds including imine bonds and borate bonds have strong molecular interactions, resulting in cellulose-based hydrogel electrolytes with good self-healing stability. Recently, chemically dynamic cross-linked self-repairing hydrogels based on imine (-C=N-) bond (also known as Schiff base) have been prepared by utilizing the reversible properties of -C=N- bonds. Researchers have constructed self-healing cellulose-based hydrogels based on dynamic Schiff base bonding using dialdehyde-modified CNCs reacted with carboxymethyl chitosan bearing amino groups (Huang et al., 2018). The boronic bond created between the boric acids and the dial-alcohols gives the hydrogel electrolyte remarkable self-healing behavior (Cho et al., 2021). For example, Chen et al. (2019) developed a borax-crosslinked polyvinyl alcohol/nanocellulose (B-PVA/NFC) hydrogel electrolyte. The NFC were entangled in the PVA chains through hydrogen bonding to form reticular structures. After addition of borax, the borax was hydrolyzed in aqueous solution to tetrahedral $B(OH)_4^-$ and trihedral $B(OH)_3$. The hydrolyzed borax formed three kinds of compounds (i.e., PVA-$B(OH)_4$-PVA, PVA-$B(OH)_4$-NFC, and NFC-$B(OH)_4$-NFC) with PVA and/or NFC to cross-link them reversibly. Therefore, a hydrogel electrolyte with high tensile properties and good self-healing (20 s) was produced through the synergistic effect of an NFC-enhanced network, borax-mediated multiple linkages, chain tangling, and interfacial/intramolecular hydrogen bonding. Self-healing was verified by assembling a flexible quasi-solid-state Zn ion hybrid supercapacitor (ZHS) with a B-PVA/NFC hydrogel electrolyte. The capacity of the ZHS immediately dropped to zero when the ZHS was cut in half. Subsequently, after healing at room temperature, most of the capacity of the ZHS was restored.

For dynamic noncovalent bonding, the self-healing of cellulose-based hydrogel reconstructed the network through noncovalent interactions, such as hydrogen bonding, electrostatic interactions, and metal coordination (Chen et al., 2021a; Fu et al., 2023). The noncovalent method was simple, inexpensive, and renewable, which was an easy way to prepare self-healing cellulose-based hydrogel electrolytes for flexible supercapacitors. Moreover, nitrocellulose could form a dynamic cross-linked network through hydrogen bonding, which served as a connecting bridge between polymer chains to help the healing properties of hydrogel electrolytes. For example, Li et al. (2018) designed a flexible double-crosslinked cellulose-based hydrogel electrolyte by connecting a rigid cellulose skeleton to a flexible PAM network with polydopamine (PDA) as the linker. Of these, the π-π stacking of catechol groups contained in PDA and the abundant hydrogen bonding contained in cellulose had favorable effects on the flexibility and self-healing properties of the hydrogel electrolyte. In addition, Lin et al. (2023) employed H^+-assisted freezing method to form CMC hydrogels by hydrogen bonding cross-linking. Next, the hydrogel was soaked in calcium chloride solution, and Ca^{2+} ions as ionic cross-linker were used capable of embedding uniformly in the CMC hydrogel to produce a strong ionic cross-linking network. Therefore, through dual physical cross-linking (hydrogen bonding and ionic coordination bonding), CMC hydrogel could be rebuilt after an instantaneous fracture. Besides, soaking in an ionic solution provides the high ionic conductivity (58.0 mS/cm) of the electrolyte. The symmetrical supercapacitor exhibited high specific capacitance (309 F/g). Finally, the supercapacitor had a healing efficiency of 93.9% after five cuttings.

6.3.4 Self-adhesion of cellulose-based hydrogel electrolytes

Self-adhesion of cellulose-based hydrogels ensures a close interfacial connection between electrolyte and electrode, which is crucial for the stability and reliability of wearable electronic devices (Zhang et al., 2022b). Several studies have found that cellulose-based hydrogels with self-adhesive properties were usually composed of catechol groups (Heidarian et al., 2019; Deng et al., 2021; Chao et al., 2022). Therefore, cellulose modified with compounds containing catechol groups (e.g., tennis acid (Fu et al., 2021), polydopamine (Yang et al., 2021b)) used as cross-linking networks or enhancers introduced into the hydrogel electrolyte structure has become a hot research issue. For example, Fu et al. (2021) constructed a self-catalytic nanoreinforced strategy by combining tannic acid−functionalized cellulose nanocrystals (TCs) with a high concentration of 7.5 M $ZnCl_2$. Subsequently, the strategy was able to catalyze the rapid polymerization (within 1 min) of [2(methacryloyloxy)ethyl] dimethyl(3-sulfopropyl) (SBMA) and acrylic acid (AA) monomers to prepare "water-in-salt" nanocellulose composite hydrogel electrolytes (TCZn/PSA). As shown in Fig. 6.7a−c, the TCZn/PSA hydrogels were able to produce robust and steady adhesion on different substrates. This was attributed to the fact that the zwitterionic and carboxyl groups in SBMA and AA synergize with the catechol molecules of TC through various interactions.

Additionally, some studies have shown that the adhesion of cellulose hydrogels mainly originated from the polymer network (Fu et al., 2023; Zhang et al., 2022b). For example, Fu et al. (2023) introduced a zwitterions (betaine) permeant into a polyacrylic acid (PAA) network reinforced with CNFs to prepare an all-round cellulose-based hydrogel electrolyte (CP/EG Zn/betaine). The prepared CP/EG Zn/betaine hydrogel electrolytes achieved the highest adhesion strength and adhesion energy (34.3 kPa and 800 J/m). The strong adhesion of CP/EG Zn/betaine hydrogel electrolyte on various electrodes (metal and carbon electrodes) was attributed to a variety of interactions of metal coordination, hydrogen bonding, and electrostatic attraction. When the assembled zinc-ion supercapacitor was subjected to successive deformations ranging from squeezing, rolling, folding to compressing, there was almost no loss of capacity. The structure of the zinc-ion supercapacitor was closely assembled without any stratification or detachment, which ensured dependability and safety in practical applications. For flexible energy storage devices,

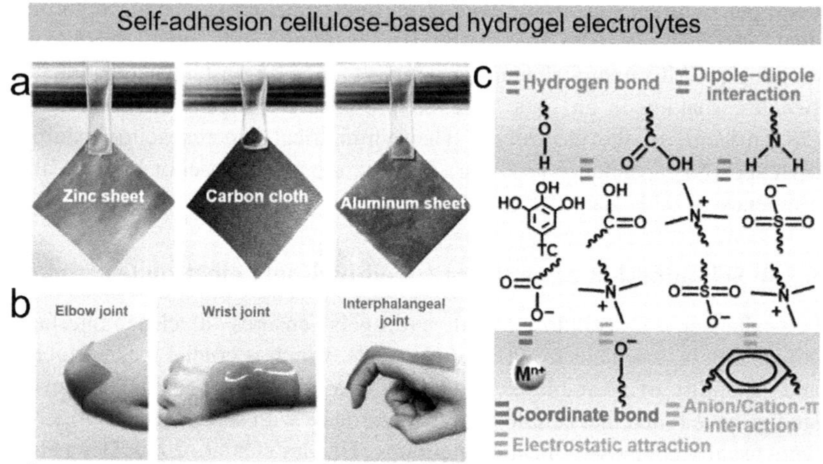

FIGURE 6.7 (a) Photographs showing the interfacial adhesion between the hydrogel and different substrate materials. (b) Photographs of hydrogel self-adhesion on different joints of the human body. (c) Schematic diagram of the adhesion mechanism of TCZn/PSA hydrogels. *Reprinted with permission from (Fu et al., 2021). Copyright 2021, American Chemical Society.*

electrochemical stability was especially important when subjected to arbitrary deformations. It requires strong interfacial adhesion between the gel electrolyte and the electrode formed under conditions of mechanical deformation.

6.4 Chapter summary

In summary, the preparation of cellulose-based supercapacitor electrodes mainly involves compounding cellulose with conductive materials to make composite electrodes, such as porous carbon, film, hydrogel, and aerogel. By adjusting the ratio of each component of the electrode, the electrochemical properties of the supercapacitor electrode are optimized, thus improving the specific capacity, cycling performance, and mechanical properties of the supercapacitor. Meanwhile, cellulose has frequently been used as a reinforcing agent in hydrogel electrolyte compositions, which enhances the hydrogel electrolyte properties in various aspects, such as structural stability, interfacial adhesion, and tensile strength. In addition, cellulose-based hydrogel electrolytes tend to form porous structures to obtain better ionic conductivity. Although cellulose-based supercapacitor electrodes and hydrogel electrolytes have been extensively studied, there are still some unresolved challenges to be addressed.

Currently, cellulose-based electrode films are limited by a disorderly/dense microstructure with low porosity, which results in hindered electron conduction and ion transfer/storage. In addition, the poor coupling between the active material and the conductive network interface leads to slow mass transfer at thc charge interface, rendering the energy density of supercapacitors difficult to further improve. Therefore, how to design nanostructures of cellulose and active components for high-performance electrochemical supercapacitors has become one of the challenging problems in the field. In addition, numerous celluloses (nanocellulose, water-soluble cellulose, and non−water-soluble cellulose) have significant differences in their physical properties, although they have the same chemical composition and better affinity to each other. Therefore, the selection of a suitable morphology of cellulose to design the desired electrode structure is the focus of the research.

The synthesized mechanism of the complex cellulose electrode materials and the interfacial reactive process with the electrolyte are to be further explored. In addition, the structure, transport properties, and ion matching of the active materials are investigated by further combining experiments with theoretical calculations to provide directions for subsequent research. With the continuous development of flexible electronic products, fiber-typed supercapacitors have attracted much attention due to their light weight, controllable volume, good bending, and stretching performance, and braid ability. At present, carbon fibers are commonly used as electrode materials for fiber-based supercapacitors. However, the narrow voltage window results in low energy density of supercapacitors. Cellulose material can rely on the unique 1D fiber structure, which ensures the high mechanical strength and good flexibility of the electrode. In the future, it is possible to construct fiber-shaped electrodes with self-stretching by wet spinning and winding processes, with cellulose as the bonding matrix and support, and pseudo-capacitor materials as reinforcing materials, and thus increase the energy density of fiber-typed supercapacitors. There is great application potential in the field of wearable e-textiles.

At present, most cellulose-based composite hydrogel electrolytes are only applicable to particular environmental situations. Therefore, there remains a great challenge to make multifunctional cellulose-based composite electrolytes with the necessary ionic conductivity, stretchable, self-healing, and low-temperature resistant properties. Cellulose-based supercapacitor as an energy storage device can only be better played with other electrical devices, sensors, photovoltaic conversion, and other power generation devices into a composite multifunctional device, which is the future design of cellulose-based supercapacitors to be considered.

Acknowledgments

The financial support from National Natural Science Foundation of China (22478036) is gratefully acknowledged.

References

Amaral, M.M., Venâncio, R., Peterlevitz, A.C., Zanin, H., 2022. Recent advances on quasi-solid-state electrolytes for supercapacitors. J. Energy Chem. 67, 697−717.

Cai, J., Niu, H., Wang, H., Shao, H., Fang, J., He, J., Xiong, H., Ma, C., Lin, T., 2016. High-performance supercapacitor electrode from cellulose-derived, inter-bonded carbon nanofibers. J. Power Sources 324, 302–308.

Chao, Y., Gao, H., Zhu, X., Wang, H., Yang, Z., Zhou, W., Li, Y., Chen, H., Yang, Y., Hu, Y., 2022. Stretchable, adhesive and antibacterial mussel-inspired cellulose nanocrystal/pectin biomass-based multifunctional tough hydrogels. Ind. Crop. Prod. 187, 115272.

Chen, D., Zhao, X., Wei, X., Zhang, J., Wang, D., Lu, H., Jia, P., 2020. Ultrastretchable, tough, antifreezing, and conductive cellulose hydrogel for wearable strain sensor. ACS Appl. Mater. Interfaces 12 (47), 53247–53256.

Chen, J., Chen, H., Chen, M., Zhou, W., Tian, Q., Wong, C.P., 2022. Nacre-inspired surface-engineered MXene/nanocellulose composite film for high-performance supercapacitors and zinc-ion capacitors. Chem. Eng. J. 428, 131380.

Chen, K., Huang, J., Yuan, J., Qin, S., Huang, P., Wan, C., You, Y., Guo, Y., Xu, Q., Xie, H., 2023. Molecularly engineered cellulose hydrogel electrolyte for highly stable zinc ion hybrid capacitors. Energy Storage Mater. 63, 102963.

Chen, L.F., Huang, Z.H., Liang, H.W., Gao, H.L., Yu, S.H., 2014. Three-dimensional heteroatom-doped carbon nanofiber networks derived from bacterial cellulose for supercapacitors. Adv. Funct. Mater. 24 (32), 5104–5111.

Chen, L.F., Huang, Z.H., Liang, H.W., Yao, W.T., Yu, Z.Y., Yu, S.H., 2013. Flexible all-solid-state high-power supercapacitor fabricated with nitrogen-doped carbon nanofiber electrode material derived from bacterial cellulose. Energy Environ. Sci. 6, 3331.

Chen, M., Chen, J., Zhou, W., Han, X., Yao, Y., Wong, C.P., 2021a. Realizing an all-round hydrogel electrolyte toward environmentally adaptive dendrite-free aqueous Zn-MnO_2 batteries. Adv. Mater. 33 (9), e2007559.

Chen, M., Chen, J., Zhou, W., Xu, J., Wong, C.P., 2019. High-performance flexible and self-healable quasi-solid-state zinc-ion hybrid supercapacitor based on borax-crosslinked polyvinyl alcohol/nanocellulose hydrogel electrolyte. J. Mater. Chem. A 7, 26524.

Chen, M., Zhang, Y., Xing, G., Chou, S.L., Tang, Y., 2021b. Electrochemical energy storage devices working in extreme conditions. Energy Environ. Sci. 14 (6), 3323–3351.

Chen, R., Ling, H., Huang, Q., Yang, Y., Wang, X., 2021c. Interface engineering on cellulose-based flexible electrode enables high mass loading wearable supercapacitor with ultrahigh capacitance and energy density. Small 18, 2106356.

Cheng, Y., Ren, X., Duan, L., Gao, G., 2020. A transparent and adhesive carboxymethyl cellulose/polypyrrole hydrogel electrode for flexible supercapacitors. J. Mater. Chem. C 8 (24), 8234–8242.

Cho, S., Hwang, S.Y., Oh, D.X., Park, J., 2021. Recent progress in self-healing polymers and hydrogels based on reversible dynamic B–O bonds: boronic/boronate esters, borax, and benzoxaborole. J. Mater. Chem. A 9 (26), 14630–14655.

Deng, X., Huang, B., Wang, Q., Wu, W., Coates, P., Sefat, F., Lu, C., Zhang, W., Zhang, X., 2021. A Mussel-inspired antibacterial hydrogel with high cell affinity, toughness, self-healing, and recycling properties for wound healing. ACS Sustain. Chem. Eng. 9 (8), 3070–3082.

Du, W., Bai, Y.L., Xu, J., Zhao, H., Zhang, L., Li, X., Zhang, J., 2018. Advanced metal-organic frameworks (MOFs) and their derived electrode materials for supercapacitors. J. Power Sources 402, 281–295.

Fan, W., Wang, Q., Rong, K., Shi, Y., Peng, W., Li, H., Guo, Z., Xu, B.B., Hou, H., Algadi, H., Ge, S., 2023. Mxene enhanced 3D needled waste denim felt for high-performance flexible supercapacitors. Nano-Micro Lett. 16, 36.

Fourmann, O., Hausmann, M.K., Neels, A., Schubert, M., Nyström, G., Zimmermann, T., Siqueira, G., 2021. 3D printing of shape-morphing and antibacterial anisotropic nanocellulose hydrogels. Carbohydr. Polym. 259, 117716.

Fu, Q., Hao, S., Meng, L., Xu, F., Yang, J., 2021. Engineering self-adhesive polyzwitterionic hydrogel electrolytes for flexible Zinc-ion hybrid capacitors with superior low-temperature adaptability. ACS Nano 15 (11), 18469–18482.

Fu, Q., Hao, S., Zhang, X., Zhao, H., Xu, F., Yang, J., 2023. All-round supramolecular zwitterionic hydrogel electrolytes enabling environmentally adaptive dendrite-free aqueous zinc ion capacitors. Energy Environ. Sci. 16 (3), 1291–1311.

Gao, J.T., Guo, F.J., Ji, C.C., He, X.J., Mi, H.y., Qiu, J.S., 2022. A flexible and stable zinc-ion hybrid capacitor with polysaccharide-reinforced cross-linked hydrogel electrolyte and binder-free carbon cathode. J. Mater. Chem. A 10, 24639–24648.

Guan, Q.F., Yang, H.B., Han, Z.M., Ling, Z.C., Yin, C.H., Yang, K.P., Zhao, Y.X., Yu, S.H., 2021. Sustainable cellulose-nanofiber-based hydrogels. ACS Nano 15 (5), 7889–7898.

Han, J., Ding, Q., Mei, C., Wu, Q., Yue, Y., Xu, X., 2019a. An intrinsically self-healing and biocompatible electroconductive hydrogel based on nanostructured nanocellulose-polyaniline complexes embedded in a viscoelastic polymer network towards flexible conductors and electrodes. Electrochim. Acta 318, 660–672.

Han, J., Wang, S., Zhu, S., Huang, C., Yue, Y., Mei, C., Xu, X., Xia, C., 2019b. Electrospun core–shell nanofibrous membranes with nanocellulose-stabilized carbon nanotubes for use as high-performance flexible supercapacitor electrodes with enhanced water resistance, thermal stability, and mechanical toughness. ACS Appl. Mater. Interfaces 11 (47), 44624–44635.

Heidarian, P., Kaynak, A., Paulino, M., Zolfagharian, A., Varley, R.J., Kouzani, A.Z., 2021. Dynamic nanocellulose hydrogels: recent advancements and future outlook. Carbohydr. Polym. 270, 118357.

Heidarian, P., Kouzani, A.Z., Kaynak, A., Paulino, M., Nasri-Nasrabadi, B., Varley, R., 2019. Double dynamic cellulose nanocomposite hydrogels with environmentally adaptive self-healing and pH-tuning properties. Cellulose 27 (3), 1407–1422.

Hsu, H.H., Khosrozadeh, A., Li, B., Luo, G., Xing, M., Zhong, W., 2019. An Eco-friendly, nanocellulose/RGO/in situ formed polyaniline for flexible and free-standing supercapacitors. ACS Sustain. Chem. Eng. 7 (5), 4766–4776.

Hsu, H.H., Liu, Y., Wang, Y., Li, B., Luo, G., Xing, M., Zhong, W., 2020. Mussel-inspired autonomously self-healable all-in-one supercapacitor with biocompatible hydrogel. ACS Sustain. Chem. Eng. 8 (18), 6935–6948.

Huang, S., Li, Z., Li, P., Du, X., Ma, M., Liang, Z., Su, Y., Xiong, L., 2023. Ultrahigh-voltage aqueous electrolyte for wide-temperature supercapacitors. J. Mater. Chem. A 11 (28), 15532–15539.

Huang, W., Wang, Y., Huang, Z., Wang, X., Chen, L., Zhang, Y., Zhang, L., 2018. On-demand dissolvable self-healing hydrogel based on carboxymethyl chitosan and cellulose nanocrystal for deep partial thickness burn wound healing. ACS Appl. Mater. Interfaces 10 (48), 41076–41088.

Huang, Y., Zhong, M., Shi, F., Liu, X., Tang, Z., Wang, Y., Huang, Y., Hou, H., Xie, X., Zhi, C., 2017. An intrinsically stretchable and compressible supercapacitor containing a polyacrylamide hydrogel electrolyte. Angew. Chem. Int. Ed. 56 (31), 9141–9145.

Jiang, D., Wang, H., Wu, S., Sun, X., Li, J., 2021. Flexible zinc–air battery with high energy efficiency and freezing tolerance enabled by DMSO-based organohydrogel electrolyte. Small Methods 6, 2101043.

Jiang, Q., Lei, Y., Liang, H., Xi, K., Xia, C., Alshareef, H.N., 2020. Review of MXene electrochemical microsupercapacitors. Energy Storage Mater. 27, 78–95.

Jiao, S., Zhou, A., Wu, M., Hu, H., 2019. Kirigami patterning of MXene/Bacterial cellulose composite paper for all-solid-state stretchable micro-supercapacitor arrays. Adv. Sci. 6, 1900529.

Juan, L.T., Lin, S.H., Wong, C.W., Jeng, U.S., Huang, C.F., Hsu, S.H., 2022. Functionalized cellulose nanofibers as crosslinkers to produce chitosan self-healing hydrogel and shape memory cryogel. ACS Appl. Mater. Interfaces 14 (32), 36353–36365.

Khadem, A.H., Hasan, T.U., Rahman, A.N.M.M., Smriti, S.A., Alimuzzaman, S., 2022. Fabrication, properties, and performance of graphene-based textile fabrics for supercapacitor applications: a review. J. Energy Storage 56, 105988.

Kim, S.W., Lee, K.H., Lee, Y.H., Youe, W.J., Gwon, J.G., Lee, S.Y., 2022. Transparent and multifoldable nanocellulose paper microsupercapacitors. Adv. Sci. 9, 2203720.

Li, L., Lu, F., Wang, C., Zhang, F., Liang, W., Kuga, S., Dong, Z., Zhao, Y., Huang, Y., Wu, M., 2018. Flexible double-cross-linked cellulose-based hydrogel and aerogel membrane for supercapacitor separator. J. Mater. Chem. A 6, 24468–24478.

Li, S., Huang, D., Zhang, B., Xu, X., Wang, M., Yang, G., Shen, Y., 2014. Flexible supercapacitors based on bacterial cellulose paper electrodes. Adv. Energy Mater. 4, 1301655.

Li, X., Yuan, L., Liu, R., He, H., Hao, J., Lu, Y., Wang, Y., Liang, G., Yuan, G., Guo, Z., 2021. Engineering textile electrode and bacterial cellulose nanofiber reinforced hydrogel electrolyte to enable high-performance flexible all-solid-state supercapacitors. Adv. Energy Mater. 11, 2003010.

Lin, X., Wang, M., Zhao, J., Wu, X., Xie, J., Yang, J., 2023. Super-tough and self-healable all-cellulose-based electrolyte for fast degradable quasi-solid-state supercapacitor. Carbohydr. Polym. 304, 120502.

Liu, H., Xu, T., Cai, C., Liu, K., Liu, W., Zhang, M., Du, H., Si, C., Zhang, K., 2022a. Multifunctional superelastic, superhydrophilic, and ultralight nanocellulose-based composite carbon aerogels for compressive supercapacitor and strain sensor. Adv. Funct. Mater. 32, 2113082.

Liu, S.L., Zhang, R.Z., Mao, J.F., Zhao, Y.L., Cai, Q., Guo, Z.P., 2022b. From room temperature to harsh temperature applications: fundamentals and perspectives on electrolytes in zinc metal batteries. Sci. Adv. 8, eabn5097.

Liu, Z., Chen, J., Zhan, Y., Liu, B., Xiong, C., Yang, Q., Hu, G.H., 2019. Fe^{3+} cross-linked polyaniline/cellulose nanofibril hydrogels for high-performance flexible solid-state supercapacitors. ACS Sustain. Chem. Eng. 7 (21), 17653–17660.

Lu, A., Wan, H., Qin, C., 2022. A flexible, robust cellulose/phytic acid/polyaniline hydrogel for all-in-one supercapacitors and strain sensors. J. Mater. Chem. A 10, 17279.

Lyu, S., Chang, H., Zhang, L., Wang, S., Li, S., Lu, Y., Li, S., 2023. High specific surface area MXene/SWCNT/cellulose nanofiber aerogel film as an electrode for flexible supercapacitors. Compos. B Eng. 264, 110888.

Ma, L., Chen, S., Wang, D., Yang, Q., Mo, F., Liang, G., Li, N., Zhang, H., Zapien, J.A., Zhi, C., 2019. Super-stretchable zinc-air batteries based on an alkaline-tolerant dual-network hydrogel electrolyte. Adv. Energy Mater. 9 (12), 1803046.

Manjakkal, L., Pullanchiyodan, A., Yogeswaran, N., Hosseini, E.S., Dahiya, R., 2020. A Wearable supercapacitor based on conductive PEDOT:PSS-coated cloth and a sweat electrolyte. Adv. Mater. 32, 1907254.

Moon, W.G., Kim, G.-P., Lee, M., Song, H.D., Yi, J., 2015. A biodegradable gel electrolyte for use in high-performance flexible supercapacitors. ACS Appl. Mater. Interfaces 7 (6), 3503–3511.

Nakayama, A., Kakugo, A., Gong, J.P., Osada, Y., Takai, M., Erata, T., Kawano, S., 2004. High mechanical strength double-network hydrogel with bacterial cellulose. Adv. Funct. Mater. 14 (11), 1124–1128.

Nele, V., Wojciechowski, J.P., Armstrong, J.P.K., Stevens, M.M., 2020. Tailoring gelation mechanisms for advanced hydrogel applications. Adv. Funct. Mater. 30 (42), 2002759.

Ni, M., Ma, C., Huang, H., Han, L., Fu, X., Yang, Z., Li, J., Pan, L., Xu, M., 2022. Flexible high-energy-density hybrid supercapacitors with alkaline hydrogel electrolytes. ACS Appl. Energy Mater. 5 (6), 6724–6733.

Park, J., Lee, J., Kim, W., 2022. Redox-active water-in-salt electrolyte for high-energy-density supercapacitors. ACS Energy Lett. 7 (4), 1266–1273.

Peng, H., Lv, Y., Wei, G., Zhou, J., Gao, X., Sun, K., Ma, G., Lei, Z., 2019. A flexible and self-healing hydrogel electrolyte for smart supercapacitor. J. Power Sources 431, 210–219.

Peng, X., Liu, H., Yin, Q., Wu, J., Chen, P., Zhang, G., Liu, G., Wu, C., Xie, Y., 2016. A zwitterionic gel electrolyte for efficient solid-state supercapacitors. Nat. Commun. 7, 11782.

Rostami, J., Benselfelt, T., Maddalena, L., Avci, C., Sellman, F.A., Cinar Ciftci, G., Larsson, P.A., Carosio, F., Akhtar, F., Tian, W., Wågberg, L., 2022. Shaping 90 wt% nano MOFs into robust multifunctional aerogels using tailored bio-based nanofibrils. Adv. Mater. 34, 2204800.

Sheberla, D., Bachman, J.C., Elias, J.S., Sun, C.J., Shao Horn, Y., Dincă, M., 2016. Conductive MOF electrodes for stable supercapacitors with high areal capacitance. Nat. Mater. 16 (2), 220–224.

Simotwo, S.K., Chinnam, P.R., Wunder, S.L., Kalra, V., 2017. Highly durable, self-standing solid-state supercapacitor based on an ionic liquid-rich ionogel and porous carbon nanofiber electrodes. ACS Appl. Mater. Interfaces 9 (39), 33749–33757.

Sun, Z., Qu, K., You, Y., Huang, Z., Liu, S., Li, J., Hu, Q., Guo, Z., 2021. Overview of cellulose-based flexible materials for supercapacitors. J. Mater. Chem. A 9, 7278.

Tang, H., Chen, R., Huang, Q., Ge, W., Zhang, X., Yang, Y., Wang, X., 2022. Scalable manufacturing of leaf-like MXene/Ag NWs/cellulose composite paper electrode for all-solid-state supercapacitor. EcoMat 4, e12247.

Tian, W., VahidMohammadi, A., Reid, M.S., Wang, Z., Ouyang, L., Erlandsson, J., Pettersson, T., Wågberg, L., Beidaghi, M., Hamedi, M.M., 2019. Multifunctional nanocomposites with high strength and capacitance using 2D MXene and 1D nanocellulose. Adv. Mater. 31, 1902977.

Wang, D., Li, Z., Yang, L., Zhang, J., Wei, Y., Feng, Q., Wei, Q., 2023. Hydrogel electrolyte based on sodium polyacrylate/KOH hydrogel reinforced with bacterial cellulose aerogel for flexible supercapacitors. Chem. Eng. J. 454, 140090.

Wang, J., Huang, Y., Liu, B., Li, Z., Zhang, J., Yang, G., Hiralal, P., Jin, S., Zhou, H., 2021. Flexible and anti-freezing zinc-ion batteries using a guar-gum/sodium-alginate/ethylene-glycol hydrogel electrolyte. Energy Storage Mater. 41, 599–605.

Wang, S., Jiang, F., Xu, X., Kuang, Y., Fu, K., Hitz, E., Hu, L., 2017a. Super-strong, super-stiff macrofibers with aligned, long bacterial cellulose nanofibers. Adv. Mater. 29 (35), 1702498.

Wang, Y., Wang, X., Li, X., Bai, Y., Xiao, H., Liu, Y., Liu, R., Yuan, G., 2019. Engineering 3D ion transport channels for flexible MXene films with superior capacitive performance. Adv. Funct. Mater. 29, 1900326.

Wang, Z., Carlsson, D.O., Tammela, P., Hua, K., Zhang, P., Nyholm, L., Strømme, M., 2015. Surface modified nanocellulose fibers yield conducting polymer-based flexible supercapacitors with enhanced capacitances. ACS Nano 9 (7), 7563–7571.

Wang, Z., Tammela, P., Strømme, M., Nyholm, L., 2017b. Cellulose-based supercapacitors: material and performance considerations. Adv. Energy Mater. 7, 1700130.

Wu, X., Shi, Z., Tjandra, R., Cousins, A.J., Sy, S., Yu, A., Berry, R.M., Tam, K.C., 2015. Nitrogen-enriched porous carbon nanorods templated by cellulose nanocrystals as high performance supercapacitor electrodes. J. Mater. Chem. A 3, 23768.

Wu, Y., Sun, S., Geng, A., Wang, L., Song, C., Xu, L., Jia, C., Shi, J., Gan, L., 2020. Using TEMPO-oxidized-nanocellulose stabilized carbon nanotubes to make pigskin hydrogel conductive as flexible sensor and supercapacitor electrode: inspired from a Chinese cuisine. Compos. Sci. Technol. 196, 108226.

Xu, T., Du, H., Liu, H., Liu, W., Zhang, X., Si, C., Liu, P., Zhang, K., 2021. Advanced nanocellulose-based composites for flexible functional energy storage devices. Adv. Mater. 33, 2101368.

Xu, Z., Xie, F., Wang, J., Au, H., Tebyetekerwa, M., Guo, Z., Yang, S., Hu, Y.S., Titirici, M.M., 2019. All-cellulose-based quasi-solid-state sodium-ion hybrid capacitors enabled by structural hierarchy. Adv. Funct. Mater. 29, 1903895.

Yan, Y., Duan, S., Liu, B., Wu, S., Alsaid, Y., Yao, B., Nandi, S., Du, Y., Wang, T.W., Li, Y., He, X., 2023. Tough hydrogel electrolytes for anti-freezing zinc-ion batteries. Adv. Mater. 35 (18), 2211673.

Yang, L., Song, L., Feng, Y., Cao, M., Zhang, P., Zhang, X.F., Yao, J., 2020. Zinc ion trapping in a cellulose hydrogel as a solid electrolyte for a safe and flexible supercapacitor. J. Mater. Chem. A 8, 12314.

Yang, P., Yang, J.L., Liu, K., Fan, H.J., 2022. Hydrogels enable future smart batteries. ACS Nano 16 (10), 15528–15536.

Yang, Y., Zhang, D., Liu, Y., Shen, L., Zhu, T., Xu, X., Zheng, J., Gong, X., 2021a. Solid-state double-network hydrogel redox electrolytes for high-performance flexible supercapacitors. ACS Appl. Mater. Interfaces 13 (29), 34168–34177.

Yang, Z., Huang, R., Zheng, B., Guo, W., Li, C., He, W., Wei, Y., Du, Y., Wang, H., Wu, D., Wang, H., 2021b. Highly stretchable, adhesive, biocompatible, and antibacterial hydrogel dressings for wound healing. Adv. Sci. 8 (8), 2003627.

Yu, W., Lin, W., Shao, X., Hu, Z., Li, R., Yuan, D., 2014. High performance supercapacitor based on Ni_3S_2/carbon nanofibers and carbon nanofibers electrodes derived from bacterial cellulose. J. Power Sources 272, 137–143.

Zhang, B., Chen, H., Hu, Q., Jiang, L., Shen, Y., Zhao, D., Zhou, Z., 2021a. CelluMOFs: green, facile, and flexible metal-organic frameworks for versatile applications. Adv. Funct. Mater. 31 (43), 2105395.

Zhang, H., Gan, X., Song, Z., Zhou, J., 2023. Amphoteric cellulose-based double-network hydrogel electrolyte toward ultra-stable Zn anode. Angew. Chem. Int. Ed. 62 (13), e202217833.

Zhang, K., Pang, Y., Chen, C., Wu, M., Liu, Y., Yu, S., Li, L., Ji, Z., Pang, J., 2022a. Stretchable and conductive cellulose hydrogel electrolytes for flexible and foldable solid-state supercapacitors. Carbohydr. Polym. 293, 119673.

Zhang, Q., Hou, X., Liu, X., Xie, X., Duan, L., Lü, W., Gao, G., 2021b. Nucleotide-tackified organohydrogel electrolyte for environmentally self-adaptive flexible supercapacitor with robust electrolyte/electrode interface. Small 17 (46), 2103091.

Zhang, T., Cheng, Q., Ye, D., Chang, C., 2017. Tunicate cellulose nanocrystals reinforced nanocomposite hydrogels comprised by hybrid cross-linked networks. Carbohydr. Polym. 169, 139–148.

Zhang, T., Yang, L., Yan, X., Ding, X., 2018. Recent advances of cellulose-based materials and their promising application in sodium-ion batteries and capacitors. Small 17, 2103091.

Zhang, X., Zhao, J., Xia, T., Li, Q., Ao, C., Wang, Q., Zhang, W., Lu, C., Deng, Y., 2020. Hollow polypyrrole/cellulose hydrogels for high-performance flexible supercapacitors. Energy Storage Mater. 31, 135–145.

Zhang, Y., Shang, Z., Shen, M., Chowdhury, S.P., Ignaszak, A., Sun, S., Ni, Y., 2019. Cellulose nanofibers/reduced graphene oxide/polypyrrole aerogel electrodes for high-capacitance flexible all-solid-state supercapacitors. ACS Sustain. Chem. Eng. 7 (13), 11175–11185.

Zhang, Y., Wu, D., Huang, F., Cai, Y., Li, Y., Ke, H., Lv, P., Wei, Q., 2022b. "Water-in-salt" nonalkaline gel polymer electrolytes enable flexible zinc-air batteries with ultra-long operating time. Adv. Funct. Mater. 32, 2203204.

Zhao, J., Gong, J., Wang, G., Zhu, K., Ye, K., Yan, J., Cao, D., 2020. A self-healing hydrogel electrolyte for flexible solid-state supercapacitors. Chem. Eng. J. 401, 125456.

Zheng, Q., Cai, Z., Ma, Z., Gong, S., 2015. Cellulose nanofibril/reduced graphene oxide/carbon nanotube hybrid aerogels for highly flexible and all-solid-state supercapacitors. ACS Appl. Mater. Interfaces 7 (5), 3263–3271.

Zheng, Y., Wang, Y., Zhao, J., Li, Y., 2023. Electrostatic interfacial cross-linking and structurally oriented fiber constructed by surface-modified 2D MXene for high-performance flexible pseudocapacitive storage. ACS Nano 17 (3), 2487–2496.

Zhong, C., Deng, Y., Hu, W., Qiao, J., Zhang, L., Zhang, J., 2015. A review of electrolyte materials and compositions for electrochemical supercapacitors. Chem. Soc. Rev. 44 (21), 7484–7539.

Zhou, G., Li, M.C., Liu, C., Wu, Q., Mei, C., 2021a. 3D printed $Ti_3C_2T_x$ MXene/cellulose nanofiber architectures for solid-state supercapacitors: ink rheology, 3D printability, and electrochemical performance. Adv. Funct. Mater. 32, 2109593.

Zhou, S., Guo, K., Bukhvalov, D., Zhu, W., Wang, J., Sun, W., He, M., 2021b. H-bond/ionic coordination switching for fabrication of highly oriented cellulose hydrogels. J. Mater. Chem. A 9 (9), 5533–5541.

Zhou, S., Kong, X., Zheng, B., Huo, F., Strømme, M., Xu, C., 2019. Cellulose nanofiber @ conductive metal–organic frameworks for high-performance flexible supercapacitors. ACS Nano 13, 9578–9586.

Zhu, K., Sun, Z., Li, Z., Liu, P., Li, H., Jiao, L., 2023a. Design strategies and recent advancements for low-temperature aqueous rechargeable energy storage. Adv. Energy Mater. 13 (8), 2203708.

Zhu, X., Ji, C., Meng, Q., Mi, H., Yang, Q., Li, Z., Yang, N., Qiu, J., 2022. Freeze-tolerant hydrogel electrolyte with high strength for stable operation of flexible zinc-ion hybrid supercapacitors. Small 18 (16), 2200055.

Zhu, Y., Ma, J., Das, P., Wang, S., Wu, Z.S., 2023b. High-voltage MXene-based supercapacitors: present status and future perspectives. Small Methods 7 (8), 2201609.

Chapter 7

Cellulose composites for optical application

Ting Xu,[1] Chuanling Si[1,*] and Mingguo Ma[2,*]

[1]*State Key Laboratory of Biobased Fiber Manufacturing Technology, Tianjin Key Laboratory of Pulp and Paper, College of Light Industry and Engineering, Tianjin University of Science and Technology, Tianjin, PR China;* [2]*MOE Engineering Research Center of Forestry Biomass Materials and Bioenergy, Research Center of Biomass Clean Utilization, Beijing Key Laboratory of Lignocellulosic Chemistry, College of Materials Science and Technology, Beijing Forestry University, Beijing, PR China*

[*]*Corresponding authors: E-mail: sichli@tust.edu.cn and mg_ma@bjfu.edu.cn*

7.1 Introduction

In nature, many living things experience colorful colors. Polarized structural color is a special structural color in chemical color. At present, a series of bionic optical materials with fine structure has been developed (Zhao et al., 2012). In the scientific research of natural/biomimetic polarization structure color, it has been found that cellulose nanocrystals (CNCs) have controllable "orderly self-assembly" performance, developing optical materials with chiral nematic structure by using their spiral self-assembly characteristics (Tardy et al., 2017).

CNCs have the characteristics of high crystallinity, high purity, high strength, high elastic modulus, good biocompatibility, biodegradability, and unique optical properties (Habibi et al., 2010). When the crystalline nanocellulose suspension is higher than the critical concentration, CNCs can be self-assembled to form chiral nematic helical structures with iridescence, which have circularly polarized light characteristics and belong to cholesteric liquid crystal materials (Yang et al., 2022). The chiral nematic spiral structure formed by the self-assembly of chiral CNCs can selectively reflect light of different wavelengths, generate structural color, and react rapidly to external light, electricity, force, heat, and other stimuli. After being stimulated, the

Surface Science and Advanced Functional Applications of Cellulose Composites
https://doi.org/10.1016/B978-0-443-13675-7.00007-5

structural color can be adjusted. This characteristic of chiral CNCs can be used to imitate the bright colors of butterfly wings and the discoloration effect of chameleon skin to achieve bionic discoloration. Chiral CNCs can realize the change of structure and color by external stimulation and have important value in the development of color-changing materials, stealth materials, and other fields (Querejeta-Fernandez et al., 2014). Using the characteristics of chiral CNCs, chiral CNCs have important research value in biology, medicine, optics, physics, catalysis, electronics, optical instruments, information storage, information encryption, optical anticounterfeiting, and other fields (Kaushik et al., 2015).

Peng et al. (2020) reviewed recent progress in the controllable arrangement of nanocellulose in polymeric matrices, including highly oriented structure, helical structure, and gradient structure, discussing the current nanotechnologies that enable the arrangement of nanocellulose in nanocomposite materials. Schütz et al. (2020) reviewed the equilibrium liquid crystal formation and kinetic arrest to photonic bandgap films using suspensions of CNCs. They summarized the development of cholesteric nuclei (tactoids) from the isotropic phase and their potential impact on the final dry films, the kinetic arrest about the cholesteric helical order, the "coffee-ring effect" active prior to kinetic arrest, and the compression of the helix during the final water evaporation, giving rise to visible structural color in the films. Qu et al. (2022) reviewed the fundamental design principles, methodologies, responsive mechanisms, and practical applications of responsive chiral photonic CNC materials. The composites can generate Bragg reflective light response under the stimulation of humidity, pH, gas, solvent, temperature, light, mechanical force, electric field, magnetic field, etc. They also summarized chiral photonic crystals with chiral response and shape memory functions, and stimulus response fluorescence and circularly polarized fluorescence materials. The stimulus-responsive chiral photonic crystals based on CNCs can produce structural and optical properties changes under a variety of external stimuli, which can be used in visual sensors, photonic crystal actuators, rewritable photonic crystal paper, anticounterfeiting labels, and other fields.

The development of alcohol detectors with excellent structural color output and reconfigurable nanostructures will certainly set off reform in colorimeter sensors. Qi et al. (2023) fabricated a portable colorimetric photonic indicator (CPI) for ethanol concentration sensing via shape memory photonic crystals. CPI shows an ultrafast color response (<1 s) to the perception of ethanol concentration and outputs structural colors from blue to red, with reflection covering $\sim$477 to $\sim$628 nm. CPI has the lower half-peak width (less than 33 nm) and low expansion index (0.03) in water for the output of stable reflection spectrum of photonic crystal. It finds stable structure color of CPI (0% concentration corresponds to the wavelength of $\sim$477 nm, which is blue). The maximum reflection peak is observed at the concentration of $\sim$75% ($\sim$628 nm, red) (Fig. 7.1). It shows green structure color at 100%

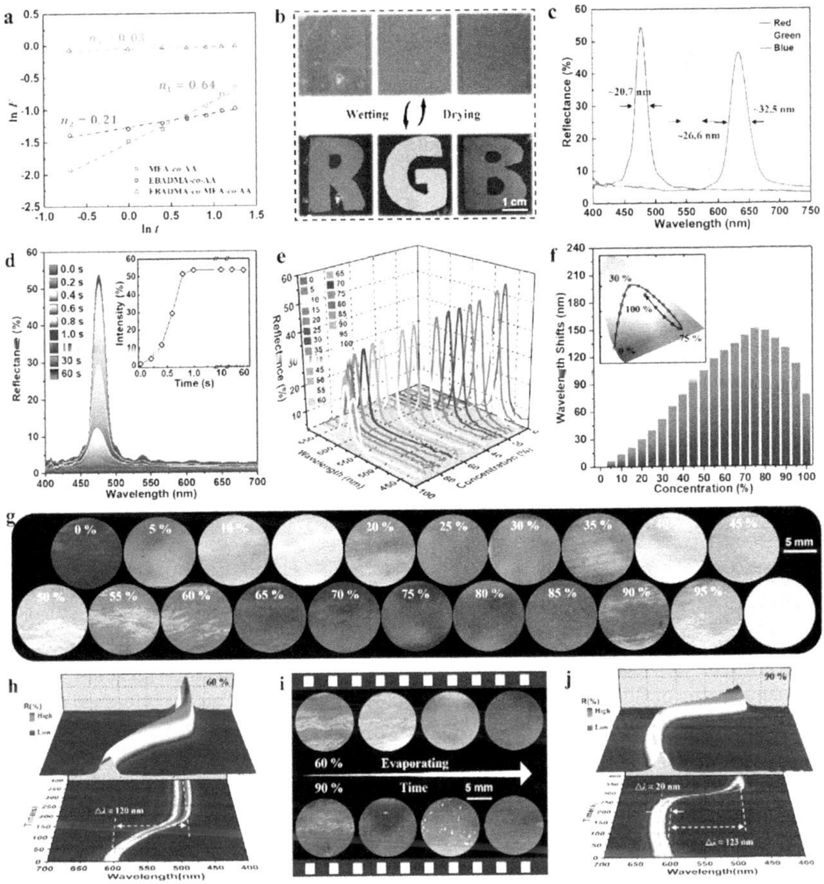

FIGURE 7.1 CPI swelling kinetics curve in water, water response process, reflection spectra/photos corresponding to different concentrations of ethanol solution, and evaporation process of mixture corresponding to similar structural colors. *Reproduced from Qi et al. (2023), with permission from* ***Elsevier.***

concentration. It has been observed that the structural color first shifts to red and then to blue at the high concentration mixture for 60% and 90% ethanol solution during the evaporation process. It finds only blueshift for the structural color at the low concentration mixture.

It further verified the practicability of CPI by tracking the ethanol concentration of distillates in the process of 30% alcohol distillation (Fig. 7.2). In the early stage of distillation (<15 min), the ethanol concentration in the distillate was high (>75%), and in the middle stage, the ethanol concentration remained stable (~75%), and then gradually decreased. Finally, 200 mL of 30% ethanol–water mixture is divided into ~70 mL of distillate and ~125 mL of tower kettle liquid (distillate, ~529 nm, green, concentration

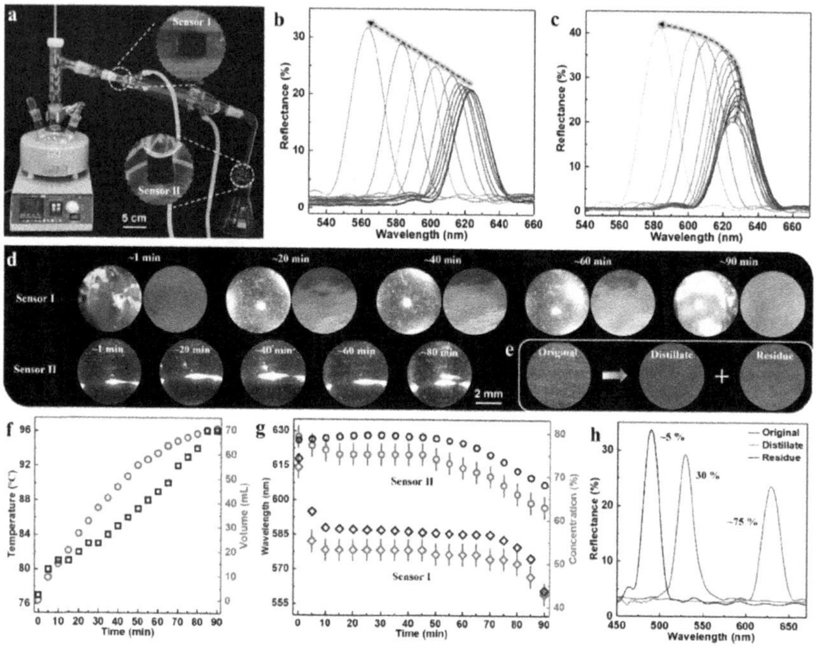

FIGURE 7.2 (a) Apparatus for 30% alcohol distillation, insets show the sensing site of CPI; reflection spectra of CPI response to the evaporation of the initial distillate (b) before and (c) after cooling; (d) typical structural colors of "sensor I" (on the right are the corresponding structural color of CPI response to the distillates before cooling) and "sensor II" during distillation; (e) structural colors of CPI corresponding to the original, total distillate, and residue liquid; (f) time-dependent temperature at branch pipe and the volume of distillate during distillation; (g) the immediate λmax of "sensors I and II" and the deduced concentration of immediate distillate during distillation; (h) reflection spectra and deduced concentrations corresponding to the original mixture, total distillate, and residue liquid. *Reproduced from Qi et al. (2023), with permission from* ***Elsevier.***

~30%; distillate, ~628 nm, red, concentration ~75%; tower kettle liquid, ~489 nm, light blue, concentration ~5%). This work provides reference for process optimization and engineering dynamics analysis of distillation/distillation-related chemical equipment, promoting the intelligent development of ethanol industry.

The drying process in preparing the films included three stages such as an equilibrium liquid crystal phase formation stage, a nonequilibrium kinetic arrest, and structural collapse as the evaporated remaining solvent. Honorato-Rios et al. (2016) established a detailed phase diagram of aqueous CNC suspensions with two different values of the surface charge in an equilibrium liquid crystal phase formation stage. It is difficult to detect macroscopically due to substantial critical fluctuations with a continuous weakening of the first-order character up to the point where phase coexistence. Chen et al. (2023)

synthesized remotely programmed and modeled reversible and complex morphing in monolithic liquid crystal elastomers kirigami encoded with pre-designed topological patterns in its microstructure by combining liquid crystal elastomer with paper-cut as multifunctional devices for soft robotics, flexible electronics, and biomedicine. It obtained a rich variety of out-of-plane shape transformations, including auxetic structures and undulating morphologies, by combining different topological microstructures and kirigami geometries. These elastomers can also achieve many other functions, such as imitating butterfly wings, flower buds, windmills that can rotate in both directions and are driven by light, and dual-mode mobile software robots (Fig. 7.3). It further realized different functionalities, such as a bioinspired fluttering butterfly, a flower bud, dual-rotation light mills, and dual-mode locomotion.

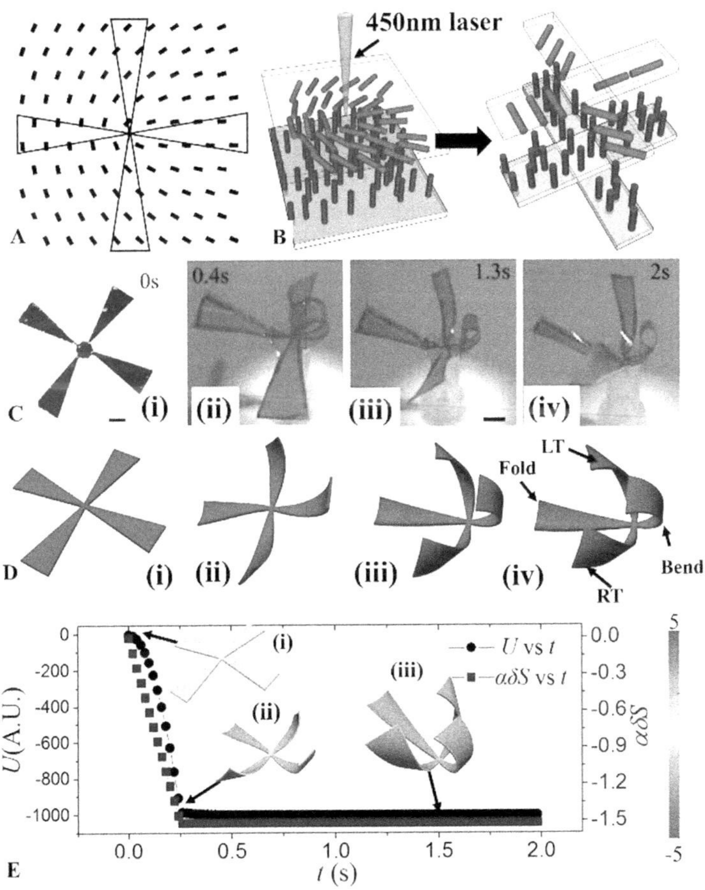

FIGURE 7.3 Liquid crystal elastomer with +1/2 defect. *Reproduced from Chen et al. (2023), with permission from* ***American Chemical Society.***

In this chapter, we introduced the synthesis and property of cellulose, nanocellulose, and advanced functional cellulose composites for optical applications. Surface chemistry of advanced functional cellulose composites for optical applications received special attention. Finally, the problem and potential directions were suggested based on our knowledge. We try to contribute great content to this book "***Surface Science and Advanced Functional Applications of Cellulose Composites.***"

7.2 Synthesis and property of cellulose for optical applications

Cellulose, especially CNC, is reported to have special property and widely optical applications of photonic applications, real-time and visual detection of shrimp and crab freshness, pigments and flash agent, etc. Various methods of self-assembly, coating technology, template method, and so on were developed for the synthesis of cellulose for optical applications. These synthetic methods had an effect on the microstructure of cellulose, further playing an important role on the property of cellulose. Therefore, it is very important to develop appropriate strategy to obtain cellulose with expected structure and designed performance, expanding its application in the optical field.

The size of the electric vector of circularly polarized light remains unchanged, but its direction changes uniformly with time, which has the characteristics of high optical sensitivity and high optical resolution, and broad application prospects in stereo optical display, circularly polarized light information encryption, biological coding, optical data storage and optical devices. In nature, circularly polarized light is a secret communication tool for signal propagation and territorial defense. Zheng et al. (2018) reported uncovering the circular polarization potential of chiral photonic cellulose films with one-dimensional photonic placket for photonic applications. The placket is adjustable in the near ultraviolet and near infrared range, which decompose the incident light matching its photonic placket into left-handed circularly polarized light and right-handed circularly polarized light. The left-handed circularly polarized light is reflected by the film, and the right-handed circularly polarized light passes through the film. Both sides of the crystalline nanocellulose film receive left and right circularly polarized light, with very high intensity and asymmetry factor of 0.87. The crystalline nanocellulose film can convert spontaneous radiation into dextral circularly polarized light, and its asymmetry factor is as high as 0.68. The circularly polarized light based on the crystalline nanocellulose membrane is chirality accurate, high intensity, and adjustable. It reported the application potential of crystalline nanocellulose film in the field of circularly polarized light anticounterfeiting.

Jia et al. (2019) synthesized amine-responsive cellulose-based ratiometric fluorescent materials with the advantages of simple preparation, good stability, good reversibility, good biodegradability, and processing performance for real-time and visual detection of shrimp and crab freshness. The cellulose-based antibacterial material was obtained by utilizing several reactive hydroxyl groups on cellulose, which is formed by chemical bonding of porphyrin and quaternary ammonium salt groups on the cellulose molecular chain at the same time (see Fig. 7.4). The cellulose-based antibacterial material is a kind of antibacterial coating material driven by white light, which has a strong killing effect on drug-resistant bacteria. They designed a type of ratio-type amine reaction cellulose fluorescent material to visually monitor the change of seafood freshness with isothiocyanate fluorescein (FITC) as the indicator and protoporphyrin (PpIX) as the internal standard. The obtained CA-FITC and CA-PpIX were mixed in a simple proportion to obtain the initial fluorescence tunable solid fluorescent material. The fluorescence of the material changes from red to green, realizing the rapid, accurate, and real-time visual monitoring of amine concentration in the environment. The proportional amine is made into an indicator label in cellulose-based fluorescent membrane, realizing the in situ visual monitoring of the freshness of marine products such as sea shrimp (see Fig. 7.5).

CNCs are renewable plant-based colloidal particles, which can form photonic films through self-assembly driven by solvent evaporation. It is not easy to the preparation method of large-area CNCs thin films, which is the

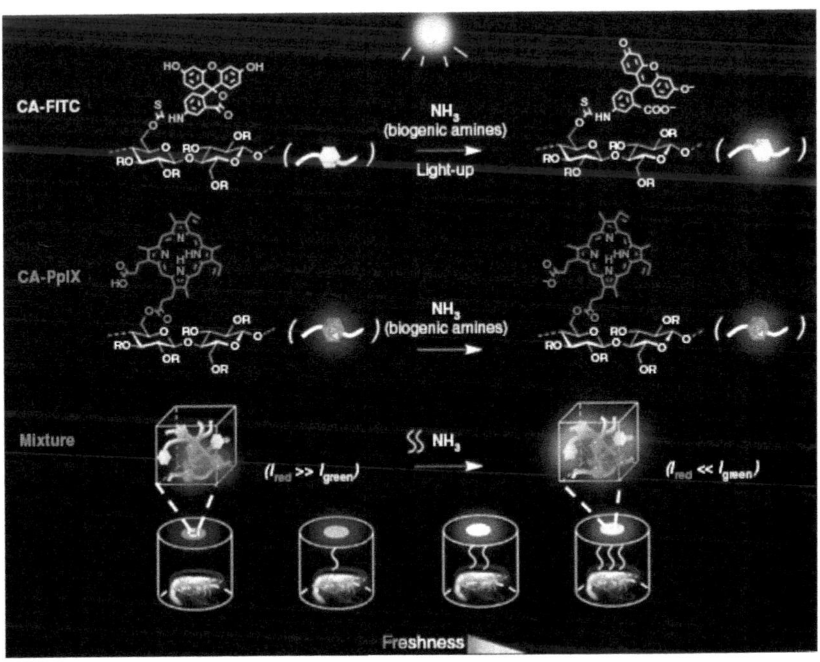

FIGURE 7.4 Schematic illustration for the mechanism of the cellulose-based ratiometric fluorescent material. *Reproduced from Jia et al. (2019), with permission from* ***Springer Nature***.

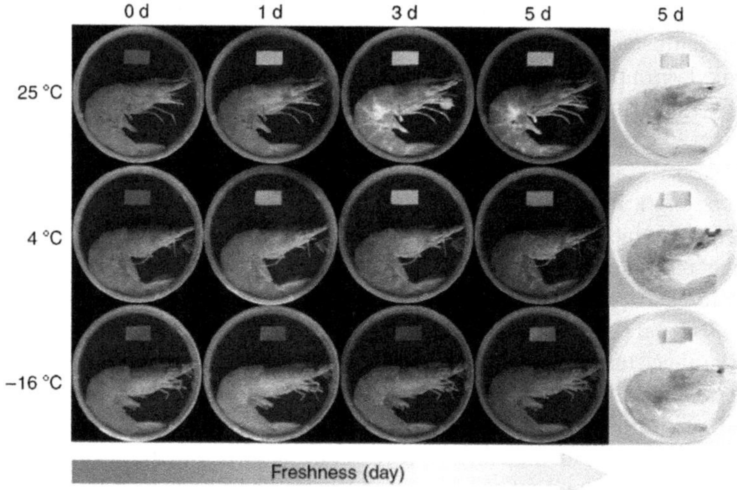

FIGURE 7.5 Monitoring of shrimp freshness by the smart trademark. Images of the cellulose-based ratiometric fluorescent material with a red (dark gray in print version) initial fluorescence as a smart trademark for monitoring the freshness of shrimps stored at different conditions. *Reproduced from Jia et al. (2019), with permission from* ***Springer Nature****.*

bottleneck restricting its commercial developments. Droguet et al. (2022) developed the continuous R2R coating technology for the large-scale fabrication of structurally colored CNC films with good optical response as effect pigments and flash agent. Importantly, these particles will not fade or redisperse after 1 year even in water, and still maintain their optical response. Corona discharge is carried out on the central part of the substrate to activate the surface, increasing the surface energy of hydrophobic polyethylene terephthalate (PET) mesh structure, thus improving the subsequent wettability of low viscosity aqueous CNC suspension. The deposition of the CNC support structure is realized through the in-line slit die to produce a continuous and controllable coating. After deposition, the coating is either statically dried under environmental conditions or passed through the heating chamber of R2R to shorten the drying time.

The same CNC suspension was used to produce a macro uniform blue blade casting film at a constant shear rate of 2.2 s^{-1}. Larger coating gaps result in higher film thickness. The film casted with gc = 700 μm and vc = 1.5 mm s^{-1} achieves the maximum achievable reflectivity, and its thickness is only 10.5 μm, meaning the membranes with well-arranged in structure. The blue, green, and red CNC films cast from the blades were prepared by applying ultrasonic waves to the same initial CNC suspension. The thin films cast in petri dishes with the same suspension in this paper show similar spectra, confirming that the optical quality will not be significantly affected by the amplification or reduction of deposition conditions. For thin cast films, the increasing total ultrasonic energy delivered to the suspension will result in a widening of the peak value and a reduction in the maximum reflectivity. The addition of polymer additives (such as hydroxypropyl cellulose) can make the reflection color of the CNC film redshift, and inhibit the edge nonuniformity, acting as a plasticizer. The interaction between the accelerated drying and the optical properties of the generated CNC film can be obtained by changing the temperature of the hot air dryer from 20 to 60°C. Blue films with good optical properties can be produced under all conditions. The shortening drying time by increasing the temperature will result in small redshift and reduced optical response. Considering the difference in thickness, the stepped film has the same reflectivity as the statically dried film. Therefore, the thin films produced by the fine step process at 60°C maintain a good optical appearance.

More recently, Wang et al. (2022) applied the template region limiting method to obtain the programmable birefringent patterns from modulating localized orientation of CNCs under the synergistic effect of Marangoni effect and capillary flow (see Fig. 7.6). The CNCs film shows the extinction birefringence optical property of "Malta Black Cross" under polarized light, achieving a dynamic change like a kaleidoscope via changing the incidence angle of linearly polarized light. It observed the transparent CNC film in sunlight, and only visible birefringence in polarized light.

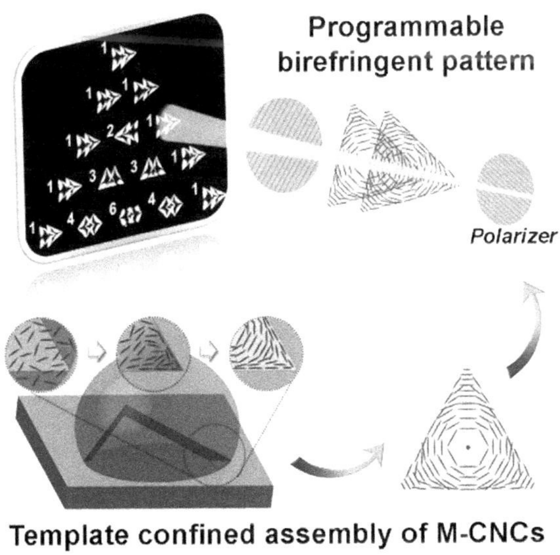

FIGURE 7.6 Template confined assembly of M-CNCs. *Reproduced from Wang et al. (2022), with permission from* ***American Chemical Society.***

Polydimethylsiloxane (PDMS) groove molds with different shapes were prepared by using photolithography and mold replica, exploring the influence of shape restriction on the drying and film forming behavior of droplets. The CNCs films with different shapes have the birefringence extinction characteristics of symmetric "Malta Black Cross" by adjusting the hydrophobic and hydrophobic properties and assembly temperature, proving the nematic arrangement. The orientation of the CNC films was investigated by the focal plane array infrared spectroscopy, presenting a symmetrical structure of upper, lower, left, and right, respectively (see Fig. 7.7). The orientation diagram and the change of the three-phase contact line during the evaporation process indicated the large orientation factor of the area of the film edge and the inscribed circle in the pattern, demonstrating that the CNCs are arranged in order in the long axis direction parallel to the template boundary and the three-phase contact line.

The control of the orientation degree of CNCs at different positions was achieved using a noncentrosymmetric template or a combination of multiple circular templates. The orientation changes with the shape of the three-phase contact line during the template and evaporation self-assembly process, so as to obtain a locally controllable orientation degree and the corresponding patterned birefringence extinction characteristics. The orientation of modified CNCs in different regions can be controlled by changing the template boundary and the contour of three-phase contact lines. The birefringent extinction patterns that are dynamically changing like a kaleidoscope were achieved by changing the different incident angles of linearly polarized light

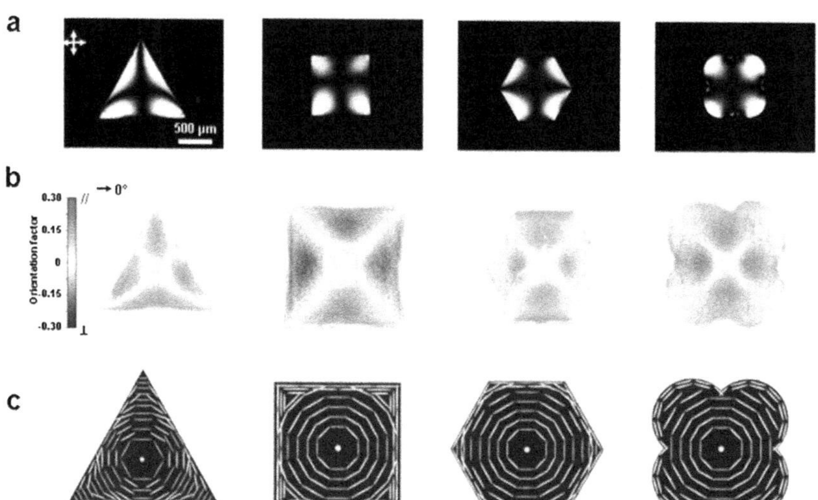

FIGURE 7.7 Analysis of assembly degree of orientation of modified CNCs. *Reproduced from Wang et al. (2022), with permission from* ***American Chemical Society.***

and films. The extinction pattern is found to change significantly after rotating the modified CNCs films for 45°. It reported that the pattern was further coded to achieve information encryption and reading.

The cholesteric liquid crystal self-assembly of water-suspended cellulose nanocrystal (CNC) into a helical arrangement was observed already more than 20 years ago, using to produce iridescent solid films by evaporating the solvent or via sol−gel processing. In addition, Parker et al. (2022) obtained the red, green, and blue cellulose photonic pigments by optimizing the composition of CNC colloidal suspension and breaking through the limitations of CNC self-assembly under spherical constraints by using the buckling phenomenon of the surface of CNC droplets. It reported that the unique multilayer structure of spherical photonic pigment particles made the structural color produced not change with the angle. The washing CNC particles with different volatile solvents would lead to different degrees of volume shrinkage and color change. CNC particles show visible red color in n-hexane solvent; meanwhile the volume of CNC particles did not change significantly after evaporating n-hexane. The volume of the CNC particles shrinks significantly after washing the CNC particles with methanol, resulting in the compression of the structure spacing, further causing the blueshift of the color. Photon particles covering full spectrum color can be obtained by cleaning CNC particles with different polarity solvents. CNC particles of different sizes showed similar blue appearance after methanol treatment, but the reflected light had different polarization. The smaller particles mainly reflect left-handed circularly polarized light, while the larger particles reflect both left-handed and right-handed circularly polarized light. Both solvent cleaning and heat treatment caused the blueshift of color by removing residual moisture. The CNC particles can be used as photon pigments in paints or coatings. These photonic pigments can produce structural colors that do not change with angle by embedding green CNC particles into transparent silane films and conducting optical characterization under two representative observation conditions.

The simpler and more scalable metal insulator metal (MIM) film has a good optical structure and is used for laser emission applications as a light absorber and an etalon. Dong et al. synthesized an optically responsive flexible soft etalon based on ultrathin double network cellulose hydrogels (Dong et al., 2019), which was deposited between two gold layers. The thin cellulose hydrogel layer can be treated with solution. The refractive index and thickness of hygroscopic gel change with the change of humidity. These soft MIM films respond to changes in the refractive index and/or thickness of the hydrogel. The obvious color changes of the structure can be observed with the change ambient humidity. The hydrogel spectrum changed continuously and reversibly in the process of expansion and contraction. The ratio of chemical cross-linking agent has a great influence on the hydration behavior of the hydrogel. The higher the ratio of chemical cross-linking agent, the greater the optical response of the hydrogel. These MIM films based on biological gel also showed good mechanical stability up to 12 months. The simpler and more scalable MIM film has a good optical structure and is used for laser emission applications as a light absorber and an etalon.

Liu and Wu (2020) reported bioinspired hierarchical liquid metacrystal fibers with hierarchical core–sheath architectures, metaperiodic cholesteric alignments, and three-dimensional (3D) topological defects for chiral optics and advanced textiles through self-assembly of CNCs within an in situ–formed hydrogel sheath using a simple microfluidic strategy. The liquid metacrystal fibers exhibit vivid, tunable interference colors, and even inverse optical activity, which have a unique ability to precisely regulate linearly and circularly polarized light in a half-sync/half-async form. Robust hydrogel sheath enables the liquid metacrystal with alignment stability and configuration programmability during drying, which endows the unprecedented freedom to tailor different optical appearances for polarization-based encryption and recognition.

Hou et al. (2022) combined simple mechanical decomposition of phosphorylated fibers, vacuum filtration, and cold press drying to synthesize mechanically robust, flame-retardant phosphorylated cellulose (PhC) films with good mechanical properties, flame retardancy, and adjustable optical properties for light management in light-emitting diodes (LEDs). When the mechanical treatment conditions were optimal (60s), the tensile strength, Young's modulus, toughness, and bending resistance of the prepared PhC film were 150 MPa, 8.5 GPa, 8.2 MJ/m^3, and 4580 times, respectively. By controlling the time of mechanical treatment, the haze of PhC film can be controlled within the range of 9%–91%, while maintaining a high light transmittance of >90%. Using PhC film for light management of LEDs can effectively improve the luminous uniformity of devices. Compared with ordinary paper, carboxymethyl cellulose (CMC) film and 2,2,6,6-tetramethylpiperidine-1-oxyl (TEMPO)-oxidized cellulose film, PhC film shows significantly improved flame retardancy.

7.3 Synthesis and property of advanced functional cellulose composites for optical application

7.3.1 Carbon-dot based phosphorescent cellulose composites

Due to the unique room temperature phosphorescence performance of carbon dot materials, as well as excellent optical stability, chemical stability, biocompatibility, and low biological toxicity, carbon dot materials have shown great application prospects in the fields of anticounterfeiting, information encryption, sensing, detection, biological labeling, and imaging. Similar to organic room-temperature phosphorescent materials and polymer room-temperature phosphorescent materials, the intersystem transition efficiency of carbon dots is low, the nonradiative relaxation rate is fast, the phosphorescent emission life is short, the efficiency is low, and the luminescent color is single (Sun et al., 2006). Dispersing carbon dots into the matrix and using the rigid environment of the matrix to inhibit the nonradiative relaxation process can effectively improve the room temperature phosphorescence life of carbon dots (Wang et al., 2023). According to the different matrix structure and its interaction with carbon point, the common interactions that can build a rigid environment include hydrogen bonding, covalent bonding, ion cluster, and space confinement in organic–inorganic hybrid systems.

Yu' group did system work about advanced functional carbon dot–based cellulose composites for optical application. For example, Xu et al. (2019) applied a coassembly strategy to prepare left handed hierarchical photonic CNC films with right-handed polarization, full-color tunable wavelengths, and tailored dissymmetry factors as optical coding labels to encode and decode information by assembling semiconductor quantum dots. The photonic bandgaps and handedness that spanned the visible spectrum were controlled by the glycerol content in the coassembled photonic films. This enabled tunable circularly polarized luminescence from the green-emitting quantum dots with a tailored maximum g(lum) from −0.48 to −0.21. In addition, Wu et al. (2019) reported an in situ green synthesis of carbon dot–based room-temperature phosphorescent materials using *Schisandra chinensis* polysaccharide as the only carbon source, which has a natural nitrogen-containing structure for endogenous nitrogen doping. The carbon dot–based room-temperature phosphorescent materials had lifetimes up to 271.2 ms under 350 nm excitation and with a smaller energy gap (0.32 eV). It exhibits adequate quenching in the presence of iron ions (Fe^{3+}). The visible room-temperature phosphorescent intensity is inversely proportional to the Fe^{3+} concentration over the range of 0.1–2 mM, with a 0.57 mu M detection limit. The prepared materials have highly stable optical and physical properties, which open a new perspective as luminescent sensors for Fe^{3+} detection with inexpensive and green raw materials.

Recently, Xu et al. (2020) applied coassembling CNCs, polyvinyl alcohol (PVA), and carbon quantum dots to prepare the hybrid chiral photonic films with circularly polarized room-temperature phosphorescence. It achieved the adjustable photonic bandgap by adjusting the ratio of CNC/PVA in the hybrid film to produce adjustable circularly polarized luminescence, which has reversible circularly polarized luminescence chirality, adjustable wavelength, and high asymmetry factor (g_{lum}) up to −0.27. It found the stable triplet exciton generated by carbon quantum dots in the chiral photonic crystal environment, resulting in an adjustable, right-handed circularly polarized room temperature phosphorescence with a lifetime of 103 ms and a room-temperature phosphorescence asymmetry factor (g_{RTP}) of −0.47. It also applied mold manufacturing technology to prepare the patterned hybrid films with multiple polarization characteristics.

More recently, Xu et al. (2022) explored the circular polarization capacity from chiral CNC films for a photocontrolled chiral helix of supramolecular polymers. It generated right-handed transmitted (T-) circularly polarized light (CPL) from self-assembled CNC films, inducing amorphous azobenzene (Azo) supramolecular polymers into chiral structures. The chiral induction effect of T-CPL is enhanced on Azo polymers with longer spacers. The absorptive dissymmetry factor (gabs) values of liquid-crystal supramolecular polymers can be amplified significantly (over 10 times) after T-CPL irradiation. By integrating carbon dots into CNC films, it achieved CPL emission with a considerable luminescence dissymmetry factor (glum) up to −0.66, which could be used for the photoalignment of Azo polymers with high chiroptical properties. Authors applied coassembly method to prepare carbon dots@CNC film and used small molecule glycerol to adjust the photonic bandgap (PBGs) of CPL films. It found bright blue light under white light irradiation and strong blue fluorescence under ultraviolet light irradiation for CPL film (G1-carbon dots). The fluorescence emission wavelength of the CNC-carbon dots film is almost the same as that of the carbon dots solution, while the fluorescence intensity of the G1-carbon dots film is related to the excitation wavelength, showing resistance to photobleaching. Besides chiral CNC films, wood-derived carbon was also developed as light-emitting material. Li et al. (2021) summarized the unique characteristics of wood-derived carbon and light-emitting materials with regard to the fabrication principles, properties, applications, challenges, and future prospects. These materials have great potential for use in bioengineering, flexible electronics, and clean energy.

Dynamic optical materials have received extensive attention, which can respond to external stimuli to provide an interesting chameleon-like visual effect and carry more information. Jin et al. (2021) reported photoswitchable, dynamic-optical, multicolored, and solid-state materials from carbon dots and cellulose for photorewritable paper, advanced anticounterfeiting, and reversible dynamic-optical properties. The materials are spontaneously coated on the surface of negatively charged carbon dots via electrostatic attraction to form a physical protective shell, inhibiting aggregation-caused quenching in carbon dots, promoting effective fluorescence resonance energy transfer. The resultant materials exhibit a phototriggering reversible dynamic-optical property, and effectively regulate fluorescence emission and fluorescence kinetics. Gan et al. (2022) synthesized carbon quantum dot with high quantum yield, multicolor emission and stable fluorescence properties by solvothermal method by controlling the dehydration and carbonization of lignin containing different nitrogen compounds, further prepare the photoluminescent transparent wood with blue, green, and orange emission colors by dispersing it into the cellulose template with porous structure. The photoluminescent transparent wood had the excellent transmittance (86.1%) and optical haze (73%). The tensile properties of photoluminescent transparent wood were enhanced to 422 MPa. By controlling the content of different carbon quantum dots, a photoluminescent transparent wood for white light conversion was prepared with the CIE color coordinates of (0.29, 0.34).

7.3.2 Photonic cellulose composites films

Information security concerns human life, social stability, and even national security. Responsive photonic crystal can change its structural color through external stimulation to achieve visual information encryption. Traditional anticounterfeiting technology is subject to simple single-channel decryption method, which is easy to be cracked. The encryption/decryption technology of multistimulus cooperation is conducive to improving information security. It is still a challenge to develop encryption materials with high security and accurate but simple decryption technology. Li et al. (2022) developed synergistically time-/temperature-resolved high security photonic inks for simple multidimensional information encryption. It found the strong interaction between propylene glycol and hydroxypropyl cellulose, improving the upper limit temperature of response of hydroxypropyl cellulose-based photonic crystals. The response temperature range and color transition rate of the photonic crystal can be adjusted by the concentration of hydroxypropyl cellulose and propylene glycol. The photonic crystal can be processed into various complex patterns as ink. It realized multidimensional information encryption of temperature dependence with high transparency and controllability of response temperature range, combined with multichannel coding technology, and the time-dependent multidimensional information encryption by using the color change speed difference in the response, combined with the Morse cipher encryption method. It provides a new idea for photonic materials to cooperate with temperature and time-resolved information coding and high-end anticounterfeiting technology.

Li et al. (2022) found that hydroxypropyl cellulose (HPC) photonic crystals achieve fast, linear, and stable structural color transition with a one-component chiral nematic mesophase following a thermally induced phase separation mechanism. It only takes 12 s for the formed photonic crystal to change from blue to red and has good reversibility of color change. The temperature dependence of the number of hydrogen bonds between hydroxypropyl cellulose and solvent molecules is the main reason for the rapid phase separation behavior and rapid and repeatable temperature response based on the molecular dynamic simulation. The photonic crystal displays wide-range (red-to-purple) colorations by controlling the mesophase concentration and external temperature (Fig. 7.8). The photonic crystal can be used in temperature sensors, intelligent decoration, and information encryption based on the reversible structure color transition.

Wang et al. (2022) fabricated photonic cellulose composite films with vivid structural colors using a dopamine-infiltration and postpolymerization approach by introducing a light-absorbing melanoid molecule, which realized the

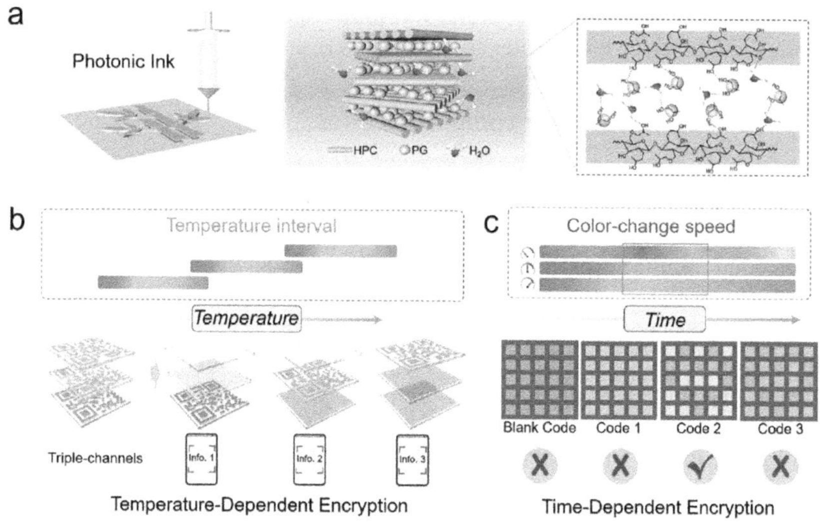

FIGURE 7.8 HPC/PG mesophase materials for advanced information encryption. (a) Schematic illustration for the preparation of cholesteric HPC/PG mesophase. (b) Temperature-resolved and (c) time-resolved information encryption and decryption. *Reproduced from Li et al. (2022) with permission from Elsevier.*

quantitative and selective detection of organic solvents, in view of the weak visibility of the structural color of CNC materials. The chiral nematic structure of CNC enables the structural coloration, while the strong light absorption of the polymeric cophase, polydopamine (PDA) enhances the color chroma and visibility. By controlling the PDA amount, the composite films can detect organic solvents quantitatively and selectively via visible color changes. It defined a critical solubility parameter distance (R-0) between PDA and "active" solvents by using three-dimensional Hansen solubility parameter calculation method from the viewpoint of the compatibility and similitude principle.

Dong et al. (2022) reported bioinspired low-/non−angle-dependent non-iridescent structural coloration enabled by self-assembled CNC composite films with balanced ordered/disordered arrays by introducing positively charged poly(dimethyl diallyl ammonium chloride), inspired by longhorn beetle having short-range ordered but long-range disordered microstructures. It aimed at the problem of visual error caused by the strong iridescence effect of the structural color of such photonic materials. It realized the controllable transition from the long-range/short-range ordered structure of the assembly to the long-range disorder/short-range ordered structure using the electrostatic interaction between CNC and cationic polyelectrolyte to regulate the self-assembly behavior of CNC.

Wang et al. (2022) prepared a hydrophobic CNC photonic sensor with ethanol response function and antiwater interference by a two-step method of coassembly and surface coating. Citric acid is used as the CNC coassembly phase for ethanol response because of the strong intermolecular affinity, and polyvinyl butyral is used as the surface coating to endow the surface hydrophobicity of the CNC/citric acid blend film. By controlling the coating thickness, the composite photonic film sensor can realize the quantitative analysis of ethanol in ethanol/water mixed solution and potentially track drunk driving as a colorimetric sensor, which has good water resistance, sensitivity, stability, reversibility, highly flexible, and controllably structurally colored.

7.3.3 Structural color of photonic cellulose film composites

Structural color is one of the strangest phenomena in nature. It is produced by the interference of light on nanostructures and often appears in animals and plants. Different from pigments, structural colors are environmentally friendly and colorfast. Some organisms can even change colors quickly and reversibly for camouflage, mating and group defense. Among them, chameleon is one of the most important representatives. By tightening or relaxing the skin, the chameleon can change the way light reflects from the surface of the guanine crystal to produce the so-called structural color. Scientists have simulated the crystal nanostructure of chameleon skin in various color changing materials. The chiral optical array structure was constructed by the uniaxial assembly of one-dimensional (1D) CNCs controlled by the force field in the assembly dynamics process (Gan et al., 2019), which improved the application value of this kind of bio-based structure color materials in the fields of information encryption. It obtained a strong blue emission with an emitting quantum efficiency as high as 13.90%. By introducing the vertical-assembly film with a specific pattern into paper materials, an anticounterfeiting image is obtained under a UV radiation. It proposed the strategy of "assembly-induced structure color monochromatic luminescence" of nanoparticles based on the nonlinear optical enhancement theory of slow photon effect and the elementary dimensional effect.

Zhang et al. (2020) reported chameleon-inspired variable coloration enabled by a highly flexible photonic cellulose film. The obtained film has high flexibility with a fracture strain of up to 39%. It realize the stretching-induced structural color change visible to the naked eye caused by stretching on CNC materials. The film show humidity- and compression-responsive properties in terms of changing apparent structural colors. By controlling the ambient humidity, it can show or hide the color traces left by the ink free writing. When the relative humidity increases to 100%, as water molecules enter the self-assembled structure of CNC, the pitch increases, and the film shows red. In this case, the mark is "hidden." When the relative humidity is reduced to 56% and 16%, the mark will appear, which is due to the reduction of pitch caused by moisture removal. Thanks to the flexibility of the CNC/polyethylene glycol diacrylate film, it has the characteristics of mechanical discoloration, because the pitch of CNC decreases during the stretching process, and the color of the film should shift blue. However, the color change is not obvious or uniform due to the poor mobility of the assembled CNC. The shape recovery is incomplete due to the formation of colloidal particles during the stretching process, hindering the recovery. When a pressure of 0–15 MPa is applied to the film, the structural color changes from red to green. The biobased photonic film has many functions such as color sensing, encryption, anticounterfeiting, etc.

Li et al. (2022) applied a ternary coassembly approach to fabricate the noniridescent and wide-color-range structural coloration enabled by CNCs with a controlled long-range and short-range ordered photonic structure and helical pitch. CNC films had weak/nonangular optical response and wide-range colors. Poly(dimethyl diallyl ammonium chloride) is introduced to influence the CNC self-assembly through electrostatic attraction, inducing the structural transition to the long-range disordered/short-range ordered structure as well as presenting the gradual change from iridescent structural colors to noniridescent colors. A hydrophilic glycerol is used to adjust the helical pitch of nanostructured CNC arrays, offering the broad-range coloration. The CNC-based composite is used as an optical coating for anticounterfeiting and realizing color patterning and angle-dependent selective color changes. Song et al. (2022) also applied a ternary coassembly and post-UV-irradiation polymerization strategy to develop flexible and elastic CNC composite films with naked-eye-visible brilliant structural colors and stretching-induced color change covering a broad wavelength region at a moderate deformation (like skin). By adjusting the stretching, the film is designed as a smart skin to adapt to surrounding environments for camouflage. This work offers a universal strategy for constructing biomimic optically functional cellulose skins.

He et al. (2018) prepared a highly flexible, controllably iridescent, and multistimuli-responsive CNC film by simply introducing a small molecule as both plasticizer and hygroscopic agent. The presence of the additive results in the enhancement of its mechanical toughness, making it possible to obtain free-standing iridescent CNC films with tunable structural colors. In response to environmental humidity and mechanical compression, such films can change structural colors smoothly by modulating their chiral nematic structures. Notably, the chromism is reversible by alternately changing relative humidity between 16% and 98%, mimicking the longhorn beetle *Tmesisternus isabellae*. This chromic effect enables various applications of the biofilms in colorimetric sensors, anticounterfeiting technology, and decorative coatings. Bai et al. (2020) employed a new coassembly strategy of CNCs and citric acid (CA) to fabricate freestanding photonic CNC films as a platform with multisensing functions. The chiral nematic structure and visible structural colors can be adjusted in a wide color range by varying the CA content. Owing to the plasticizing effect, the resulting CNC-CA films display high flexibility and can be folded freely. Notably, such films can sense different external signals, including compression, ethanol, and alkali, by changing apparent structural colors. Additionally, for volatile chemicals, the color changes are reversible, ensuring repeating applications. Wang et al. (2021) reported stimuli-responsive flexible membrane in biosensor and wearable devices via coassembling sodium alginate into assembly membranes of rod-like CNCs with an achiral array. The stretchability increased from 0.027% to 37% when 33% sodium alginate was introduced. The assembly achirality was controlled by tuning CNC concentration in suspension, and the coassembly could further control the wavelength of the assembly-induced emission from 420 to 440 nm. The improved stretchability made assembly membrane an optical sensor, whose excitation wavelength blueshifted about 30 nm under a 30% strain. Liu et al. developed a binary coassembly of 1D nanoparticles and long-chain transparent poly(vinyl alcohol) (PVA) to synthesize quantum-efficiency enhancement and mechanical responsiveness of solid-state photoluminescent flexible materials in optical sensors containing uniaxial cellulose nanocrystal arrays (Liu et al., 2022). The coassembling with PVA could improve the uniaxial orientation of CNC arrays, which played a crucial role in enhancing the emission quantum efficiency. Stretching the PVA/CNC membrane could further induce an enhancement in emission quantum efficiency together with a blueshift in excitation wavelength. The stimulation-responsive enhancement in emission quantum efficiency (from ca. 40%−60%) can even be observed by naked eyes.

They constructed a new structural toner material and a personalized writing and painting ink based on CNC uniaxial assembly of nanoscale arrays in collaboration with MOF, carbon nanoparticles, and other substances by using the plasmon superposition mechanism (Zheng et al., 2021). Via controlling the CNC assembly in the nanoscale space of metal−organic frameworks, the emission quantum yield of CNC array primarily increased by 50%. When carbon nanodots were further incorporated in MOF, their plasmonic resonance could enhance the emission coupling rate and offset the loss in emission intensity from the extinction of MOF particles, which has a high luminous quantum efficiency of 64.84%. Shi et al. (2022) developed an excitation-dependent photochromic long afterglow luminescent material for triple information anticounterfeiting based on the single axis−oriented CNC array using the coassembly strategy and optimizing the material transformation control method. It obtained a photoluminescent CNC/SiO_2 coassembled film using the hydrolysis and condensation of small molecular silicone precursor in the solvent evaporation process, and inducing the uniaxial orientation of rod-like particles in the assembly process. During the process of transforming CNC into carbon quantum dots, it controlled and maintained the uniaxial orientation array structure, optimizing the conditions to ensure that CNCs can be converted into carbon quantum dots with excellent photoluminescence properties in situ in the mesoporous structure. The emission band of carbon quantum dots is at the edge of the photonic bandgap of the CNC uniaxial array, and its slow photon effect significantly increases the luminous intensity of carbon quantum dots by about 6.7 times. The CNC uniaxial array effectively protects the triplet emission of carbon quantum dots. The material shows blue and cyan long afterglow luminescence after ultraviolet irradiation at 254 and 365 nm, respectively, with a lifetime of 1.574 and 0.492 s. The material shows excellent photoluminescence and excitation dependent long afterglow luminescence behavior and has application prospects in the field of multiple information anticounterfeiting in different optical environments such as natural light/ultraviolet light/long afterglow luminescence.

7.3.4 The other advanced functional cellulose composites

Xu et al. (2018) developed coassembly of CNCs and small molecule glycerol to prepare multifunctional chiral nematic CNCs/glycerol structural colored nanocomposites for intelligent responsive films, photonic inks, and iridescent coatings. The redshifted structural colors were obtained from the increase in the helical pitch of the chiral nematic structures according to the microstructure analysis. The glycerol promoted the crystallinity of the films to different degrees and improved the mechanical properties of the composite films. The nanocomposite films exhibited reversible reflection colors at different relative humidity because of the strong water absorption capability of glycerol, which were used as photonic inks to obtain photonic writing with unique fingerprint textures and used to make iridescent coatings on different substrates. The photonic nanocomposites can potentially be developed as optical sensors, security markings, and functional coatings. Moreover, Li et al. (2019) achieved right-handed, tunable upconverted circularly polarized luminescence (UC-CPL) emission by integrating multiple emissive, upconverting nanoparticles into CNC-based chiral photonic films that had tunable PBGs. Glycerol was also used to tune the PBGs of the chiral photonic films, which yielded tunable UC-CPL emission at 450 and 620 nm with a tailored g(lum). It obtained humidity responsive UC-CPL at blue wavelength from glycerol-composite photonic film, with a g(lum) that ranged from -0.156 to -0.033.

Zhang et al. (2021) used an evaporation-induced self-assembly technique to fabricate highly flexible, water-stable, and chiral nematic CNC photonic composite film with exceptional solvent-resistant abilities for optical coatings via poly(vinyl alcohol) (PVA). The film remains a highly ordered chiral nematic organization, being able to act as a highly transparent substrate for selective reflection of left-handed circularly polarized light, preparing fully covered and patterned full-color coatings on various substrates. They reported the composite system of polyvinyl alcohol (PVA) and CNC as an example to achieve water stability, introduces the cross-linking reaction of glutaraldehyde (GA), and fixes two orderly assembled hard/soft nanocomposites (Fig. 7.9). This chemical reaction consumes a lot of hydroxyl groups on the surface of cellulose and polyvinyl alcohol, greatly improving the solvent tolerance of

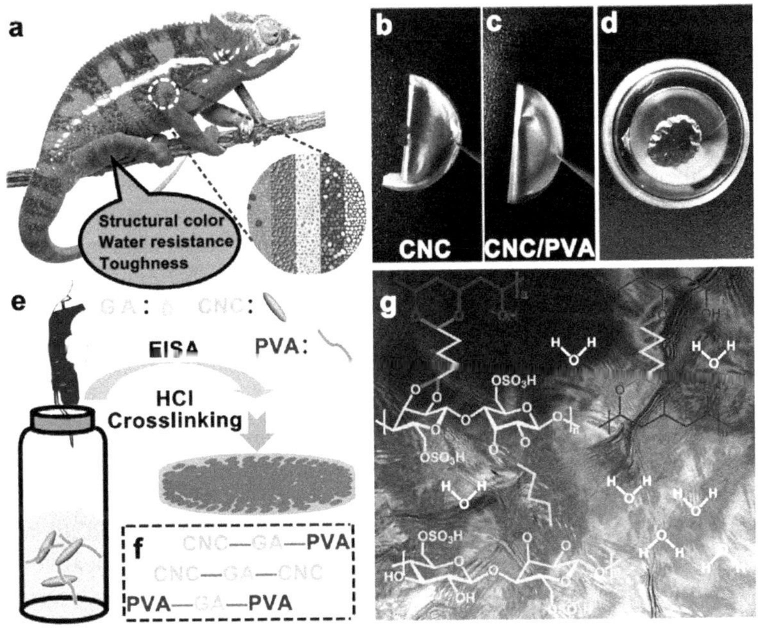

FIGURE 7.9 Design concept of a durable and solvent-resistant cellulose nanocrystal (CNC) film. (a) Chameleon skins are composed of close-packed guanine nanocrystals embedded in multifunctional superimposed iridophores that form an advanced photonic nanostructure, which makes chameleons generate highly flexible, durable, and reversible structural colors. (b) Brittle CNC film. (c, d) Inspired by the mechanism of chameleons, a highly flexible CNC film was prepared through the introduction of 30 wt% soft poly(vinyl alcohol) (PVA); however, this composite film is unstable in water and swells easily (d). (e) Schematic illustration of the preparation process of a glutaraldehyde (GA)-cross-linked CNC/PVA film for the improvement of mechanical strength and water stability through an evaporation-induced self-assembly (EISA) technique. (f) Multiple complexations between CNC and PVA cross-linking through GA. (g) Possible formation mechanism of the polymer network within CNC/PVA-GA. *Reproduced from Zhang et al. (2021) with permission from American Chemical Society.*

optical composite membranes. A 1.5 wt% glutaraldehyde cross-linked 70 wt% CNCs-loaded membrane, treated with harsh hydrophilic exposure (immersion in strong acid, strong alkali and seawater) and various organic solvents, all of which exhibit excellent structural stability. The film is still a highly ordered chiral nematic structure, which can selectively reflect left-handed circularly polarized light, and shows application potential in adjustable structure color coating. This work will pave the way for low-cost and durable photonic cellulose coating applications and high-end bioelectronics development.

Merindol et al. (2020) reported the spin-assisted layer-by layer assembly of anisotropic wood-inspired films composed of anionic cellulose nanofibrils (CNFs) and cationic poly(vinyl amine) possessing a tensile strength that exceeds the one of the wood from which the fibers originate. The thickness of the multilayer films increases linearly with the increased number of deposition cycles up to 200 layer pairs. The interference colors of films with increasing thickness are characteristic structure colors of films with uniform thickness and refractive index. It observed the obvious uniformity of the film on the micrometer scale via atomic force microscope (AFM) image. The orientation of nanofibers was characterized by AFM image and Orientation J. CNFs are randomly distributed in the center, while they are located at the edge. The direction of cellulose nanofibers increases from the center to the edge. Most images taken at the center of the rotating substrate have some degree of orientation. The order parameter increases from center to edge, and the maximum orientation of CNF in the films also increases with the rotation speed. The CNF is a reinforcing component originated from excellent mechanical properties of wood. Its tensile modulus is estimated to be 100 GPa, and its breaking strength exceeds 1 GPa. The Young's modulus of the oriented film measured perpendicular to the fiber direction was maintained at 9 GPa. When the nanofibers are parallel to the orientation, they bear most of the stress, and the Young's modulus of the films reaches 33 GPa and the tensile strength is 490 MPa. More recently, Tan et al. (2022) reported flexible functional luminescent composites from aggregation-induced-emission luminogen-based fluorescent MOF nanosheets and seaweed CNFs for humidity sensing and UV shielding. The fluorescence emission of the composite films is enhanced because of the favored affinity between MOF nanosheets and CNFs. The composite films demonstrate excellent UV-shielding capacity and high optical transmittance at the visible wavelength range, showing reversible changes in fluorescence emission intensity in response to ambient humidity. The tensile strength and modulus of the composite films are also enhanced owing to the increased adhesion between CNFs through the adsorbed MOF nanosheets.

Zhang et al. (2022) report uniformly aligned flexible magnetic films from bacterial nanocelluloses for fast actuating optical materials, enabling transformation from helicoidal cholesteric (chiral nematic) to uniaxial nematic phase with near-perfect orientation order parameter of 0.98 across large areas. These translucent magnetic films are flexible, robust, and possess anisotropic birefringence and light scattering combined with relatively high optical transparency reaching 75%. The optical birefringence and anisotropic optical texture of all films have different appearances under the cross-polarizer. Characteristic random birefringent crystalline-like aggregates with fingerprint patterns and clear boundaries were observed in pure bCNC films. For magnetic bCNC-MNP1 films with the lowest content of magnetic nanoparticles, similar but more distorted optical textures were observed. The magnetic film with higher MNP load shows unidirectional aligned optical texture on a large area (up to 1 mm in diameter), with highly anisotropic crystalline aggregates. The magnetic film containing 12.6% Fe_3O_4 nanoparticles has uniformly arranged texture and high optical anisotropy, and there is no single random crystalline aggregates. The azimuthal distribution of transmitted light collected by a hyperspectral microscope with polarization angles ranging from −90 degrees to 90 degrees shows the maximum at 0°, which represents unidirectional optical birefringence. Uniaxial assembly results in highly anisotropic light diffraction, which is contrary to the isotropic light scattering of the original bCNC thin film with random multidomain crystal-like aggregates. It observed photothermal response under light. This fast triggered reaction may be caused by the single-sided light absorption and the corresponding thermally induced twin stress, which is due to the relatively strong light absorption in the near side of the film within the visible light wavelength range. The photothermal stress induced by light is strong enough to induce shape change under higher illumination, but due to the depth-dependent light absorption gradient, it will not affect the other side of the film. The extremely low doping ratio allows the films to preserve relatively high transparency, optical birefringence, and other optical properties and induce uniaxial optical anisotropy and anisotropic light diffraction.

Nawaz et al. (2023) reported a highly fluorescent and multifunctional biopolymer-based cellulose acetate with excellent aqua blue fluorescence for extreme pH sensing and smart printing applications. A biopolymer-based structure containing responsive functional groups such as −C=O and −NH is constructed by chemically bonding 5-amino-2,3-dihydrophthalazine-1,4-dione (luminol) onto cellulose acetate using 4,4′-diphenylmethane diisocyanate as a cross-linking agent. The ratiometric system clearly differentiates the extreme acidic pH 1 from pH 2 and extreme alkaline pH 12, 13, and 14 by visual and fluorescence color change response under a narrow pH range. The transparent flexible fluorescent films prepared with 1%, 5%, and 10% Lum-4,4′-diphenylmethane diisocyanate-cellulose acetate had an outstanding UV shielding, security printing, and haze properties for smart food packaging and printing applications. Lum-4,4′-diphenylmethane diisocyanate-cellulose acetate/DMSO solution (20 mg/mL) can also be prepared into fluorescent ink to draw safety patterns, which can also be effectively used to prepare secure ink solutions for developing security codes for information encryption applications. The material has good UV protection (>90%) and haze characteristics due to many 4,4′-diphenylmethane diisocyanate, Lumino groups and aromatic ring.

7.4 Chapter summary

In summary, great advancements in advanced functional cellulose composites for optical applications have been made in recent years. Advanced functional cellulose composites have been widely used in visual sensors, photonic crystal actuators, intelligent responsive films, photonic inks, iridescent coatings, and other fields. Although these applications make the considerable progress, there are still many aspects that are needed to consummate. Obviously, CNC is mainly used in various types of cellulose and nanocellulose due to its circularly polarized light characteristics and chiral nematic helical structures with iridescence. It expects that other types of cellulose can be developed for optical applications. As for advanced functional cellulose composites, carbon dot materials have been widely used to improve the property of composites due to its unique room temperature phosphorescence performance and excellent optical stability. It is particularly important to prepare high-quality carbon dots, and then prepare high-performance advanced functional cellulose composites for optical applications. Development of simple, fast, and low-cost methods such as ATR method to explore the structural of solid, thin film, and other surface or interface layers is of great importance for broadening and improving the optical applications of advanced functional cellulose composites. It expects that advanced functional cellulose composites have a bright future like bright colors of butterfly wings.

Acknowledgments

The financial support from Beijing Nova Program (20230484431) and National Natural Science Foundation of China (22478036) is gratefully acknowledged.

References

Bai, L., Wang, Z.L., He, Y.D., Song, F., Wang, X.L., Wang, Y.Z., 2020. Flexible photonic cellulose nanocrystal films as a platform with multisensing functions. ACS Sustain. Chem. Eng. 8, 18484–18491.

Chen, J., Jiang, J., Weber, J., Gimenez-Pinto, V., Peng, C., 2023. Shape morphing by topological patterns and profiles in laser-cut liquid crystal elastomer kirigami. ACS Appl. Mater. Interfaces 15, 4538–4548.

Dong, X., Zhang, Z.L., Zhao, Y.Y., Li, D., Wang, Z.L., Wang, C., Song, F., Wang, X.L., Wang, Y.Z., 2022a. Bio-inspired non-iridescent structural coloration enabled by self-assembled cellulose nanocrystal composite films with balanced ordered/disordered arrays. Compos. B Eng. 229, 109456.

Dong, X., Li, D., Wu, J.M., Zhang, Z.L., Wang, Z.L., Song, F., Wang, X.L., Wang, Y.Z., 2022b. Non-iridescent and wide-color-range structural coloration enabled by cellulose nanocrystals with a controlled long-range photonic structure and helical pitch. ACS Sustain. Chem. Eng. 10, 10641–10648.

Dong, Y., Akinoglu, E.M., Zhang, H., Maasoumi, F., Zhou, J., Mulvaney, P., 2019. An optically responsive soft etalon based on ultrathin cellulose hydrogels. Adv. Funct. Mater. 29, 1904290.

Droguet, B.E., Liang, H.L., Frka-Petesic, B., Parker, R.M., De Volder, M.F.L., Baumberg, J.J., Vignolini, S., 2022. Large-scale fabrication of structurally coloured cellulose nanocrystal films and effect pigments. Nat. Mater. 21, 352.

Gan, J., Wu, Y., Yang, F., Zhang, H., Wu, X., Wang, Y., Xu, R., 2022. Wood-cellulose photoluminescence material based on carbon quantum dot for light conversion. Carbohydr. Polym. 290, 119429.

Gan, L., Feng, N., Liu, S.Y., Zheng, S.Y., Li, Z., Huang, J., 2019. Assembly-induced emission of cellulose nanocrystals for hiding information. Part. Part. Syst. Char. 36, 1800412.

Habibi, Y., Lucia, L.A., Rojas, O.J., 2010. Cellulose nanocrystals: chemistry, self-assembly, and applications. Chem. Rev. 110, 3479–3500.

He, Y.D., Zhang, Z.L., Xue, J., Wang, X.H., Song, F., Wang, X.L., Zhu, L.L., Wang, Y.Z., 2018. Biomimetic optical cellulose nanocrystal films with controllable iridescent color and environmental stimuli-responsive chromism. ACS Appl. Mater. Interfaces 10, 5805–5811.

Honorato-Rios, C., Kuhnhold, A., Bruckner, J.R., Dannert, R., Schilling, T., Lagerwall, J.P.F., 2016. Equilibrium liquid crystal phase diagrams and detection of kinetic arrest in cellulose nanocrystal suspensions. Front. Mater. 3, 21.

Hou, G., Zhao, S., Li, Y., Fang, Z., Isogai, A., 2022. Mechanically robust, flame-retardant phosphorylated cellulose films with tunable optical properties for light management in LEDs. Carbohydr. Polym. 298, 120129.

Jin, B.N., Tian, W., Bai, H., Zhang, J., Wang, S., Zhang, J., 2019. Amine-responsive cellulose-based ratiometric fluorescent materials for real-time and visual detection of shrimp and crab freshness. Nat. Commun. 10, 795.

Jin, K.F., Ji, X., Yang, T., Zhang, J., Tian, W., Yu, J., Zhang, X., Chen, Z., Zhang, J., 2021. Facile access to photo-switchable, dynamic-optical, multi-colored and solid-state materials from carbon dots and cellulose for photo-rewritable paper and advanced anti-counterfeiting. Chem. Eng. J. 406, 126794.

Kaushik, M., Basu, K., Benoit, C., Cirtiu, C.M., Vali, H., Moores, A., 2015. Cellulose nanocrystals as chiral inducers: enantioselective catalysis and transmission electron microscopy 3D characterization. J. Am. Chem. Soc. 137, 6124–6127.

Li, D., Wu, J.M., Liang, Z.H., Li, L.Y., Dong, X., Chen, S.K., Fu, T., Wang, X.L., Wang, Y.Z., Song, F., 2022. Sophisticated yet Convenient information encryption/decryption based on synergistically time-/temperature-resolved photonic inks. Adv. Sci. 10, 2206290.

Li, W., Chen, Z.J., Yu, H.P., Li, J., Liu, S.X., 2021. Wood-derived carbon materials and light-emitting materials. Adv. Mater. 33, 2000596.

Li, W., Xu, M.C., Ma, C.H., Liu, Y.S., Zhou, J., Chen, Z.J., Wang, Y.G., Yu, H.P., Li, J., Liu, S.X., 2019. Tunable upconverted circularly polarized luminescence in cellulose nanocrystal based chiral photonic films. ACS Appl. Mater. Interfaces 11, 23512–23519.

Liu, S.Y., Shi, Z.X., Wang, X.H., Gong, Y.B., Li, X.J., Jia, X., Gan, L., Huang, J., 2022. Quantum-efficiency enhancement and mechanical responsiveness of solid-state photoluminescent flexible materials containing uniaxial cellulose nanocrystal arrays. Cellulose 29, 1393–1403.

Liu, Y.J., Wu, P.Y., 2020. Bioinspired hierarchical liquid-metacrystal fibers for chiral optics and advanced textiles. Adv. Funct. Mater. 30, 2002193.

Merindol, R., Diabang, S., Mujica, R., Houerou, V.L., Roland, T., Gauthier, C., Decher, G., Felix, O., 2020. Assembly of anisotropic nanocellulose films stronger than the original tree. ACS Nano 14, 16525–16534.

Nawaz, H., Chen, S., Zhang, X., Li, X., You, T., Zhang, J., Xu, F., 2023. Cellulose-based fluorescent material for extreme pH sensing and smart printing applications. ACS Nano 17, 3996–4008.

Parker, R.M., Zhao, T.H., Frka-Petesic, B., 2022. Cellulose photonic pigments. Nat. Commun. 13, 3378.

Peng, N., Huang, D., Gong, C., Wang, Y., Zhou, J., Chang, C., 2020. Controlled arrangement of nanocellulose in polymeric matrix: from reinforcement to functionality. ACS Nano 14, 16169–16179.

Qi, Y., Kou, D., Sun, Y., Hu, T., Yuan, H., Zhou, C., Li, C., Lu, A.H., Wu, S., Zhang, S., 2023. Portable colorimetric photonic indicator for ethanol concentration sensing. Chem. Eng. J. 457, 141184.

Qu, D., Rojas, O.J., Wei, B., Zussman, E., 2022. Responsive chiral photonic cellulose nanocrystal materials. Adv. Opt. Mater. 10, 2201201.

Querejeta-Fernandez, A., Chauve, G., Methot, M., Bouchard, J., Kumacheva, E., 2014. Chiral plasmonic films formed by gold nanorods and cellulose nanocrystals. J. Am. Chem. Soc. 136, 4788–4793.

Schütz, C., Bruckner, J.R., Honorato-Rios, C., Tosheva, Z., Anyfantakis, M., Lagerwall, J.P.F., 2020. From equilibrium liquid crystal formation and kinetic arrest to photonic bandgap films using suspensions of cellulose nanocrystals. Crystals 10, 199.

Shi, Z.X., Zhao, W.W., Zhang, Y., Yang, D.M., Gan, L., Huang, J., 2022. Triply hiding optical information via excitation-dependent allochroic photoluminescence based on cellulose derivates. Small 19, 2205697.

Song, F., Wang, Y.Z., Zhang, Z.L., Dong, X., Zhao, Y.Y., Wang, X.L., 2022. Bioinspired optical flexible cellulose nanocrystal films with strain-adaptive structural coloration. Biomacromolecules 23, 4110–4117.

Sun, Y.P., Zhou, B., Lin, Y., Wang, W., Fernando, K.A.S., Pathak, P., Meziani, M.J., Harruff, B.A., Wang, X., Wang, H.F., 2006. Quantum-sized carbon dots for bright and colorful photoluminescence. J. Am. Chem. Soc. 128, 7756–7757.

Tan, F.C., Zha, L., Zhou, Q., 2022. Assembly of aiegen-based fluorescent metal-organic framework nanosheets and seaweed cellulose nanofibrils for humidity sensing and UV-shielding. Adv. Mater., 2201470

Tardy, B.L., Ago, M., Guo, J.Q., Borghei, M., Kamarainen, T., Rojas, O.J., 2017. Optical properties of self-assembled cellulose nanocrystals films suspended at planar-symmetrical interfaces. Small 13, 1702084.

Wang, H., Shao, R., Meng, X., He, Y., Shi, Z., Guo, Z., Ye, C., 2022. Programmable birefringent patterns from modulating the localized orientation of cellulose nanocrystals. ACS Appl. Mater. Interfaces 14, 36277–36286.

Wang, K.T., Qu, L.J., Yang, C.L., 2023. Long-lived dynamic room temperature phosphorescence from carbon dots based materials. Small 19. https://doi.org/10.1002/smll.202206429.

Wang, X.H., Feng, N., Shi, Z.X., Zhou, N., Lu, J., Huang, J., Gan, L., 2021. Stimuli-responsive flexible membrane via co-assembling sodium alginate into assembly membranes of rod-like cellulose nanocrystals with an achiral array. Carbohydr. Polym. 262, 117949.

Wang, Z.L., Wang, Q., Dong, X., Li, D., Bai, L., Wang, X.L., Wang, Y.Z., Song, F., 2022. Photonic cellulose films with vivid structural colors: fabrication and selectively chemical response. Biomacromolecules 23, 1662–1671.

Wang, Z.L., Deng, Z.P., Dong, X., Bai, L., Wang, X.L., Wang, Y.Z., Song, F., 2022. A surface diffusion barrier strategy toward water-resistant photonic materials for accurate detection of ethanol. ACS Appl. Mater. Interfaces 14, 30352–30361.

Wu, X.Y., Ma, C.H., Liu, J.C., Liu, Y.S., Luo, S., Xu, M.C., Wu, P., Li, W., Liu, S.X., 2019. In situ green synthesis of nitrogen-doped carbon-dot-based room-temperature phosphorescent materials for visual iron ion detection. ACS Sustain. Chem. Eng. 7, 18801.

Xu, M.C., Ma, C.H., Zhou, J., Liu, Y.S., Wu, X.Y., Luo, S., Li, W., Yu, H.P., Wang, Y.G., Chen, Z.J., Li, J., Liu, S.X., 2019. Assembling semiconductor quantum dots in hierarchical photonic cellulose nanocrystal films: circularly polarized luminescent nanomaterials as optical coding labels. J. Mater. Chem. C 7, 13794–13802.

Xu, M.C., Wu, X.Y., Yang, Y., Ma, C., Li, W., Yu, H.P., Chen, Z., Li, J., Zhang, K., Liu, S., 2020. Designing hybrid chiral photonic films with circularly polarized room-temperature phosphorescence. ACS Nano 14, 11130–11139.

Xu, M.C., Li, G., Li, W., An, B., Sun, J., Chen, Z., Yu, H., Li, J., Yang, G., Liu, S., 2022. Exploring the circular polarization capacity from chiral cellulose nanocrystal films for a photo-controlled chiral helix of supramolecular polymers. Angew. Chem. Int. Ed. 61, e202117042.

Xu, M.C., Li, W., Ma, C.H., Yu, H.P., Wu, Y.Q., Wang, Y.G., Chen, Z.J., Li, J., Liu, S.X., 2018. Multifunctional chiral nematic cellulose nanocrystals/glycerol structural colored nanocomposites for intelligent responsive films, photonic inks and iridescent coatings. J. Mater. Chem. C 6, 5391–5400.

Yang, T.J., Yuan, D., Liu, W., Zhang, Z., Wang, K.Y., You, Y.X., Ye, H.P., de Haan, L.T., Zhang, Z., Zhou, G.F., 2022. Thermochromic cholesteric liquid crystal microcapsules with cellulose nanocrystals and a melamine resin hybrid shell. ACS Appl. Mater. Interfaces 14, 4588–4597.

Zheng, H.Z., Li, W., Li, W., Wang, X., Tang, Z., Zhang, S.X.A., Xu, Y., 2018. Uncovering the circular polarization potential of chiral photonic cellulose films for photonic applications. Adv. Mater., 1705948

Zheng, S.Y., Liu, S.Y., Xiao, B., Liu, L., Wan, X.J., Gong, Y.B., Wei, S.Q., Luo, C.L., Gan, L., Huang, J., 2021. Integrate nanoscale assembly and plasmonic resonance to enhance photoluminescence of cellulose nanocrystals for optical information hiding and reading. Carbohydr. Polym. 253, 117260.

Zhao, Y.J., Xie, Z.Y., Gu, H.C., Zhu, C., Gu, Z.Z., 2012. Bio-inspired variable structural color materials. Chem. Soc. Rev. 41, 3297–3317.

Zhang, F.S., Ge, W., Wang, C., Zheng, X., Wang, D., Zhang, X., Wang, X., Xue, X., Qing, G., 2021. Highly strong and solvent-resistant cellulose nanocrystal photonic films for optical coatings. ACS Appl. Mater. Interfaces 13, 17118–17128.

Zhang, X.F., Kang, S., Adstedt, K., Kim, M., Xiong, R., Yu, J., Chen, X., Zhao, X., Ye, C., Tsukruk, V.V., 2022. Uniformly aligned flexible magnetic films from bacterial nanocelluloses for fast actuating optical materials. Nat. Commun. 13, 5804.

Zhang, Z.L., Dong, X., Fan, Y.N., Yang, L.M., He, L., Song, F., Wang, X.L., Wang, Y.Z., 2020. Chameleon-inspired variable coloration enabled by a highly flexible photonic cellulose film. ACS Appl. Mater. Interfaces 12, 46710–46718.

Chapter 8

Conductive nanocomposite hydrogels for flexible wearable sensor

Wenyan Guo and Mingguo Ma*

MOE Engineering Research Center of Forestry Biomass Materials and Bioenergy, Research Center of Biomass Clean Utilization, Beijing Key Laboratory of Lignocellulosic Chemistry, College of Materials Science and Technology, Beijing Forestry University, Beijing, PR China

Corresponding author: E-mail: mg_ma@bjfu.edu.cn

8.1 Introduction

In the 21st century of digitalization, information, and intelligence, many electronic frontier technologies have been derived, and smart electronics products are widely used due to their multimedia, interactive, integrated, and efficient features (Chen et al., 2022; Li et al., 2022a). Currently, most electronic devices are based on electrically conducting inorganic materials, whose chemical and mechanical properties are not suitable for applications, such as electronic skin and human–computer interaction (Yu et al., 2017). To adapt to the human body, flexible wearable sensors are gradually coming into the spotlight (Zhao et al., 2021; He et al., 2022a). Unlike conventional electronics, flexible wearable electronics have long attracted intense interest from many researchers due to their ability to continue to function in states of stretching, compression, bending, and other apparent deformations, demonstrating lightweight, portability, and superior biocompatibility (You et al., 2022; Ahmed et al., 2022; Zhang et al., 2022a). In general, conventional conducting materials suffer from the drawbacks of poor biocompatibility, weak interfacial bonding, and poor mechanical properties. How to incorporate these rigid materials into flexible wearable sensors through clever design and thus exert more favorable advantages is a key issue in the research field.

Surface Science and Advanced Functional Applications of Cellulose Composites
https://doi.org/10.1016/B978-0-443-13675-7.00008-7

Among them, hydrogels, 3D polymer networks formed from natural or synthetic materials (Liu et al., 2023), stand out among conducting materials due to their high water content, tunable mechanical properties, good biocompatibility, and excellent electrical conductivity and are widely used in sensors (Heidarian et al., 2022), actuators (Ionov, 2014), and hydrogel machines (Shen et al., 2022). Unlike rigid elastomers, the high water content of hydrogels allows for the transport of chemical molecules and conductive materials, giving the material a better electrical conductivity (Rahmani and Shojaei, 2021). The structure of the polymer network can be tuned to match the elastic modulus of biological tissues, similar to that of real human skin, minimizing the mismatch (Pang et al., 2020). Biocompatibility is designed to meet the needs of hydrogels for applications, such as health monitoring and human—computer interaction by improving their healthy contact with human skin or biological tissues and avoiding inflammatory reactions (Naahidi et al., 2017). As a result, hydrogel-based strain and pressure sensors have shown great potential in smart skin and implantable devices. Of course, the main challenge for conducting hydrogels is to achieve high conductivity without losing the aforementioned properties for immediate, accurate, and efficient signal capture and transfer. Second, the mechanical properties of conventional hydrogels are not ideal, and reasonable designs are still needed to meet the applications of flexible wearable sensors in practice. As the strength and toughness of pure hydrogel does not meet the requirements of practical applications, scholars often use the development of dual-network structure (Ling et al., 2021) or the introduction of nanomaterials (Li et al., 2023a; Liu et al., 2018a) to achieve the purpose of enhanced toughening. And most of the hydrogel is not conductive by nature, how to properly and effectively build conductive hydrogel to meet the needs of flexible electronic products is worth thinking about. Depending on the conductance mechanism, conductive hydrogels can be divided into ionic and electronic hydrogels (Zhu et al., 2023), which can be used to transport ions or electrons by constructing conductive pathways. To date, the most common and effective strategy to endow hydrogels with electrical conductivity is through the nanocomposite construction strategy. For example, Karimzadeh et al. reviewed the performance, strain sensing efficiency, and potential applications of PVA nanocomposite hydrogels based on PVA for wearable sensors and discussed the development and challenges of hydrogels for strain sensors (Karimzadeh et al., 2022). Park et al. reviewed the progress of nanocomposites constructed on the basis of different polymers and nanofillers for wearable sensors and focused on the characteristics, strategies, applications, and limitations based on different nanocomposite fillers (Park et al., 2022a).

From a material design point of view, a common strategy to improve the conductivity of hydrogels is the addition of various nanomaterials, including carbon-based nanofillers (carbon black (CB) (Xu et al., 2023a), carbon nanotube (CNT) (Zhang et al., 2021a), graphene oxide (GO) (Wang et al.,

2017), and reduced graphene oxide (rGO)) (Zhang et al., 2020a), polymer-based nanofillers (nanocellulose bonded with conductive materials (Lu et al., 2022), polyaniline (PANI) (Ma et al., 2020), polypyrrole (PPy)) (Qin et al., 2023), metallic nanofillers (including nanoparticles (Han et al., 2021), nanowires (Wang et al., 2020), and so on), and two-dimensional (2D) materials MXenes that have emerged in the past few years (Wang et al., 2022a; Yin et al., 2023a). These nanofillers can be physically or chemically added to confer higher conductivity to the hydrogel. However, high levels of conductive fillings can lead to impaired mechanical properties and poor fatigue resistance of the hydrogels. Therefore, how to balance mechanical properties with high conductivity is one of the main goals of conducting hydrogels.

From a functional design perspective, the development of conductive hydrogels with multiple functions is also crucial for flexible wearable sensors (Fang et al., 2018). To better apply conductive hydrogels to flexible wearable electronics, wear resistance and stability are essential characteristics of hydrogels, and self-healing properties are also beneficial for extending their lifetime and operation time (Pei et al., 2020; Xu et al., 2023b; Huang et al., 2022). Moreover, good compliance, biocompatibility, and reproducible adhesion to skin tissue are also essential to better capture signal changes (Cao et al., 2022; Zhang et al., 2019; Li et al., 2021a). In addition, to expand the potential of conductive hydrogels in practical applications and in extreme environments (e.g., extreme cold and heat), endowing conductive hydrogels with antifreezing and moisturizing properties is also worthy of in-depth research (Nie et al., 2022; Zou et al., 2023; Guo et al., 2022). There are ongoing efforts to design and tune polymer networks and nanofillers of hydrogels to impart various properties, such as self-healing, adhesion, antifreezing, and multisensing properties. In general, the properties of conductive nanocomposite hydrogels are determined by a combination of polymer networks and nanofillers, even with superior properties. There has been a lot of work and effort by academics to design flexible wearable sensors based on conductive nanocomposite hydrogels. The number of publications based on nanocomposite and hydrogel sensors has been increasing in recent years (Fig. 8.1).

At present, the search for high-performance nanofillers and suitable hydrogel polymer networks is crucial for the use of conductive nanocomposite hydrogels for flexible wearable sensors, which not only needs to maintain the properties of the nanofillers and polymer networks and solve the incompatibility, but also needs to consider the other properties that are expanded by the combination, as well as the multisensing performance of the sensors. Herein, this chapter presents a review of recent studies and the applications of these nanofillers in flexible wearable devices based mainly on the categories of nanofillers and introduces the multifunctional properties, including mechanical

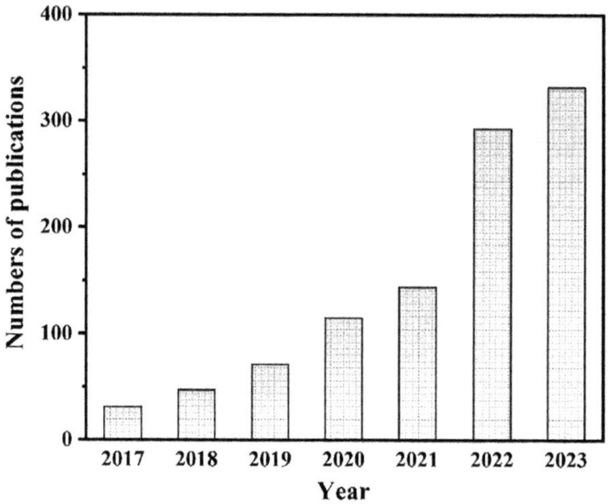

FIGURE 8.1 Number of published papers based on nanocomposites and hydrogels sensor from the Web of Science.

properties, self-healing properties, adhesion properties, and antifreeze properties (Fig. 8.2). And finally, we suggest perspectives and challenges for the future design of conductive nanocomposite hydrogels for use as flexible wearable sensors.

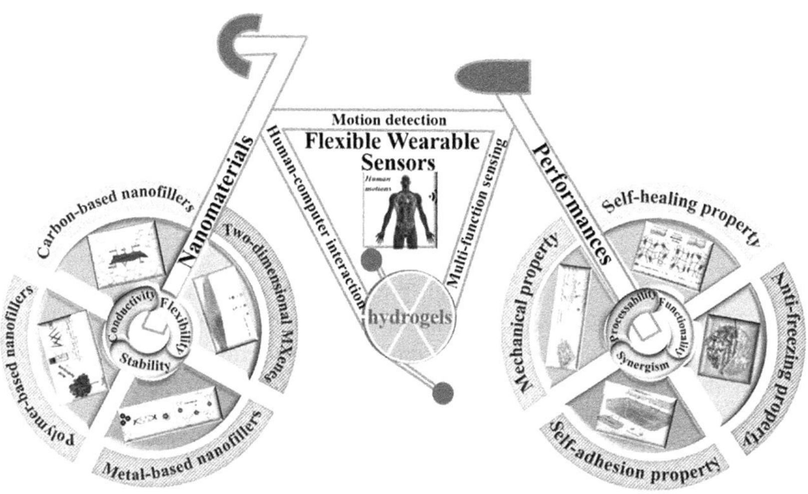

FIGURE 8.2 Schematic representation of composite hydrogels based on different nanofillers and properties for sensors.

8.2 Design of conductive nanocomposite hydrogels and classification of nanofillers

Stable hydrogels consisting of nanofillers with polymer matrices have been extensively investigated for flexible wearable sensors as promising candidates to replace traditional rigid sensors (Chen et al., 2021a; Lin et al., 2019). To better enhance the properties of conductive nanocomposite hydrogels and their potential for use in flexible wearable sensors, nanofillers dispersed in polymer substrates have a crucial role to play. On the one hand, the combination of nanofillers and hydrophilic polymer networks can substantially improve the mechanical properties of conductive nanocomposite hydrogels, and on the other hand, different nanofillers can endow conductive hydrogels with various properties, which further enhances their potential for practical applications (Chen et al., 2021b; Cheng et al., 2022). Various conductive nanomaterials and other nanofillers in combination with conductive polymers have been extensively investigated. In this section, we review the fabrication strategies for conductive nanocomposite hydrogels and summarize and outline the nanofillers into the following four categories, including carbon-based nanofillers, polymer-based nanofillers, metal-based nanofillers, and other emerging nanofillers based on MXenes.

8.2.1 Design of conductive nanocomposite hydrogels

Hydrogels are cross-networks of monomers or polymers synthesized by chemical or physical interactions in an aqueous medium. Physical cross-linking networks usually include physical entanglement, hydrophobic interactions, electrostatic interactions, metal coordination, crystallization, and π-π stacking. Chemical cross-linking networking methods typically include aldehyde-mediated, free radical polymerization, click chemistry, and condensation reactions (Zhang et al., 2023a). Conductive hydrogels are composites formed by the combination between a conventional hydrogel, which provides structural support, and various conductive compositions, which provide electrical conductivity (Fang et al., 2020). Since the conductivity of stretchable conductors is closely related to the performance of flexible wearable sensor devices, it is significant to dope the combination of conductive nanofillers, polymeric nanofillers, and conductive components into polymer matrix to prepare conductive nanocomposite hydrogels with excellent electrochemical properties to endow them with the desired conductivity. Conductive hydrogels can be broadly divided into electronically conductive hydrogels and ionically conductive hydrogels through the conductive matrix (Han et al., 2020). For electronically conductive hydrogels, the conductive mesh pyrrole can be a conductive filler or a conductive polymer formed by in situ polymerization. In common with the strategy of preparing nanocomposite hydrogels, such as the introduction of carbon nanomaterials (carbon nanotubes, graphene oxide, etc.) with metal nanoparticles and nanowires into the hydrogel precursor solution, the direct cross-linking into a three-dimensional network can give the hydrogel not only excellent electrical conductivity, but also enhance its mechanical and sensing properties (Zhang et al., 2023b). In general, chemical bonding, self-assembly, and ionic cross-linking are the main methods for the preparation of such conductive nanocomposites. Ionic conductivity via free ion transport is prevalent in ion-conducting hydrogels, and the three-dimensional network has flexibility and softness similar to biological systems by virtue of the large number of free ions present. For ion-conductive hydrogels, the directional transport of free ions determines their electrical conductivity (Guo et al., 2023). However, ion-conducting hydrogels usually require a high salt concentration to meet the electrical conductivity, which is not friendly to the biocompatibility of composite hydrogels. To further enhance the biocompatibility and electrical conductivity of ionic hydrogels, introducing highly ordered nanostructures as ion transport channels into polymer matrices is an effective way to prepare hydrogels with excellent properties. Among them, cellulose, as a rich renewable resource, has been explored and researched by many scholars, especially the micro- and nanoforms with different shapes and crystallinity (Zhang et al., 2023a). Strikingly, cellulose nanocrystals (CNCs) and cellulose nanofibers (CNFs) with high hydrophilicity have become a common strategy to endow conductive nanocomposite hydrogels with excellent electrical conductivity by virtue of their excellent biocompatibility, high Young's modulus, and mechanical properties, etc., which have been used as nanofillers in conjunction with ionic salts and fine-designed to optimize and adjust the network structure.

8.2.2 Carbon-based nanofillers

Conductive hydrogels based on carbon-based nanofillers have been the subject of intense research by scholars over the past decades. Due to their inherent properties, carbon-based nanofillers can be combined with a wide range of polymer substrates and exhibit high performance, such as excellent stability, good mechanical properties, and high electrical conductivity (Long et al., 2023; Yue et al., 2023a). Based on these versatile properties, hydrogels fabricated with carbon nanofillers can be widely used in flexible wearable electronics and sensors. Next, the main focus is on the development of conductive nanocomposite hydrogels based on the widest range of carbon-based nanofillers (including CB, CNTs, GO, etc.) for flexible wearable sensors.

8.2.2.1 Carbon black

CB fillers have been shown to provide excellent synergies due to their ability to impart good electrical conductivity and mechanical property to nanocomposites (Jamatia et al., 2023). For example, Hardman et al. prepared printable, yet elastic and soft sensing structures using gelatin and glycerol (Gly) based on 3D printing method, highlighting the simplicity and low cost of the preparation process (Hardman et al., 2021). Through the introduction of conducting CB particles, an electrically conducting hydrogel with higher robustness, stability, and strain coefficient is formed, which is promising in terms of sensor nonlinearity, low drift, and fast response. Finally, six printed hydrogel layers are formed into a 3D finger for actuation and sensing, and the underlying layer for sensing enables proprioception, providing a simple way to bend by controlling tendon tension. As illustrated in Fig. 8.3a, Gu et al. prepared organohydrogels with 3D honeycomb structure by integrating CB and CNTs into conductive poly(vinyl alcohol)/Gly hydrogels. Since no polymerization reaction takes place, there are no residual monomers in the preparation process (Gu et al., 2020). However, CB and CNT fillers can significantly improve the mechanical properties of organohydrogels through synergistic effects and exhibit toughness up to 15.93 MJ/m^3. The introduction of fillers has been successful in constructing conductive pathways, and the introduction of Gly has also given conductive nanocomposite hydrogels good thermoplastics, self-healing, antifreezing, and moisture retention properties. It was further assembled into a flexible versatile sensor, which exhibited a linear response at 600% strain, as well as good linearity and high resistive temperature coefficient when used as a temperature sensor. Huang et al. introduced CB and carboxylated multiwalled CNTs (c-MWCNTs) modified by the Hummers method into bacterial cellulose (BC)–reinforced basically cross-linked hydrogels by freeze-thawing process and calcium-ion cross-linking, which showed relatively stable mechanical properties (Huang et al., 2019). Moreover, due to the introduction of conductive fillers, the nanocomposite hydrogel can exhibit synchronization characteristics with good repeatability and stability when assembled into sensing strain sensors and capacitive pressure sensors, both at the input of the mechanical load and at the output of the electrical signal. Capacitive pressure sensors have a wide pressure range of 337 kPa and exhibit reliable repeatability and stability during compression/release. Capacitive-based sensors or dual-signal sensors can perform size detection of human motion, including grip weight. Zhang et al. were inspired by human neurons to introduce modified CB and silver nanowires into PVA/tannic acid (TA)/polyacrylamide (PAM) nanocomposite hydrogel networks, where modified CB mimicked the structure of somatic cells (Zhang et al., 2022b). And silver nanowires mimicked the structure of neural synapses, which resulted in flexible conductive nanocomposite hydrogels with higher sensitivity and fast response, where the strain sensing was in the range

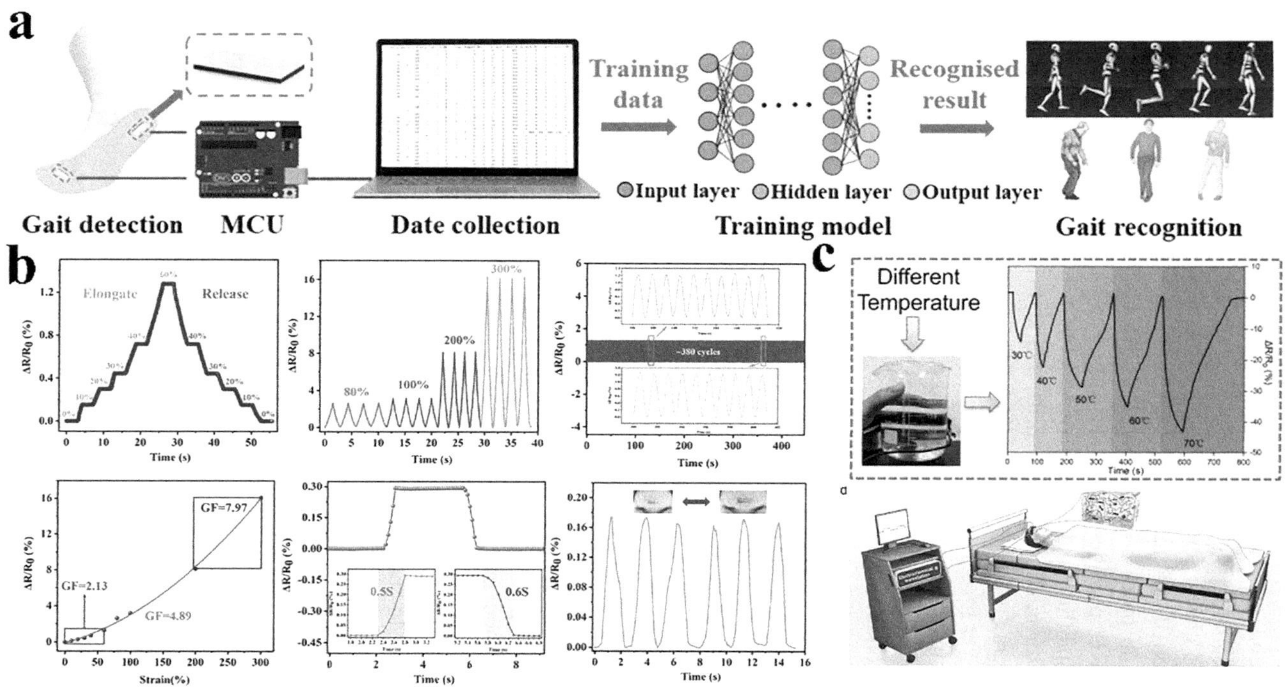

FIGURE 8.3 Composite hydrogels based on carbon-based nanofillers for flexible wearable sensors. (a and b) The process of human gait recognition based on GO-PAM hydrogel. (c) Application of GO-PVA hydrogel as temperature sensor for real-time monitoring of different temperatures. *(a and b) Reproduced from Wang et al. (2022c) with permission from Elsevier. (c) Reproduced from Li et al. (2023a) with permission from Elsevier.*

of 0%–160%, 160%–300%, 300%–355% with GFs of 5.17, 9.21, and 68.64, respectively, and response/recovery times of about 100 and 200 ms, respectively. Their assembly into flexible wearable sensors can provide excellent stretchability and wide-area sensing with strain and pressure sensitivity, flexibility, malleability, and stability.

8.2.2.2 Carbon nanotubes

CNTs nanocomposite hydrogels have attracted a lot of attention in biomedical engineering and wearable electronics due to their excellent electrical and mechanical properties (Jiao et al., 2021). Wang et al. developed conductive hydrogels with high transparency, good mechanical properties, adhesion, and multiresponsive repair capabilities by combining the strategies of in situ polymerization. And sparse conductive networks by introducing c-MWCNT with catechol nanoparticles into PAM networks (Wang et al., 2022b). The commercial c-MWCNT acts as a linear conductance unit, and the catechol nanoparticles better disperse and modify the CNTs through conjugation effects, thus balancing their water dispersibility and conductivity. Due to the large number of dynamical interactions in conducting hydrogels, damaged hydrogels can be repaired by microwave irradiation for 15 s and by pH--mediated mechanism, and the multiple stimulus-induced repair behavior gives hydrogels an advantage under different operating conditions. The conductive nanocomposite hydrogels assembled into strain sensors also have high sensitivity and fast response, with a GF of 1.57 in the 0%–50% strain range, 2.86 in the 50%–200% strain range, 4.97 in the 200%–400% strain range, and 7.8 in the 400%–900% strain range, and the response and relaxation times are 0.4 and 0.2 s, respectively. The excellent electromechanical performance allows the monitoring of human body movements with varying degrees of deformation. Moreover, Deng et al. designed multifunctional conductive hydrogels through β-cyclodextrin (β-CD), *N*-isopropylacrylamide (NIPAM), CNT, and PPy, with β-CD and NIPAM as the subject and guest molecules, respectively, and endowed the nanocomposite hydrogels with better self-repairing properties through the subject–guest interactions (Deng et al., 2018). In addition, the hydrogel network consists of hard CNT and soft PINPAM, forming a helical chain to form a good elastic network, which gives the hydrogel better resilience and fast recovery. Conductive nanofillers, such as CNT and PPy, also have the ability to absorb NIR light and convert it into heat, and their potential application as NIR light-responsive valves was further demonstrated by designing a simple device to illustrate their formal behavior. The introduction of nanofillers also makes flexible conductive hydrogels promising for pressure sensors, human motion monitoring sensors, and self-healing devices. Additionally, Liu et al. constructed stretchable, tough, and self-healing gelatin/multiwalled CNT (MWCNT)-PPy conductive hydrogels by template degradation (Liu et al., 2021). And then the gels were immersed into ammonium sulfate solution to make folds on the surface of the material, which led to the enhancement of mechanical and electrical properties of the hydrogels by a simple method of immersion–heating-template coating–reimmersion. The resultant hydrogels possessed high mechanical properties, high strain, electrical conductivity, self-healing, and biodegradability. Furthermore, the addition of MWCNTs can further improve the electrochemical properties of the conductive nanocomposite hydrogels, which can be used as sensors to monitor deformation strains, such as tensile, bending, and torsion. Meanwhile, as supercapacitors, the capacity retention rate of more than 98% after self-healing, and capacity retention rate after self-healing is over 98%. These excellent properties enable nanocomposite hydrogels to be used in many fields, such as human health monitoring and intelligent robotics.

8.2.2.3 Graphene oxide

GO is widely used in conductive hydrogels to enhance the mechanical and electrical properties of hydrogels due to its layered structure and the abundance of functional groups, such as epoxy, carboxyl, and hydroxyl groups, which provide a wide range of excellent properties of high electrical conductivity, hydrophilicity, and large specific surface area (Salimiyan et al., 2023). Wang et al. developed a self-powered hydrogel sensor based on GO and PAM that can be used to monitor and recognize subtle human movements, and the hydrogel sensor can also be used as a frictionless nanogenerator to harvest mechanical energy (Fig. 8.3a and b). (Wang et al., 2022c) Performance was optimized by dispersing GO into PAM with different mass ratios, resulting in frictionless nanogenerators based on nanocomposite hydrogel films with high output voltage and current. The optimal sample is a 0.02 wt% GO-PAM hydrogel film, which has an output voltage of 990 V, a current of 63.84 μA, and an output power of 26 mW after assembled into a TENG. The mechanical and electrical properties of the optimal sample, as well as the strain sensing performance of the assembled strain sensor, are also investigated. In addition, a wearable in-shoe monitoring system is designed, which consists of a flexible insole, a data processing board, and a PC interface developed using Python. Different algorithms were used to recognize the gait of daily human actions and the gait of patients, and concluded that the algorithm using artificial neural networks had the highest recognition rate for different gaits, which could provide a more convenient option in human gait monitoring and recognition for patients, with potential for medical applications. Furthermore, Chen et al. reported a porous cross-linked and honeycomb dense polyvinyl alcohol (PVA) hydrogel doped with GO and rGO nanosheets with dimethyl sulfoxide/water binary solvent, which resulted in nanocomposite organohydrogels with good mechanical properties, high toughness (3.1 MPa), and reliable mechanical and electrical properties over a wide range of temperatures (−30–60°C) (Chen et al., 2021b). By embedding highly conductive rGO into the structural framework of GO, the weak noncovalent interactions between graphene sheets and polymer substrate can be effectively avoided. And the homogenous dispersion of graphene as well as the enhancement of van der Waals forces between the polymer substrate can be further promoted, which leads to an effective improvement in the overall mechanical and electrical properties of the conductive nanocomposite hydrogel. RGO and GO nanosheets can interconnect and construct a conductive network in the organohydrogels, thus ensuring that the organohydrogels are assembled into a flexible strain/pressure sensor. It exhibits a fast and sensitive response to the applied strain, with a GF value of 2.1 and linearity of 0.99 over the stretching in the range of 0%–600%. The linear relationship between the sensor strain and the resistance ensures the stability of the output signal. In addition, the organohydrogels are able to be constructed as a temperature sensor, exhibiting excellent linearity, stable thermal response, and accurate sensing with a temperature sensitivity of (3.81%/°C). The application of this work to flexible strain, pressure, and temperature sensors also provides a new paradigm for wearable versatile sensors or electronic skins. Li et al. introduced GO and ionic liquid (IL) into starch/PVA network to fabricate nanocomposite hydrogel by green physical cross-linking method (Fig. 8.3c) (Li et al., 2023a). The nanocomposite hydrogel has better mechanical properties (tensile strength of 0.64 MPa) due to different molecular structures and abundant noncovalent interactions in the polymer network, such as hydrogen bonding, electrostatic interactions, etc. In addition, GO is rich in oxygen-containing groups, which provide electrical conductivity and frost resistance. The multimodal wearable sensors assembled from the nanocomposite hydrogels can be applied to sensing and monitoring the changes of strain, pressure, and temperature, and they are proved to have good fatigue resistance after 2000 cycles. This work explores starch-based hydrogels with high performance for wearable flexible sensors. Besides, Zhou et al. fabricated a biomimetic hybrid hydrogel with a layered structure by introducing GO, sodium casein, PDA, and calcium chloride into a PAM network (Zhou et al., 2022). The hybrid hydrogel exhibits excellent elongation (up to 10,300%) due to the presence of a large number of covalent and noncovalent bonds in the polymer network with multiple cross-links. Furthermore, the presence of PDA not only provides stable adhesion of the hybrid hydrogel but also converts a portion of the GO into rGO during reduction process, which improves the conductivity and electrical properties of the hybrid hydrogel. Moreover, the flexible wearable sensors based on hybridized hydrogels have high strain sensitivity (GF of 13.45), which can be used to monitor nonstop types of human activities in real time in a stable and fast manner. Gloves composed of hydrogel-based sensor arrays can also be used to monitor complex hand gestures and finger movements. The integration of sensor systems into a cloud-based platform allows for reliable remote monitoring across continents, exhibiting high sensitivity and fast response.

8.2.2.4 Reduced graphene oxide

RGO and GO belong to the graphene family and are widely used in flexible composites due to various excellent properties, such as high specific surface area, extraordinary electrical conductivity, high mechanical flexibility, and high electronic mobility (Yin et al., 2023b). The rGO can be derived chemically or thermally from the GO, and the electrical properties can be improved by restoring the sp2 carbon atoms in the GO. However, rGO is poorly dispersed in water or hydrophilic materials, so it is crucial to employ strategies that allow it to be uniformly dispersed in polymer substrates and enhance the mechanical and electrical properties. For example, Park et al. prepared a self-assembled particulate hydrogel by using GO-coated agarose microbeads as a precursor material and thermally annealed it at 70°C, accompanied by the reduction of GO to rGO and the rearrangement of rGO, to develop a nanocomposite hydrogel consisting of rGO as the main conductive component (Park et al., 2022b). The GO and rGO enhanced overall the macroporous structure, in addition to helping to dissipate external energy and preventing crack expansion to improve the mechanical properties of the conductive hydrogel, is also capable of providing high electrical conductivity (20 mS/cm) and some in vitro–in vivo biocompatibility. That can be used for biomedical applications, such as implantable bioelectrodes, wearable flexible sensors, and 3D-printed tissue scaffolds. Since no chemical cross-linking agents, initiators, are required in the preparation process, the conductive hydrogel-based electrodes are soft and can be used to monitor EMG signals as if they were in direct contact with muscle tissue. Alternatively, conductive hydrogel-based electrodes can be assembled into flexible pressure and strain sensors that can monitor external stimuli. Pan et al. innovatively introduced rGO and proanthocyanin composites with neural-like nanonetworks into Gly/water binary solvent-plasticized PVA-borax hydrogel networks to prepare e-skin with biomimetic haptics (Pan et al., 2020). The introduction of Gly calculated with borax and the formation of borax-Gly-water network through abundant hydrogen bonding enabled the PVA hydrogel to exhibit remarkable stretchability (>5000%) and self-healing properties (3s, 95.73%). In addition, the reduction of GO with the green reducing agent PC and the formation of nerve-like nanonetworks in the polymer network endowed the PVA hydrogel with excellent electrical properties. And the flexible strain sensor based on the composite hydrogel also possessed a high sensitivity (GF = 14.14), which can be used to monitor and mimic a number of real skin motions, including facial expressions and throat vocalizations. PVA hydrogels can also be used as adhesive electrodes for detecting EMG and ECG signals. Moreover, Zhang et al. proposed a spongy porous hydrogel with both excellent mechanical and self-repairing properties by rGO using liquid metal (LM) (Zhang et al., 2021b). Acrylic acid (AA) was employed as the reactive monomer to generate acidic conditions through the ionization of carboxyl groups, under which the LM triggered the deoxygenation reaction of GO to form rGO, leading to the formation of a conductive network of LM/rGO hybrids and thc formation of a microporous structure in the hydrogel. In the polymer network, LM and rGO act as conductive fillers, resulting in nanocomposite hydrogels with good mechanical properties ($\sim$2800%), elasticity, and self-healing properties. In addition, flexible strain wearable sensors based on nanocomposite hydrogels also possess high strain sensitivity and wide sensing range, with a GF value of 1.77 in the 0%–100% strain range and increasing to 9.86 for 600% strain increase. When nanocomposite hydrogels are applied to pressure sensors, they can be used as capacitive and resistive types, respectively. The sensitivity and effective dielectric constant of capacitive pressure sensors increases with the gradual deformation of the macropore under external pressure. For nanocomposite hydrogels, the synthesis temperature affects the microporous structure of the hydrogel, and the sensitivity of the sensor increases with the synthesis temperature. When used as resistive pressure sensing, the GF is also stable at 1.2 over a wide range of compressibility. As well, Yin et al. constructed a conductive printable hydrogel by a gentle two-step reduction of GO using ascorbic acid and sodium citrate (Yin et al., 2023b). In particular, the integration of the rGO dissipates the mechanical strength and enhances the mechanical properties and electrical conductivity of the hydrogel through the rich noncovalent interactions. Beyond that the reduction product can induce the effect of saltation, which can synergistically enhance the mechanical properties. The rGO can form a 3D conductive pathway to effectively shield electromagnetic waves with an optimal EM shielding efficiency of 29.4 dB. Flexible wearable sensors based on nanocomposite hydrogels can also monitor external pressure changes (204.17 kPa) and strain (0%–35%) in either resistive or capacitive modes, with a certain stability and repeatability at 60 kPa, and can be recycled more than 20,000 cycles times. Besides, the composite hydrogel can degrade in soil within 32 days, which indicates its environmental stability.

8.2.3 Polymer-based nanofillers

Polymer-based nanofillers, such as CNCs, CNFs, and nanoconducting polymers including PPy, and PANI are also commonly used for the construction of conductive nanocomposite hydrogels due to their unique properties (Geng et al., 2023; Zhang et al., 2023c). Nanocellulose is usually obtained by top-down methods, such as acid hydrolysis and mechanical stripping, as well as bottom-up methods, such as bacterial synthesis, which allows the processing of natural cellulose into CNCs, CNFs, and bacterial nanocellulose (BNCs) of different sizes and structures. Of interest, nanocellulose has a surface area of a few hundred square meters per gram and a surprisingly high Young's modulus, roughly in the range of 20–50 GPa. As a new green, renewable, and degradable nanomaterial, nanocellulose can be used to build toughened materials or combined with other conductive fillers to develop highly efficient and sustainable flexible wearable electronics and sensors (Li et al., 2023b). Conductive polymers, on the other hand, are of interest for their π-conjugated polymers with alternating single and double covalent bonds for conducting electrons, and for obtaining tunable mechanical properties and excellent electrical conductivity by introducing conductive polymers into hydrogel systems.

8.2.3.1 Cellulose nanocrystals

CNCs are promising candidates for the construction of nanocomposite conductive hydrogels due to their wide range of sources, high crystallinity, high mechanical strength, light weight, high stiffness, and biodegradability. Tang et al. proposed a green method to oxidize CNCs and sodium alginate under mild conditions and introduce aldehyde-based functional groups, followed by the introduction of amine-containing vinyl functional groups onto a polymer substrate via Schiff-bonding reaction and prepared based hydrogels via a one-pot method (Tang et al., 2020). The mechanical properties and elastic limitations of nanocomposite hydrogels have been increased by the introduction of functional CNCs, which can act as reinforcing and cross-linking agents for hydrogels and can dissipate energy efficiently. Synergistic dynamical noncovalent interactions such as physical cross-linking, hydrogen bonding, and Schiff bonding in the adjoint system also confer certain self-healing properties on nanocomposite hydrogels. Stimulus response properties were increased by incorporating ethylene glycol (EG) migration into the hydrogel. When externally triggered, it can exhibit a response within 20 s, demonstrating that it is also chemically stable as a thermally triggered brake. Moreover, Nie et al. introduced CNCs modified with poly (diallyl dimethylammonium chloride) and phytic acid with antifreeze properties into a PVA substrate in a water-Gly binary system by a simple freeze-thaw cycling method to prepare a nanocomposite hydrogel with antifreeze and self-healing properties (Nie et al., 2022). And with the increase of the amount of physical acid, the antifreeze property of the nanocomposite hydrogels was also improved, which could maintain the elasticity and mechanical properties at −30°C. Besides, the nanocomposite hydrogels were endowed with certain self-healing properties through the hydrogen-bonding interactions between phytic acid, CNCs, and PVA, and the abundant hydroxyl groups in the binary solvent, which maintained a good tensile strength (1.0 MPa) and a self-healing efficiency of 85.0% even after healing for 4 h at −30°C. The nanocomposite hydrogel was assembled into a flexible wearable sensor, which also possessed a wide strain detection range up to 600%, fast response time (310.0 ms), and good fatigue resistance, and the flexible strain sensor could maintain stable and repeatable signal monitoring even at the low temperature of −30°C. Liu et al. prepared multistimulated hydrogels with variously anisotropic optical properties and thermal sensitivity by introducing self-assembled CNCs with chiral nematic ordered structures into a polymer network of two poly(ethylene glycol) derivatives and PAM with different side-chain lengths via in situ photopolymerization (Liu et al., 2022a). The cross-linking density was varied mainly by changing the content of AM and the ratio of CNCs to polymer, and it was found that the arrangement of CNCs strongly depended on the angle of self-assembly as well as on the residence time, which could be exploited to tailor the CNCs and prepare them into multistimulated hydrogels. Increasing the self-assembly angle from 0 to 90 degrees shifted the arrangement of CNCs from chiral to symmetric nematic columns, thus providing a way to produce tunably anisotropic CNCs hydrogels, which can be used to study the self-assembly process from longitudinal and transverse cross-sectional views of the samples. Moreover, the introduction of the thermosensitive polymer matrix not only responds to compression but also exhibits a reversible response to temperature. With its volume and ESR decreasing with increasing temperature, the responses to pressure and temperature are independent of each other, which can be applied to strain and temperature sensors. As shown in Fig. 8.4a and b, Cui et al. inspired by mussel proteins and gecko bionics introduced

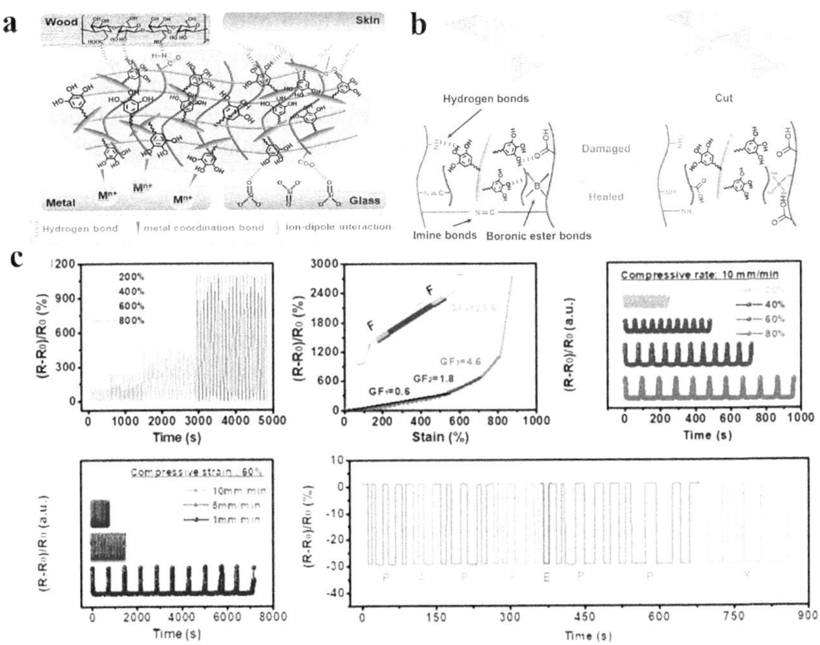

FIGURE 8.4 Nanocellulose-based composite hydrogels for flexible wearable sensors. (a and b) Schematic of the adhesion mechanism of TA-coated CNCs hydrogels between various substrates and UV blocking. (c) Electromechanical properties of CNFs and PPy-reinforced hydrogel sensors and use in Morse code. *(a and b) Reproduced from Cui et al. (2024) with permission from Elsevier. (c) Reproduced from He et al. (2023) with permission from Elsevier.*

TA-coated CNCs with UV filtration into a polymer substrate composed of oxidized alginate, gelatin by a simple one-step method. The solvent employed a water-Gly binary reagent to successfully construct an all-biomass-based hydrogel that was self-adhesive, self-repairing, transparent, frost-resistant, biocompatible, environmentally stable, and antimicrobial (Cui et al., 2024). Biomass hydrogels possess some self-adhesion and self-healing properties due to the synergistic bonding of borate and hydrogen due to the presence of amines in the substrate catechol. The use of binary reagents also gives it some elasticity and conductivity even at low temperatures. Flexible wearable sensors with biomass hydrogel assemblies can also achieve fast response in real-time monitoring.

8.2.3.2 Cellulose nanofibers

CNFs are widely used in flexible wearable sensors due to their reliability, biocompatibility, biodegradability, high aspect ratio, and excellent nanosize effects. For example, Yang et al. introduced dopamine-modified CNFs into the reaction solution synthesized by thermally induced ring-opening polymerization of the natural compound TA through a simple one-step strategy and obtained conductive composite hydrogels with both better stretchability and self-adhesion after cooling (Yang et al., 2023). The TA was lithiated to give it electrical conductivity and hydrophobicity, with a maximum conductivity of 17.36 mS/cm. The combination of the high mechanical strength of the CNFs, the carboxyl functional groups contained in the molecules, and the dynamic disulfide bonding contained in the TA leads to a degree of self-healing, adhesion, and excellent stretchability of the conductive nanocomposite hydrogels. Flexible wearable sensors based on conductive nanocomposite hydrogels also exhibit a wide sensing range ($\sim$1500) and stable, fast response signals. When the hydrogel sensors are attached to the body, they can detect small and large body movements, simple emotions, and speech recognition. As well, Lu et al. developed a conductive nanocomposite hydrogel with tunable strain response by introducing CNFs capped by TA and trivalent iron ions into a PAM network (Lu et al., 2020). A variety of physical interactions, such as ligand and hydrogen bonding as well as the synergistic effect of covalent cross-linked networks, were constructed to confer the nanocomposite hydrogel with excellent mechanical properties (energy storage modulus of more than 14 KPa and fracture stress of up to 108 KPa), certain adhesive properties, and good electrical conductivity (conductivity of up to 3.12 S/m). However, the rich catechol group enables nanocomposite conductive hydrogels to adhere to different substrates. Finally, nanocomposite bulk hydrogels are assembled into flexible wearable sensors for monitoring human motion. Li et al. prepared conductive ionic hydrogels with certain stretchability (up to 696%), water retention, and environmental stability (30 days) by introducing CNFs and alumina hexahydrate into a PVA network via a simple one-pot method, and the solvent used was a water-dimethyl sulfoxide binary solvent (Li et al., 2022b). The use of binary solvents formed strong hydrogen bonds, which enabled the conductive nanocomposite hydrogels to maintain elasticity and conductivity over a wide temperature range of −50–50°C, and to retain more than 85% of their weight even when stored in the environment for 30 days. Besides, the flexible wearable sensors based on conductive nanocomposite hydrogels also have a certain sensitivity (GF of 1.57), fast response (130 ms), and good durability (up to 1000 cycles), which allow them to monitor human movement even in harsh environments or after 30 days of storage. It also exhibits sensitivity to pressure and temperature, allowing for versatile sensing. To achieve the challenge of balancing electrical conductivity and mechanical properties. He et al. prepared high mechanical properties (tensile strength of 2.54 MPa, tensile strain of 925%) and high toughness (17.71 MJ/m) by introducing phosphorylated CNFs, trivalent ferric ions, and PPy into a covalently cross-linked network system of PAM and PAA nanocomposite conductive hydrogels (Fig. 8.4c) (He et al., 2023). Due to the multiple physical interactions, including metal coordination and hydrogen bonding, the PPy chains are entangled in the system, and the PPy can be further modified by in situ polymerization induced by trivalent ferric ions, as well as the synergistic covalent cross-linking between polymers, the hydrogel can be effectively strengthened and toughened. Moreover, the system constitutes a conductive pathway that mixes the ionic conductive mechanism of trivalent iron ions with the electronic conductive mechanism of PPy, which leads to the enhancement of conductivity (995 mS/m) and sensing performance of the conductive nanocomposite hydrogels. Based on the combination of excellent mechanical properties and strain response, flexible strain sensors assembled with conductive nanocomposite hydrogels have a wide operating range (~873%) and high sensitivity (maximum GF value of 25.6), which can be used for fast and instantaneous monitoring of human motion. Yu et al. prepared a highly ionic organohydrogel reinforced by cellulose nanofibers with excellent antifreeze and antidehydration properties by introducing sorbitol and $CaCl_2$ into the hydrogel via solvent substitution (Yu et al., 2022a). As CNFs act as dynamic link bridges in polymer networks, they provide hierarchical honeycomb structures in nanocomposite conductive hydrogels, which are capable of achieving enhancement and toughening purposes. In addition, the strong hydrogen bonds formed between sorbitol and water enabled the nanocomposite hydrogels to be elastic and conductive at −50°C and to combine over 90% resistance to dehydration. Due to the complete dissociation of $CaCl_2$ ions and the excellent antifreezing properties, the conductive nanocomposite hydrogels based on nanocomposite can be assembled not only into strain sensors for the detection of human physiological activities, but also into stretchable humidity sensors, and exhibit high humidity sensitivity over a wide range of humidity detection (23~97% relative humidity).

8.2.3.3 Polyaniline

PANI, as a conductive conjugated polymer, has been favored by many scholars for its ability to form interactions with specific groups in the hydrogel such as −COOH, $-NH_2$, or metal ions when introduced into the polymer network, resulting in denser cross-linking of the hydrogel network and enhancing its mechanical properties (Qian et al., 2023). Ge et al. inspired by human muscles and mimicking their microstructure and multifunctionality, PANI nanofibers have been introduced into polymer substrates consisting of polyacrylic acid (PAA) to take advantage of their fiber-enhanced mechanical pliability and chemosensitivity and to develop nanocomposite conductive hydrogels, which are highly stretchable as well as having self-healing properties (Ge et al., 2020). The nanocomposite conductive hydrogels exhibit surprising area expansion (~1500%) and satisfactory self-healing properties due to the presence of multiscale nanostructures in the nanocomposite hydrogels, the coordination of metal ions (Fe^{3+}), the strong hydrogen bonding introduced by the binary reagent of water-Gly, and electrostatic interactions (Fig. 8.5a). The presence of trivalent iron ions chelates the carboxyl groups on PAA, and the formation of reversible tridentate coordination can further enhance the self-healing performance. In addition, the presence of amino groups on PANI NF can also form hydrogen bonds between the chains, resulting in a dynamic zipper-like structure, which helps conductive hydrogels to recover quickly after rupture, with healing efficiency increasing with recovery time. Gly can inhibit ice crystallization. The use of binary reagents also makes the conductive hydrogel have a certain degree of antifreezing properties and water retention, even at subzero temperatures of −26°C, conductive hydrogel can also show close to 800% of the strain and conductivity of the hydrogel containing Gly can still be left at 96.8% of the weight after 7 days. Notably, based on the piezoresistive and thermosensitive properties of PANI NF, conductive hydrogels with nanocomposites can be assembled into strain and temperature sensors. And it has a high strain coefficient (GF value of 18.28) in the strain range of 169%−268.9%, with a detection limit as low as 5% strain, and possesses excellent thermal sensing (−0.016/C) and discernible temperature resolution (2.7°C). The nanocomposite conductive hydrogel-based electronic device can be used for a flexible touch keyboard for signature recognition and a "fever indicator" for human forehead temperature detection. To balance the mechanical and conductive properties of conductive hydrogels, Su et al. successfully developed conductive hydrogels with strong mechanical and self-healing properties by prepenetrating conductive polyethylene (PANI) precursors into hydrophobically bonded PAA hydrogel matrices using a prepenetration method (Su et al., 2021). Due to the dynamic interfacial interactions such as hydrogen bonding and electrostatic interactions formed between PANI and the polymer substrate, a variety of energy dissipation mechanisms were constructed in the conductive hydrogel, which resulted in

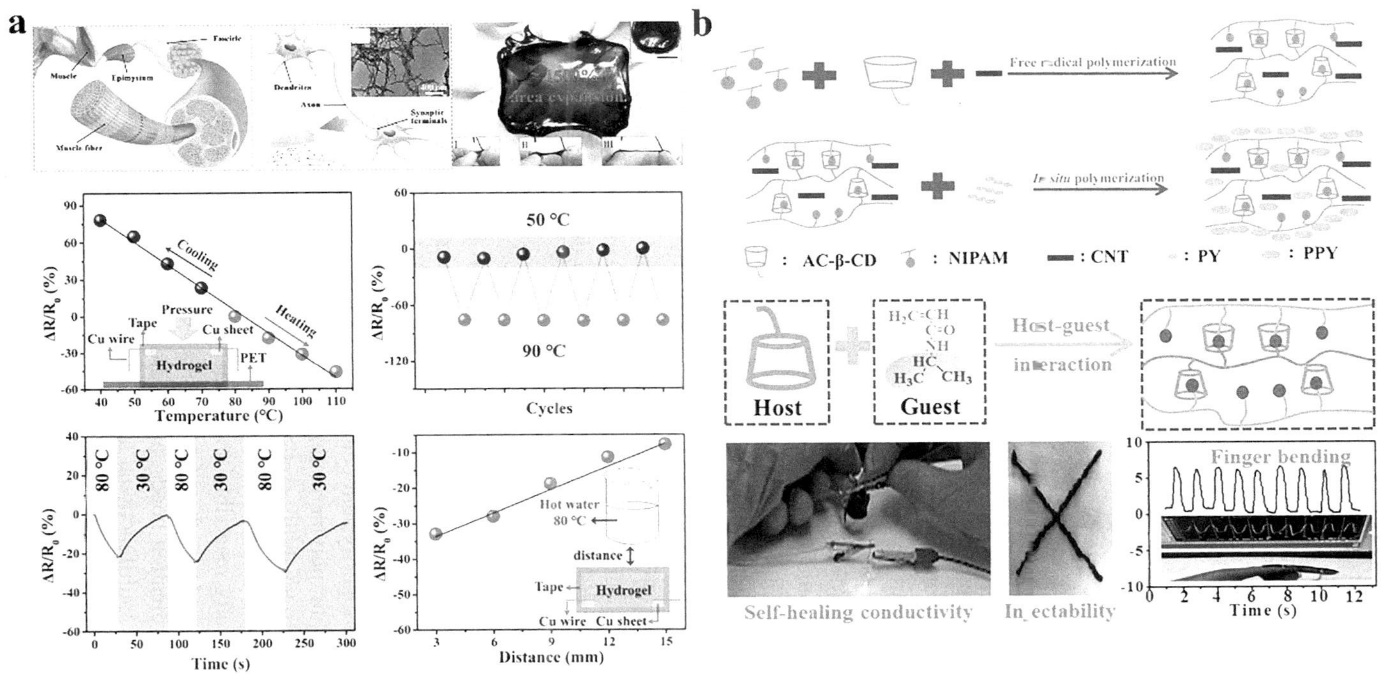

FIGURE 8.5 PANI/PPy-based nanocomposite hydrogels for flexible wearable sensors. (a) The mechanism of "sensory transmission process" inspired by human muscle and the biaxial area expansion and uniaxial stretchability of PANI-PAA hydrogel. (b) β-cyclodextrin- and *N*-isopropylacrylamide-based hydrogel with self-repairing property through the interaction of host and guest, and the introduction of PPy also allows hydrogel sensors to easily monitor finger movements. *(a) Reproduced from Ge et al. (2020) with permission from American Chemical Society. (b) Reproduced from Deng et al. (2018) with permission from American Chemical Society.*

excellent mechanical properties of the conductive hydrogel. In addition, the interconnected PANI network also endowed the composite hydrogels with high electrical conductivity (3.35 S/m) and good sensory properties. Assembling the conductive hydrogels into flexible wearable sensors also provides excellent sensing performance, even after self-healing, the hydrogel sensors have a GF value as high as 17.9 and a strain as low as 0.05% as well as a fast response time (80 ms), which can also be used to monitor the human body's motion and physiological signals and, based on the conductive hydrogels, can be used in touchscreens for handwriting monitoring and in electronic skins for pressure monitoring, too. Successfully provides a strategy to enhance both mechanical and electrical properties, Chen et al. by introducing pyrimidinone (UPy) groups into a brittle PANI/poly-4-styrenesulfonic acid network developed a novel polymer hydrogel possessing stretchability, instability, electrical conductivity, and rapid self-healing properties (Chen et al., 2019). The formation of an interpenetrating conductive network by cross-linking of UPy groups combined with in situ polymerization and dynamic interactions such as electrostatic interactions as well as hydrogen bonding in the polymer network contribute to energy dissipation and network reconstruction of the conductive hydrogel, endowing the hydrogel with mechanical and self-healing properties (self-healing within 30 s). The construction of the interpenetrating PANI/PSS network also provides ion transport–assisted electron conduction with a conductivity of up to 13 S/m. Flexible strain sensors based on conductive hydrogels also have excellent sensing performance, showing linear response at 300% strain with a GF value of 3.4, allowing fast and accurate monitoring of human motion. In addition, by exploiting the reversibility of noncovalent cross-linking, hydrogels can be easily prepared into different shapes.

8.2.3.4 Polypyrrole

The conductive polymer PPy is often used to build hydrogel-based flexible electronic devices due to its good electrical properties and more flexible mechanical properties than metals. To design flexible wearable sensors for underwater environments. For example, Zheng et al. introduced the conducting polymer PPy into a gel system containing the natural polymer filipin protein and TA via in situ polymerization and successfully prepared conductive hydrogels with stretchability, self-repairing, and self-adhesion without the need to use additional cross-linking agents (Zheng et al., 2022). Due to the multiple dynamically reversible noncovalent bonds present in hydrogel systems, including hydrophobic interactions, electrostatic interactions, catechol–metal coordination bonds, and hydrogen bonds, conducting hydrogels are endowed with multifunctional and mechanical properties, as well as some antimicrobial and biocompatible properties. At the same time, the ability of TA to penetrate the hydration layer can be exploited to achieve underwater self-healing and self-adhesion, showing good adhesion to PTFE, glass, and biological tissues, which breaks the traditional self-adhesion and self-healing properties of hydrogels. In addition, the electronic conductive network constructed by PPy also endows the hydrogel with excellent sensing performance, which can be directly adhered to the human body surface when it is assembled into flexible wearable sensors and can quickly sense and monitor the movement of human body size strain, physiological signals of microexpressions, such as smile, cough, and breathing, and also realize the transmission of underwater information through Morse code. To address the problems of low electrical conductivity and poor mechanical strength commonly found in protein hydrogels, Xu et al. developed a fully physically cross-linked conductive hydrogel by introducing conductive conjugated polymers and metal ligands into the polymer chain (Xu et al., 2023c). Firstly, bovine serum albumin modified with methacrylic anhydride and PPy was introduced into the PAM and PAA polymer chains, and through hydrogen-bonding complexation as well as metal–ligand coordination between the metal ion Fe^{3+} and the polymer chains. The composite conductive hydrogel was made to have excellent mechanical properties and certain self-healing properties, with a tensile strength of 5.36 MPa, a toughness of 17.66 MJ/m^3, the modulus of elasticity is 1.61 MPa. The presence of PPy and Fe^{3+} gives the hydrogel excellent electrical conductivity and sensing properties, where the conductivity can be as high as 1.13 S/m and the GF value is 4.98. Protein-based hydrogel sensors can monitor fine and large body movements, and they can demonstrate reliable stability and reproducibility. It can also be used as a soft electrode in an ECG device to monitor the ECG signal. Zhang et al. inspired by the ultrathin and flexible nature of human biofilms; a hydrogel film based on an ordered crystal structure was fabricated using a film casting method using a water-Gly binary reagent (Zhang et al., 2021c). The introduction of PPy, a conductive polymer, was used to form an interpenetrating conductive network, and by constructing noncovalent interactions between the conductive networks. Including hydrogen bonding and π-π stacking, the conductive hydrogel film was made to have an ultrathin structure, good flexibility, and antiexpansion properties. While UV resistance is provided by the presence of PPy. The ultrathin structure of the hydrogel film has an elongation of 523.3% at break, a tensile stress of 3.5 MPa, and good friction resistance. In addition, assembling the hydrogel film into a flexible wearable sensor has a fast response time (200 ms) and excellent sensing performance (GF value of 2.1). And the sensor not only monitors human motion but also recognizes the surface texture and roughness of objects, such as glass, brushes, wood, and sandpaper, which can be further expanded to the biosensing and e-skin neighborhood. Han et al. successfully prepared transparent, stretchable, and self-adhesive nanocomposite conductive hydrogels by interweaving in situ–formed poly(dopamine)-doped PPy nanofibrils in an elastic PAM polymer network (Han et al., 2018). As the PPy no-frills form an interpenetrating network in the PAM network, they form a complete conductive pathway with high electrical conductivity (12 S/m), which also gives the hydrogel excellent mechanical properties (tensile strain can be more than 2000%) and allows the passage of visible light for transparency, as well as UV shielding capability. The catechol group in PDA-PPy nanofibrils also endows conductive hydrogels with a certain degree of adhesion, allowing them to adhere freely to the body. Sensors based on nanocomposite conductive hydrogels can be directly adhered to human skin to monitor human movement without additional assistance and can also be used as self-adhesive electrodes to monitor physiological signals, including electrocardiograms and magnetocardiograms. Deng et al. prepared nanocomposite hydrogels with self-healing, flexibility, and high conductivity based on subject−object interactions by introducing PPy and multiwalled CNT as conductive components (Deng et al., 2018). And it has fast stimulation response performance for temperature and near-infrared light, and assembled into flexible sensors capable of monitoring human movement (Fig. 8.5b).

8.2.4 Metal-based nanofillers

Metal-based nanofillers, including silver nanowires, silver nanoparticles, gold nanoparticles, etc., have shown great potential in flexible wearable devices due to their excellent intrinsic properties, such as mechanical strength, chemical stability, and excellent electrical conductivity. To realize conductive hydrogels for multifunctional sensing, Yu et al. prepared a conductive nanocomposite hydrogel with antimicrobial functionality sensitive to strain and temperature stimuli by introducing silver nanoparticle (AgNP)-modified CNTs into a PVA/borax polymer network (Yu et al., 2022b). The mechanical and electrical conductivity of the conductive nanocomposite hydrogel is further enhanced by the conductive network composed of two fillers, AgNP and CNTs, and it also possesses a certain degree of ductility (maximum ductility of 261%) with an elasticity modulus of 187.83 kPa. The conductive nanocomposite hydrogel-based sensor also has a linearity of 0.989, a strain factor of 1.78, a 250 ms fast response time, and a 261% response range. The synergistic effect of AgNP and CNTs enabled the conductive nanocomposite hydrogel to respond to changes in water temperature with excellent response (TCR = −2.99%/°C), high sensitivity (0.1°C), and good linearity (0.998) in the range of 20−40°C. The conductive nanocomposite hydrogel is used as a sensor to measure the water temperature. Inspired by mussel proteins, Li et al. introduced CNCs modified with PDA and gold nanoparticles (AuNPs) into a PVA network to prepare conductive nanocomposite hydrogels with stretchable, biocompatible, and self-healing properties (Li et al., 2023c). Due to the synergistic effect of AuNPs, PDAs, and CNCs, composite hydrogels have enhanced mechanical and conductive properties. The conductive nanocomposite hydrogel is assembled into a flexible wearable sensor with a fast response and stable instantaneous output signal with a response time of 160 ms and a GF value of

0.99, enabling rapid capture and monitoring of human body movements. Besides, using lactate oxidase as a model enzyme, the resulting sweat sensor also achieves satisfactory sensitivity and linearity, with a detection limit as low as 0.31 mM and a linearity of 0.9987, enabling versatile applications in sweat sensing and motion monitoring. As illustrated in Fig. 8.6a, Huang et al. prepared nanocomposite hydrogels with stretchability, self-healing, and EMI shielding properties by constructing silver nanowires (AgNWs) aerogels by directional freezing, followed by in situ polymerization of acrylamide and acrylic-11-aminoundecanoic acid in the aerogels (Huang et al., 2022). Due to the introduction of AgNWs to construct a three-dimensional (3D) conductive network, the nanocomposite hydrogels showed a strain of more than 1000% and a conductivity of up to 83 S/cm and exhibited surprising EMI shielding performance (66 dB in the X-band). Based on hydrophobic interactions such

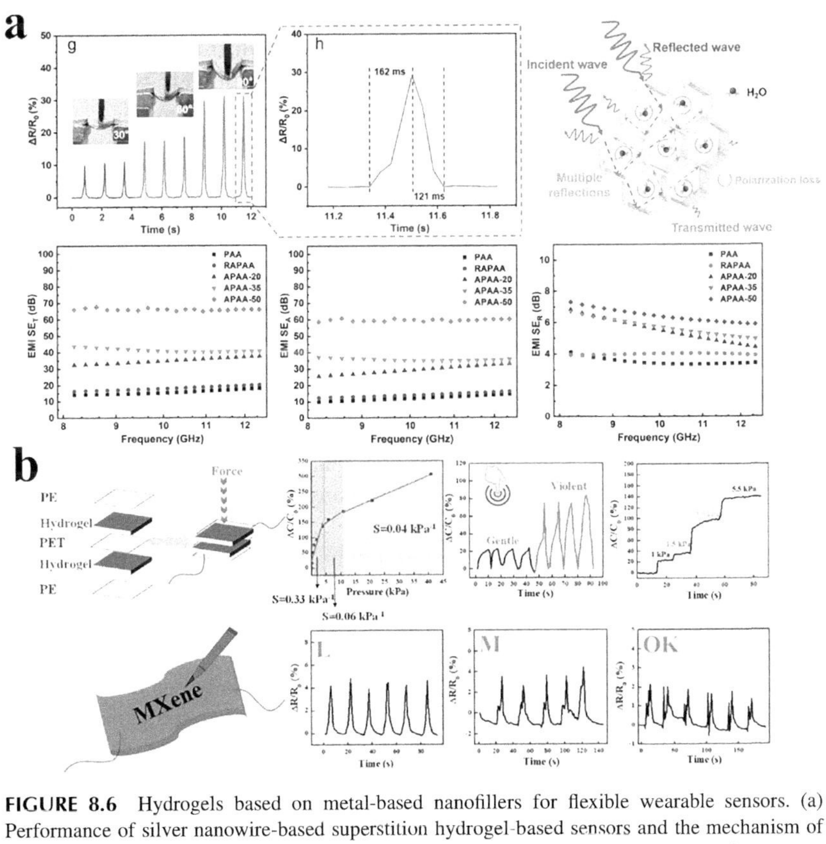

FIGURE 8.6 Hydrogels based on metal-based nanofillers for flexible wearable sensors. (a) Performance of silver nanowire-based superstition hydrogel-based sensors and the mechanism of EMI shielding. (b) Multifunctional conductive hydrogels based on lignin amine silver nanoparticles and MXenes as pressure sensors. *(a) Reproduced from Huang et al. (2022) with permission from Elsevier. (b) Reproduced from Jiang et al. (2022) with permission from Elsevier.*

as hydrophobic linkages and hydrogen bonding in polymer substrates, conductive hydrogels also confer a certain self-healing property, with 90% EMI repair efficiency for the repaired hydrogels. Finally, nanocomposite hydrogels are assembled into flexible strain sensors that can stabilize, instantaneously and quickly monitor human body movements, and remain stable after repair. Furthermore, Zhang et al. were the first to develop a tough hydrogel with Janus adhesion through the strategy of medium guidance (Zhang et al., 2021d). Firstly, the filter paper absorbed $HAuCl_4$ diffused to one side of the hydrogel, and then it was reduced to AuNPs by gelatin, and the proportion of AuNPs in the nanocomposite hydrogel was adjusted by the diffusion time and the diffusion concentration, and the AuNPs generated by the adhesive surface were able to enhance the adhesion to the skin and a variety of materials. Lithium chloride was added to the hydrogel to enhance its electrical conductivity, and the conductive nanocomposite hydrogel was assembled into a flexible wearable sensor that can accurately monitor body size and movement. Wang et al. designed one-dimensional (0D) SF@AgNPs, then introduced them and MXenes sheets into a copolymer network of AM and betaine, and fabricated conductive nanocomposite hydrogels with excellent antimicrobial properties, mechanical properties, adhesive properties, and role in promoting wound healing (Wang et al., 2023a). Moreover, the introduction of SF@AgNPs and MXenes nanosheets has also given composite hydrogels excellent mechanical properties and self-recovery capabilities, and amphiphilic catechol moieties have also allowed hydrogels to repeatedly adhere to human skin and various materials. Finally, sensors based on conductive nanocomposite hydrogels also have a wide sensing range, excellent reproducibility, allowing for the monitoring of large and subtle human movements. To enhance the mechanical properties, balanced conductivity, and self-repairing properties of hydrogels, Jiang et al. prepared conductive nanocomposite hydrogels with stretchability, strong adhesion, and self-repairing properties by introducing lignosamine silver nanoparticles (LA-AgNPs) and MXenes into the acrylic network system to construct a dynamic redox system (Fig. 8.6b). (Jiang et al., 2022)

Specifically, the amine group on the lignin amine as well as the catechol group reduced the silver ions to AgNPs, while the catechol was converted to a quinone or semiquinone structure, and the presence of photoelectrons generated by plasmon resonance on the surface of the AgNPs converted both of them back. Moreover, the presence of MXenes can also act as a cross-linker for the acrylic network, allowing rapid gelation of nanocomposite hydrogels without the need for the presence of other cross-linkers through hydrogen bonding and electrostatic interactions. The introduction of MXene and LA-AgNPs also imparts strong adhesion, antimicrobial, self-healing, and high electrical conductivity to the hydrogels. The conductive nanocomposite hydrogels can be assembled into strain and pressure sensors based on nanocomposites, exhibiting high sensitivity for fast and instantaneous monitoring of human motion, with GF values as high as 9.17 for strain sensors and S of 0.33 k/Pa for pressure sensors.

8.2.5 Novel two-dimensional (2D) material-MXenes

Recently, 2D materials such as MXenes have been widely used in flexible wearable electronics due to their unique properties, such as high electrical and thermal conductivity and large specific surface area (Zhao et al., 2023). MXenes are mainly prepared by etching the A layer in the MAX phase of the precursor, which is a 13–16 group element. Due to their abundant terminal functional groups, MXenes exhibit excellent hydrophilicity, and in combination with their high electrical conductivity, large specific surface area, and strong mechanical properties. Functional composites based on MXenes are also widely used in energy storage, electromagnetic shielding, and sensors (Zhuo et al., 2023). This section focuses on the application of MXenes-based nanocomposite conducting hydrogels to sensors. Inspired by the biomineralization process, Li et al. introduced MXenes into a polymer substrate composed of AA and amorphous calcium carbonate to produce nano conducting hydrogels with high stretchability, self-healing, and biocompatibility (Fig. 8.7). (Li et al., 2021b) Due to the rich functional groups (F, OH, O, etc.)

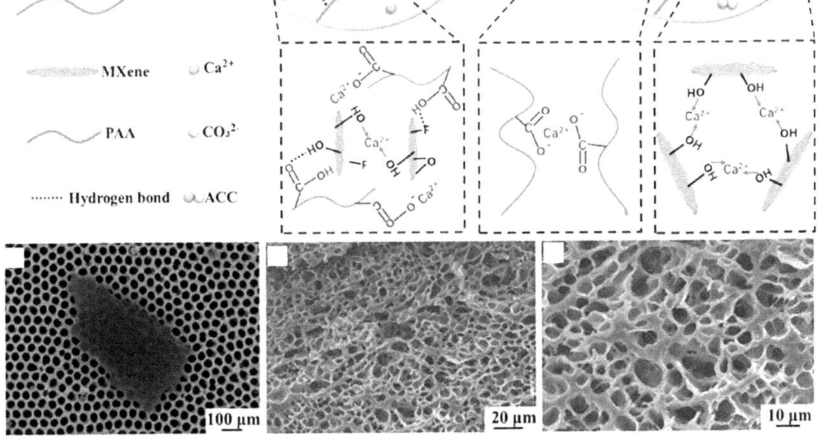

FIGURE 8.7 Hydrogels based on MXenes for flexible wearable sensors. Schematic of the preparation of MXenes-PAA hydrogel and monitoring of EMG signals as an e-skin. *Reproduced from Li et al. (2021b) with permission from American Chemical Society.*

on the surface of MXenes nanosheets, on the one hand, they can form supramolecular interactions with metal ions and carboxyl groups present in the hydrogel network, which gives the conductive hydrogel excellent self-repairing performance (0.2 s). On the other hand, it also confers the flexible wearable sensors based on the conductive hydrogel with high sensitivity and fast response time (20 ms). The skin sensors enable real-time monitoring of human movement and wireless monitoring of electrophysiological signals, including EMG and ECG signals. More significantly, conductive nanocomposite hydrogels exhibit biocompatibility, and the ability of epidermal sensors to degrade in phosphate-buffered reagents brings its potential for applications in wearable skin epidermal sensors. Liao et al. introduced MXenes into a polymer substrate consisting of PAM and PVA and prepared conductive nanocomposite hydrogels, which were then immersed in EG solution to replace some water molecules by solvent substitution to develop a nanocomposite organic hydrogel with good mechanical properties, long-lasting moisturization (8 d), self-healing, antifreezing (−40°C), and conductivity (Liao et al., 2019). The nanocomposite organohydrogel is then assembled into a flexible strain sensor. It is worth mentioning that the strain range is up to 350% even at low temperatures. The GF value of up to 44.85 indicates the high sensitivity of strain sensing and fast and instantaneous monitoring of human body movements, etc. Feng et al., inspired by the strong mechanical strength and contraction capability of muscles, fabricated anisotropic MXenes as conductive hydrogels by directional freezing (Feng et al., 2021). Due to the anisotropy of the MXenes conductive hydrogel, it has the ability to enhance mechanical properties and electrical conductivity in the direction of the properties. Through the replacement of the Gly-water binary solvent, the conductive nanocomposite hydrogel has a wide range of temperature resistance from −36 to 25°C. It was also tested for conductivity at low temperatures and maintains 89% conductivity even at −25°C. It still maintains a certain elasticity and electrical conductivity at low temperatures. It is worth mentioning that this is the first time for MXenes hydrogels with anisotropic and ordered orientations as flexible wearable sensors for instantaneous, accurate, and stable monitoring of human movement and speech recognition. Flexible sensors based on conductive hydrogels can also be used to monitor motion via wireless devices. Multiple flexible sensors can also be assembled into a 3D sensor array that can be used to detect the magnitude and spatial distribution of forces or strains. Zhang et al. used a one-pot method to introduce MXenes into a polymer network containing TA, sodium alginate, and PAM to prepare conductive hydrogels with excellent tensile properties (strain >4600%), good adhesion, and self-healing properties (Zhang et al., 2023d). The rich functional groups and high hydrophilicity of the MXenes nanosheets led to their uniform distribution in the hydrogel and the construction of the conductive network, which led to the good sensitivity (GF value of 6.6) of the conductive hydrogel-based sensor. Moreover, based on the excellent tensile properties, adhesion, and self-repairing, the conductive hydrogel can also be used as the electrode of the friction nanogenerator. And the use of silicone rubber as the friction layer of the TENG can have high output performance and prevent the electrode from dehydrating, which affects its conductivity. The conductive hydrogel-based TENG has environmental stability and better energy harvesting efficiency. Using this TENG, a human–computer interaction system is constructed that enables gesture visualization, multidimensional signal acquisition, and robotic gesture control. Since MXenes will inevitably interact with water and dissolved oxygen in the hydrogel, affecting the use of the hydrogel over a long period of time, the oxidation and transformation of MXenes also deplete their own mechanical and electrical properties. It is also crucial to inhibit their oxidation and prolong the lifespan of the hydrogel. Chae et al. proposed to inhibit the oxidation of MXenes by increasing the interaction with the functional groups on the surface of MXenes based on catechol-functionalized PVA to develop a new functional hydrogel with a combination of mechanical, self-repairing, and electrical properties (Chae et al., 2023). Due to the presence of the hydrophobic part of the benzene ring in the catechols, the MXenes are functions such that their exposure to water is reduced, and intensive studies of them have demonstrated that the grafting of catechols does indeed have the function of inhibiting the oxidation of MXenes. The resulting conductive nanocomposite hydrogel is used for flexible wearable strain and pressure sensors, due to the presence of catechol groups and the large number of noncovalent bonds between the polymer substrate. The hydrogel has a certain degree of adhesion, which can be directly adhered to the surface of the skin for the monitoring of human movement, and can be mounted on the throat to monitor the subtle human movement, such as recognition of voice, recognition of handwriting, and so on. Surprisingly, it was finally shown that signals based on conductive hydrogel sensors can be classified using deep learning models. He et al. prepared an organohydrogel with strong underwater adhesion properties by introducing MXenes into a polymer substrate copolymerized with the hydrophobic monomers 2-phenoxyethyl acrylate, 2-methoxyethyl acrylate, and the hydrophilic monomer N-(2-hydroxyethyl) acrylamide, and the binary solvent dimethylsulfoxide, inspired by barnacle proteins (He et al., 2022b). The underwater adhesion properties are important for soft electronic devices with underwater interfaces, and the presence of a large number of noncovalent interactions in organohydrogels gives hydrogels the ability to adhere to a wide variety of substrates. The presence of hydrophobic monomers also gives the organohydrogel resistance to wetting and swelling, effectively preventing water molecules from disrupting the adhesion interface, in combination with dynamical interactions, gives the hydrogel immediate and reproducible underwater adhesion. In addition, the introduction of MXenes nanosheets not only enhances the mechanical properties of the organohydrogel but also confers electrical conductivity to the organohydrogel, which can be assembled into a flexible wearable strain (strain coefficient of 7.17 at 200%–500% strain) and pressure sensor (0.63/kPa at 0–70 kPa). And the sensors were investigated for underwater monitoring, including human motion and pressure sensing, and also to further expand the application of hydrogel sensor devices underwater. The ability to perceive small mechanical changes is due to the unique electromechanical properties exhibited by MXenes hydrogels. Herein, Lee et al. found that the mechanical vibrational energy can be sensed as a voltage in MXenes-PVA hydrogels (Lee et al., 2021). Based on this, MXenes sheets were introduced into PAA/PVA hydrogels and the effect of internal ions on the electromechanical properties of conducting hydrogels was investigated. In particular, under external stimuli, such as strain and pH, the orientation of the MXenes nanosheets can be altered when a stimulus is applied to the conducting hydrogel, and the strain-induced orientation of the surface charge-guided ionic pathway can enable modulation of ion transport in the conducting hydrogel. In general, people sweat during excessive exercise to change pH values, and a reasonably designed conductive hydrogel sensor can track the pH value of sweat. It has been confirmed that both external strain stimuli and pH can regulate electron and ion transport in nanocomposite hydrogels. Based on this, a strain sensor is prepared for monitoring muscle fatigue during exercise, and changes in the signal can be monitored by a Bluetooth device.

Here, we present four different types of nanofillers. A common strategy to improve the conductivity of composite hydrogels is to introduce conductive nanofillers into the hydrogel matrix. However, this approach generally requires a high filler content, which may lead to separation of the nanofillers from the matrix, resulting in low tensile properties as well as weak fatigue resistance of the hydrogel (Karimzadeh et al., 2022). For carbon-based nanofillers, although they have been popular in recent years due to their relatively low production cost and mechanical properties, the low intrinsic conductivity is still a major drawback for the application of flexible wearable sensors and other electronic devices, and how to make up for this shortcoming through rational design is still worth exploring by scholars. In addition, metal-based nanofillers with various morphologies have become hot conductive materials by virtue of their intrinsic high conductivity (Zhang et al., 2023b). Although silver nanofillers have the advantages of easy large-scale synthesis and high conductivity as potential nanofillers for flexible wearable sensors, the oxidation of silver and the leaching of toxic silver ions still need to be solved. For gold nanofillers, with excellent chemical stability, biocompatibility, and high conductivity, they are suitable not only for flexible wearable sensor devices but also for implantable devices (Park et al., 2022a). However, the high cost and large-scale production of composites are still challenging, and it is worthwhile to continue to explore the rational introduction of other materials to maximize the advantages of gold nanocomposite fillers. MXenes are favored by many scholars because of their high hydrophilicity and conductivity, and their introduction into hydrogel matrices significantly improves the performance of the hydrogel, but their presence is susceptible to oxidization and mechanical stress, which can affect the long-term performance of the electronic devices. Although some studies have explored strategies to inhibit their oxidation, there is still a lack of long-term monitoring of electronic devices and the degree of oxidation and damage of MXenes, as a hot 2D material, its research fever remains high, still has a bright application prospect, and is worth further exploration.

Therefore, in the future research, not only the performance of various nanofillers and the comprehensive performance of the prepared conductive nanocomposite hydrogels, but also the long-term stability and durability of the nanocomposite hydrogels, as well as the stable operation and long-lasting work as a flexible wearable sensor device need to be considered, for example, how to reasonably modify the nanofillers or coat the protective layer so as to avoid direct contact with human tissues or exposure to air; how to collaborate different nanofillers to exploit their strengths and avoid their weaknesses, so as to make a breakthrough in the performance of conductive nanocomposite hydrogels. Although there are still some difficult challenges and tasks to overcome, the huge demand for high-performance hydrogel flexible wearable sensors and electronic devices will drive scholars to work tirelessly to climb the peak, and at the same time, surprising opportunities and bright paths are also introduced to the eyes.

8.3 Properties of conductive hydrogel

From the point of view of functional design as well as practical applications, electrical conductivity alone is no longer sufficient for the operating environment and long-term service life of sensors. Typically, the performance of conductive nanocomposite hydrogel sensors depend on the nanofillers as well as the polymeric network of the hydrogel, which gives both the properties of the nanofillers and the properties and advantages of the polymeric network (Xu et al., 2023d). The introduction of nanofillers is sufficient to further improve the mechanical properties of hydrogels and give them higher electrical conductivity and other functionalities, such as mechanical, self-healing, and adhesion properties (Zhang et al., 2023e). In practical applications, however, water-rich hydrogels inevitably freeze and evaporate at zero temperature, which greatly affects the practical use of hydrogel sensors. Therefore, it is important to endow nanocomposite hydrogels with certain antifreeze properties by modulating their fabrication strategies.

8.3.1 Mechanical properties

As the field of wearable electronics continues to grow, flexible nanocomposites have become a promising alternative to traditional rigid electronic materials due to their soft and stretchable properties. Not only can they improve the mismatched mechanical properties between rigid electronics and soft human skin, but also facilitate the reduction of impedance and skin irritation through conformal contact, which is crucial for the fabrication of flexible strain sensors. However, most conventional polymer materials tend to be hard and brittle, especially conventional polymer-based materials, and even some elastomers with an increased proportion of conductive fillings to improve conductivity have only a limited amount of strain at break-up (Xia et al., 2023). Compared with conventional conductive composites, conductive hydrogels stand out in flexible strain sensors and electronic skins due to their tunable mechanical properties and excellent tensile properties, which have led many scholars to explore them (Zhu et al., 2023). Among them, the most prominent is the fabrication strategy of stretchable conductive nanocomposite hydrogels by combining various conducting nanomaterials with excellent electrical and mechanical properties with hydrogels. These conductive nanomaterials form a permeable network in the polymer network, which preserves the mechanical and electrical properties of the conductive nanocomposite hydrogel well even under mechanical deformations, such as repeated stretching and compression.

Liu et al. proposed a "doping-then-gelation" strategy to synthesize conductive nanocomposite hydrogels with superstretchability (2830%), fast self-repairing properties, and high breaking strength (120 kPa) by doping PANI with AA monomers, which avoids the aggregation of PANI on the one hand and uses high-density electrostatic interactions between chains to trigger specific chain-entangled networks on the other hand (Liu et al., 2022b). The high density of electrostatic interactions between chains triggers a specific chain entangled network. Finally, this conductive nanocomposite hydrogel was assembled into a flexible strain sensor with reproducible stability, high sensitivity (GF value of 12.63), and fast response time (222 ms). Superior mechanical properties can be achieved by chemical- and physical-based intermolecular interactions. A strategy combining supramolecular interactions, including hydrogen bonding and cation-π interactions, was reported by Sun et al. to develop conductive nanocomposite hydrogels with high strength (1.63 MPa), high robustness, and swelling resistance (Sun et al., 2023). The combination of multiple supramolecular interactions endowed the composite hydrogel with reliable mechanical properties (toughness of 5.5 MJ/m^3) and excellent electrical conductivity (conductivity of 2.16 S/m) by combining the rigid conductive polymer, PPy, with the soft PAM-alginate. The conductive nanocomposite hydrogel was assembled into a flexible wearable sensor, which also possesses high sensitivity (GF value of 4.1) and linear detection range (up to 400%). The conductive nanocomposite hydrogel can be used as an underwater sensor for underwater communication by virtue of its resistance to swelling, in addition to monitoring human motion. To develop hydrogels with excellent mechanical properties and resistance to extreme environments, Liu et al. fabricated a two-network hydrogel with stable electrical conductivity, excellent adhesion, and effective flame retardation by introducing pyruvate-doped PPy and SiO_2 nanoparticles into a PAM polymer network (Liu et al., 2022c). The dual-network hydrogels exhibited excellent stretchability (1896%) and conductivity at both high and low temperatures, which were still effectively flame retardant when heated at 200°C for 1200 s. The hydrogel is also able to act as a flame retardant at ambient and cryogenic temperatures. The flexible strain sensors based on the dual-network hydrogel have a wide sensing range (up to 500%), good sensitivity (GF of 2.9), and stable sensing performance even under low temperature and heat treatment, which provides a good opportunity for the development of wearable devices in extreme environments. Cellulose has a supramolecular fiber structure due to its inter- and intramolecular hydrogen bonding. The use of cellulose and its derivatives to enhance the mechanical properties of hydrogels has also been favored by many academics. Zhang et al. fabricated a biomimetic hydrogel with skin features based on a porous cellulose network with adjustable crystallinity and chemically cross-linked elastin, where the cellulose supramolecular network was able to significantly enhance the mechanical properties and the toughness was increased by 45-fold (up to 4.3 MJ/m^3) (Zhang et al., 2022c). The in situ polymerization of aniline further gives it excellent mechanical properties and electrical conductivity, with tensile strains up to 900% and an intensity of 0.8 MPa. With a GF value of 1.7 in the 90%–600% strain range, the bionic hydrogel-based strain sensor is not only stable and sensitive to monitor human movement, but also has adjustable swelling and can still operate in a water environment. Nanogenic fibrillated cellulose (NFC) derived from natural cellulose fibers is also a common strategy to enhance the mechanical properties of materials by introducing NFC into covalently cross-linked networks, thanks to its outstanding properties, such as high aspect ratio and excellent mechanical properties. Wang et al. fabricated conducting nanocomposite hydrogels with excellent mechanical and antimicrobial properties by introducing silver nanoparticle-doped PDA@NFC into a PAM network using radical polymerization of the interpenetrating network (Fig. 8.8) (Wang et al., 2021). Due to the presence of NFC, the conductive nanocomposite hydrogel has a tensile strain of more than 1100%, a tensile strength of more than 700 kPa, a toughness of more than 7.5 kJ/m^3, and a good self-healing property with a recovery rate of 94.8%. Moreover, flexible sensors based on conductive nanocomposite hydrogels also have the advantages of wide sensing range, high compressive sensitivity, multimodal sensing, and sensing performance after 10,000 cycles.

8.3.2 Self-healing properties

In addition to superior mechanical properties, there is an urgent need to incorporate new functionalities into materials, such as self-healing properties, as flexible electronics inevitably suffer from friction and damage during use,

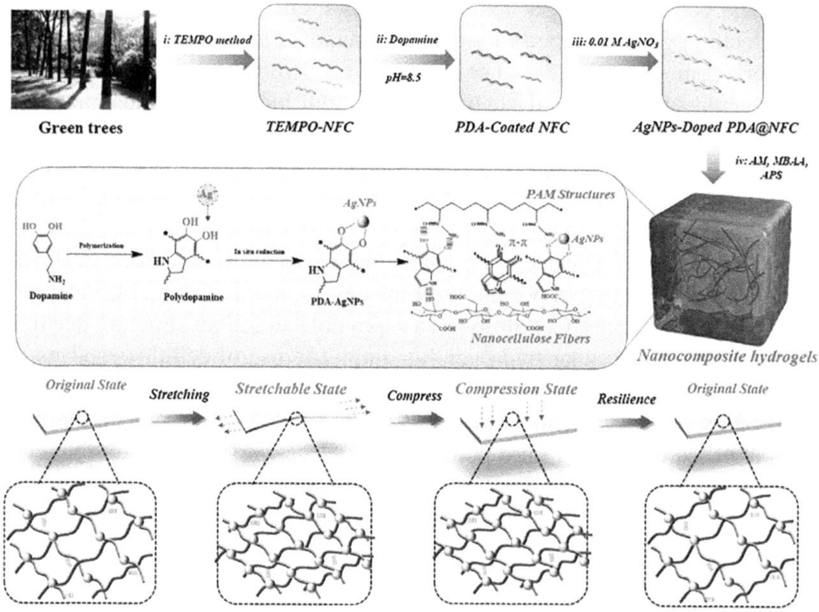

FIGURE 8.8 Mechanical properties of nanocomposite conductive hydrogels. Schematic of the preparation of composite hydrogels based on silver nanoparticles and when deformation occurs. *Reproduced from Wang et al. (2021) with permission from Elsevier.*

leading to malfunctions and thus limiting their lifespan (Wang et al., 2022d). The self-healing property allows spontaneous repair without external stimuli, which is essential to maintain the original mechanical and electrical properties for a short period of time after damage, thus extending the working life of flexible devices and significantly reducing costs. Based on these advantages, self-healing hydrogels are also increasingly being used in the sensor field. Conductive self-healing hydrogels can achieve the purpose of self-healing either through external stimuli (e.g., light, heat, etc.) or through the interactions of the hydrogel itself, which are usually constructed as dynamic covalent interactions and dynamic noncovalent interactions in self-healing hydrogel systems. Of course, there are some hydrogel systems in which multiple dynamic interactions are employed (Wang et al., 2023b).

8.3.2.1 Dynamic covalent interaction

Dynamic covalent interactions mainly include the Diels–Alder (DA) reaction, the Schiff base reaction, and borate bonding. The DA reaction is a representative dynamic reversible reaction between a dienophilic reagent and a diene via addition, which has been used to construct self-healing hydrogels due to its remarkable reversibility and high efficiency. Fang et al. fabricated stretchable and self-healing conductive composite hydrogels based on dynamic polyurethane substrates with DA chemistry and synergistic interpenetration with conductive polyethylene (Fang et al., 2021). Multiple dynamic junctions in the dual interpenetrating and synergistic nanocomposite conducting hydrogels, including DA covalent bonding, ionic interactions, and hydrogen bonding, give the hydrogels a 500% elongation at breakup, a tensile strength of 1.1 MPa, and a high conductivity of 7.9 S/m. Furthermore, the conductive nanocomposite hydrogel is assembled into a flexible strain sensor for immediate, stable, and accurate monitoring of human motion with a GF of 2.9. One of the well-known reversible chemical bonds is the Schiff base reaction, which is a dynamical interaction through the formation of an amine bond between the amine and aldehyde groups. Hydrogels based on the Schiff base reaction with certain network structures and good biocompatibility have been widely used to construct dynamic networks. Peng et al. demonstrated fast self-healing performance based on multiple reversible dynamic covalent bonds and supramolecular interactions between polymer networks and MXenes nanosheets, with a healing time of only 0.06 s (Peng et al., 2020). The surprisingly fast healing ability and electrical conductivity of the nanocomposite hydrogel is achieved mainly through the polymer network formed by cross-linking of PVA, poly(vinyl imine), and 4-formylphenylboronic acid. The dynamic covalent bonding is provided by the imine and boronic acid bonds, and the abundance of functional groups on the MXenes nanosheets, e.g., −OH, −F, and −O. The electronic skin sensor based on this nanocomposite hydrogel also has orientation sensing and fast response performance (0.12 s), which can recognize not only the movement of different parts of the prosthetic body, but also the direction of movement of the prosthetic limb. Fu et al. fabricated conductive nanocomposite hydrogels with stretchability, self-healing, degradability, and reliability based on multikinetic interactions, such as Schiff bond reactions, hydrogen bonding, and metal−ligand bonding (Fu et al., 2022). Reversible fracture reconstruction using molecular-level multidynamic interactions (mainly using groups such as−OH, −COOH, −CHO, and Fe^{3+} coordination, etc.), mainly through sol−gel and iontophoresis methods, resulted in conductive nanocomposite hydrogels with mechanical properties much better than those of pure gelatine hydrogels, and significant self-healing properties. The sensors based on this conductive nanocomposite hydrogel also have excellent mechanical properties and high sensitivity. As a pressure sensor, the compressive stress is up to 1310 kPa, the compression response time is 200 ms, and the compression sensitivity is up to 1.41/kPa at 15 kPa. The GF value can be up to 2.24 at 6% of tensile strain, which can be applied to the monitoring of omnidirectional monitoring of human movement, handwriting recognition, and personal signatures. The borate bond is a dynamic covalent bond created by the reaction of diols with boric acid, a covalent bond that can be broken reversibly and formed in a gentle environment. Polymers with hydroxyl groups, such as PVA, are favored by scholars for the fabrication of self-healing hydrogels based on borate bonding. Song et al. prepared conductive nanocomposite hydrogels with stretchability, thermal stability, and self-healing properties by grafting PANI onto the surface of polycarboxylic acid multiparty CNCs (multi-CNC-PANI) via template polymerization into polymer matrices containing borate and PVA (Fig. 8.9) (Song et al., 2020). The rod-shaped multi-CNC-PANI in the hydrogel substrate acted as a dynamic bridge, which conferred a hierarchical structure to the PVA/borax-added conductive hydrogel, and the dynamic interactions in the system. Both dynamic borate and hydrogen bonding led to the nanocomposite organohydrogel's excellent mechanical properties (tensile strain of 1085%, fracture strength of 171.52 KPa), fast self-healing properties (healing within 120 s in both air and underwater, with an efficiency of up to 99.56%), good biocompatibility, and adhesion. In addition, flexible strain sensors based on conductive nanocomposite hydrogels were assembled into flexible strain sensors; both the original hydrogels and the self-healing ones were able to sense and differentiate the human body movement. And the GF value increased with the increase of strain, up to 18.79 in the strain range of 550%−700%, which were able to detect and differentiate the signal of the pulse rate, with instantaneous, fast, and stable performance. For example, Chai et al. constructed conductive

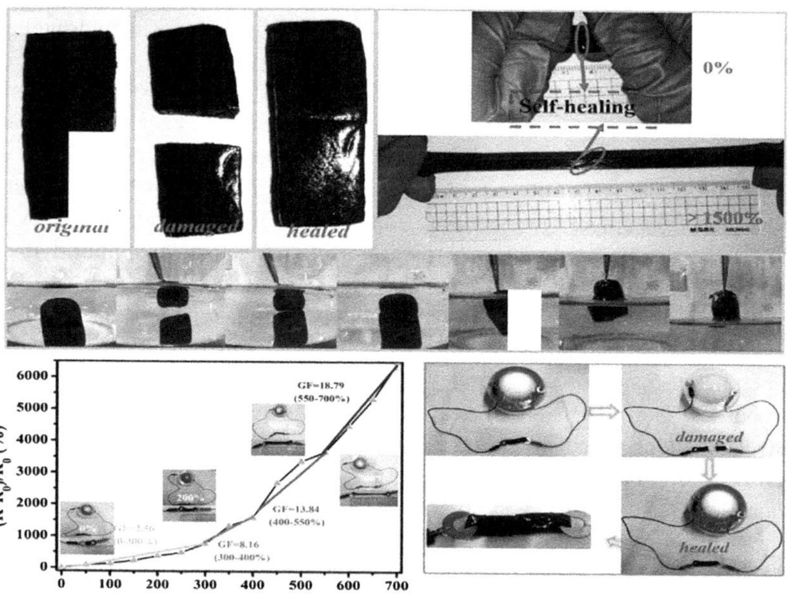

FIGURE 8.9 Self-healing properties of nanocomposite conductive hydrogels. Schematic of the preparation process of hydrogels based on borate bonding and the sensitivity. *Reproduced from Song et al. (2020) with permission from Elsevier.*

nanocomposite hydrogels with stretchable, self-adhesive, and self-healing properties by introducing PEDOT: PSS-modified CNC into a PVA/borax system (Chai et al., 2023). Based on the large number of hydrogen bonds as well as dynamic borate bonding and electrostatic interactions in the polymer network, the conductive nanocomposite hydrogels can self-heal without external stimuli in 60 s and recover to 92.57% with the increase of the self-healing time up to 12 min, both in air and water. The self-healing hydrogel still has good tensile properties, and the presence of dynamic reversible bonds gives the hydrogel a degree of self-adhesion that can adhere to different substrates. In addition, the conductive nanocomposite hydrogels have excellent thermoelectric properties, which can generate repeatable and stable voltages with a power factor of 8.12 μW/m K^2. At a temperature difference of 25 K, the output voltage reaches 31.72 mV.

8.3.2.2 Dynamic noncovalent interaction

Compared with dynamic covalent interactions, dynamic noncovalent interactions have the advantage of high reversibility, strong interaction forces, and simple fabrication strategies, which are more favorable for the construction of self-healing hydrogels (Zhang et al., 2022d). Dynamic noncovalent interactions mainly include hydrogen bonding, metal–ion coordination interactions, and host–guest interactions (Wang et al., 2023c). Hydrogen bonding is a force between a hydrogen atom and two other atoms of high electronegativity, which is weaker than conventional ionic, covalent, and metallic bonds, but stronger than electrostatic attraction and van der Waals forces. As a result, hydrogen bonding has been favored by many scholars as a mainstay of dynamic noncovalent interactions due to its adjustable strength, excellent recombination reversibility, and excellent biocompatibility (Firdous et al., 2023). For example, Cai et al. utilized hydrogen bonding between tetrafunctional borate ions and hydroxyl groups to provide self-repairing properties, where free polymer chains and the mobility of borate ions enable hydrogen bonding to rapidly trigger self-repairing of the polymer network without the need for external stimuli (e.g., heat, pH, light, catalysts, etc.). And a variety of self-healing conductive hydrogels based on single-walled CNT (SWCNT), graphene, and silver nanowires were demonstrated under ambient conditions (Cai et al., 2017). The SWCNT hydrogel recovered 98% of the original conductivity within a short self-healing time of 3.2 s, and the resistance was also effectively restored after five cutoff healing processes at the same location with high self-healing efficiency.

The strain sensor based on SWCNT conductive hydrogel has a wide response range up to 1000%, good stability, repeatability, and fast signal response (GF value of 1.51) during human movement monitoring, which still has inherent properties after 1000 cycles at 700% strain. To develop self-healing sensors that restore sensing function after damage healing, Su et al.

used a prepenetration method to introduce conductive PANI into a self-healing hydrophobically bonded PAA polymer network and enabled the conductive nanocomposite hydrogels to have excellent mechanical properties through various dynamic interactions between the conductive polymers and the hydrogel network (Su et al., 2021). Multiple dynamic interactions such as hydrogen bonding and electrostatic interactions give nanocomposite conductive hydrogels excellent mechanical properties (tensile up to 2590%, strength up to 0.9 MPa, toughness of 7.85 MJ/m^3, and fracture energy of 4200 J/m^2), self-healing properties, high electrical conductivity, and sensing properties. Notably, the presence of dynamical interactions also compensates the negative effect of enhanced mechanical properties on self-healing properties, achieving a simultaneous enhancement of mechanical, self-healing, and electrical properties.

Metallic—ion coordination interactions, a type of ionic bonding that occurs between anions and cations, can be used as an alternative to intermolecular interactions, and metal—ion coordination has also been widely used to construct dynamical networks through excellent break—recombination reversibility and tunable strength. Inspired by the multilevel structure of biological soft tissues, Liu et al. combined dynamic Fe^{3+}-crosslinked CNC as a "hard" network with covalently cross-linked PVA and polyvinylpyrrolidone as a "soft" network, and prepared CNC with excellent mechanical strength, toughness, and self-healing properties (Liu et al., 2017). Conductive nanocomposite hydrogels have been fabricated with excellent mechanical strength, toughness, and self-healing properties. The combination of "soft" and "hard" layered network structures with reversible coordination bonds resulted in nanocomposite hydrogels with a tensile strength of 2.1 MPa, a toughness of 9.0 MJ/m^3, and a good tensile strain (up to 830%). Moreover, CNC and Fe^{3+} were reconstituted by ionic coordination, which enabled the composite hydrogels to exhibit self-healing and good self-recovery properties within 5 min (30 min) without external stimuli. Based on the tunable electromechanical properties exhibited by nanocomposite hydrogels, they can be used as strain sensors when monitoring human movement, respiration, and even pulse rate changes in different states of motion. For example, inspired by mussels, Heidarian et al. obtained hydrogels with electrical conductivity and self-healing properties without external stimuli by first forming images from carboxyl chitosan, oxidized cellulose nanofibers, and chitin nanofibers, which were then immersed into TA and FeIII solutions to form various dynamic interactions, such as catechol—metal ligand bonds and hydrogen bonds (Heidarian et al., 2022). Respectively, the conductive hydrogel achieved 89% self-healing and 100% self-resuscitation at room temperature due to the synergistic action of dynamic chemical and physical bonds. And the self-healing properties were verified with the aid of macroscopic observations, compression measurements, and rheological resummation. Based on the printed activity, biocompatibility, and self-healing properties of conductive nanocomposite hydrogels, assembled into flexible strain sensors, motion signals can also be monitored.

Host—host interactions are another class of intermolecular interactions valued by scholars, usually arising between host-host parts, which are selectively regulated in the main body part through noncovalent interactions such as π-π stacking and hydrophobic interactions to form inclusion complexes. Chen et al. constructed cyclodextrin topological nanoparticles based on natural and renewable β-cyclodextrin and prepared tough, stretchable as well as self-healing properties hydrogels by means of its use as a physical cross-linking agent through subject-—guest interactions and chain entanglement (Chen et al., 2021c). The mechanical properties and electrical conductivity of the composite hydrogels are further enhanced by the combination of prestretching and salt/PPy, with excellent tensile properties (up to 1860%), tensile strength of 180 kPa, and good fatigue resistance. Based on the reversible host—guest interaction and the irreversible chain entanglement, the composite hydrogel exhibits self-healing property, which is observed after 30 s. And the conductive nanocomposite hydrogel was assembled into a flexible strain sensor with good fatigue resistance as well as a wide sensing range (up to 1800%), which can be used for human activity monitoring. Ren et al. obtained conductive nanocomposite hydrogels with good flexibility, stretchability, self-adhesion, and self-healing properties based on the primary-object complexation of poly(β-cyclodextrin) with adamantane and cross-linking of O-carboxymethyl chitosan via Schiff bonding, cross-linking of PVA via borate bonding with borax as the second cross-linker, and by introducing CNTs into the hydrogel substrate (Ren et al., 2021). Due to the synergistic effects of host—guest interactions, borate bonding, and hydrogen bonding, the composite hydrogel is endowed with the ability of rapid self-healing, and the healing efficiency is as high as 97%—103% (15 s). Sensors based on this composite hydrogel can also be used to monitor the beating of rat heart and the respiratory movement of pig lungs.

8.3.3 Self-adhesion properties

Self-adhesive properties are indispensable for flexible strain sensors and electronic skins, which are important for interfacial connections. The excellent adhesion properties allow for reliable conformational contact with soft human skin, which reduces interfacial resistance and contributes to stable and continuous signal output from portable and wearable flexible sensors (Zheng et al., 2023). In the past, hydrogel needs to stick to the surface of the desired substrate with the help of external adhesives (such as transparent tape, bandage, etc.). This material is strong but bulky, does not fit well on uneven surfaces or human skin with significant deformations, and is poorly ventilated, making it unfavorable for prolonged adhesion to the skin surface. In addition, the detection of certain physiological signals with the help of hydrogel sensors with external adhesives can lose accuracy, stability, and durability to some extent. Therefore, the development of hydrogels with self-adhesive capability that can adhere firmly to various substrates without the aid of external forces is of great practical importance (Zong et al., 2023). To date, a variety of catechol-functionalized adhesion hydrogels have been developed, inspired by natural mussels, with dopamine and TA being the most prominent.

8.3.3.1 Dopamine

Dopamine is the most typical example because its polymer polydopamine (PDA), produced by covalent bonding and noncovalent self-assembly during oxidation, which is structurally similar to the adhesive proteins secreted by mussel-like organisms and has good stability and biocompatibility. And its phenolic and quinone groups are highly reactive toward amine and sulfhydryl groups, which can be reacted by Schiff base reaction, Michael addition, or boric acid–catechol complexation. Li et al. fabricated conductive nanocomposite hydrogels with self-adhesion, self-healing, and biocompatibility by introducing PDA-modified CNTs into an in situ–polymerized AM polymer network (Li et al., 2023d). Due to the π-π stacking generated by PDA and the presence of hydrogen bonding in the system, the conductive hydrogels confer good self-adhesion, which can be directly adhered to human skin without biotoxicity and provide self-healing efficiency of up to 97.3%. In addition, the conductive nanocomposite hydrogel was assembled into a flexible piezoresistive sensor with a GF value of 3.93 at 400% strain, which can be used to monitor human motion by collecting changes in electrical resistance and has a fast response time (76 ms). Owing to contact initiation and electrostatic induction, pressure sensors can linearly sense pressure and thus generate voltage. Using the generated voltage to detect pressure changes, smart rings and gloves can be developed based on multifunctional pressure sensors for signal transmission and pressure distribution sensing. For example, Li et al. inspired by mussel proteins, CNCs, a green nanomaterial with high mechanical strength, were modified by PDA and gold nanoparticles and obtained nanocomposite conductive hydrogels combining biocompatibility, self-repairing properties, and stretchability in the introduction of PVA mesh pyrrole 94. Because of the dynamic interaction between the PVA and PDA and the hydrogen bonding between the PVA molecular chains, the conductive nanocomposite hydrogel is further endowed with a certain self-healing property, which can increase the healing efficiency to 87.6% with time. The sensors based on conductive nanocomposite hydrogels also have good sensing performance with a strain factor GF of 0.99 and a fast response (160 ms). In addition to its application to the monitoring of human motion, the obtained sweat sensor, using lactate oxidase as a model enzyme, exhibited a low detection limit (0.31 mM) and a good linear correlation (0.9987) in the range of 0.5–30 mM, providing a versatile sensing strategy for the monitoring of human motion and sweat sensing detection. Bai et al. proposed an "in situ polymerization and freeze-thaw" approach, in which dopamine was polymerized in situ in aqueous PVA to form an interpenetrating network (Fig. 8.10a) (Bai et al., 2023), and then liquid conductive phytate was introduced to prepare conductive, water-retaining, and self-adhesive nanocomposite hydrogels by cyclic freeze-thawing, exhibiting enhanced ductility, biocompatibility, and homogeneity. The introduction of phytic acid enhances the intermolecular forces with the polymer substrate, and the combination of the catechol group on the PDA and the hydrogen bonding in the system can give good adhesion properties to the hydrogel, which can be directly and stably adhered to the skin surface. Moreover, the hydroxyl groups of phytic acid and water form a large number of hydrogen bonds, which reduces the evaporation of water and has a degree of water retention that can still be used after 4 months. Flexible sensors based on conductive nanocomposite hydrogels can also instantly and accurately monitor small signal changes, such as smile and handwriting recognition. Jing et al.

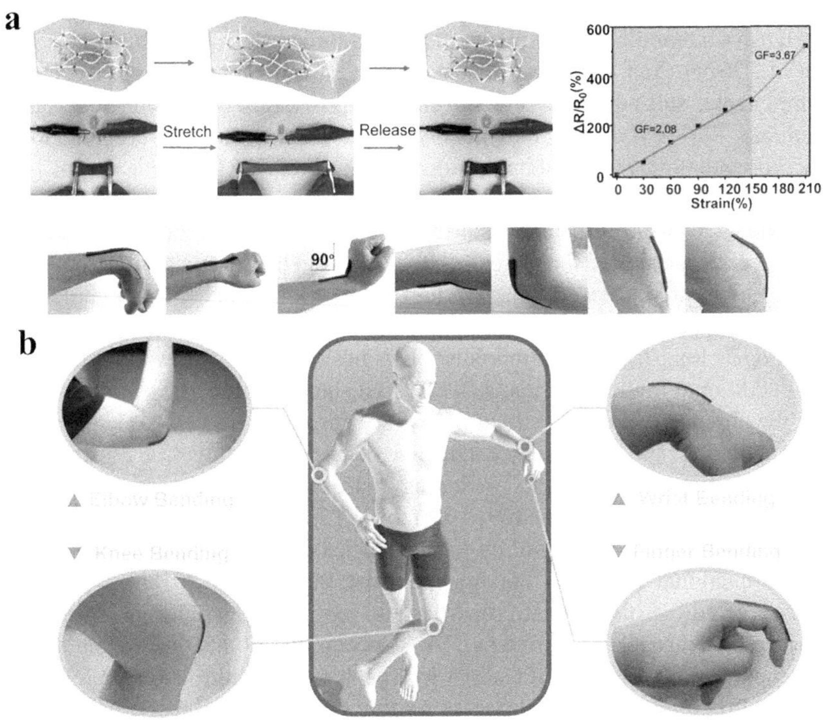

FIGURE 8.10 Self-adhesive properties of nanocomposite conductive hydrogels. (a) Adhesion mechanism of PDA/PA/PVA nanocomposite conductive hydrogels. (b) Preparation process and use as a sensor of TA/PANI@CNCs nanocomposite hydrogels for the monitoring of human body movement. *(a) Reproduced from Bai et al. (2023) with permission from American Chemical Society. (b) Reproduced from Yan et al. (2020) with permission from Elsevier.*

synthesized a mussel-inspired strain-sensitive self-healing hydrogel by preparing a physically and chemically cross-linked network in a two-step process (Jing et al., 2018). First, GO was reduced by dopamine self-polymerization, whereas later acrylic monomers were polymerized in the presence of chemical and physical cross-linkers to form double cross-linked hydrogels. The presence of noncovalent bonds, such as reversible ionic interactions, hydrogen bonding, and metal–ligand bonding in the polymer system, gives the conductive nanocomposite hydrogels better mechanical properties (tensile higher than 600%), adhesion, self-healing properties, and biocompatibility. Among these, biocompatibility is a key metric for evaluating direct contact biosensors, thus demonstrating the feasibility of nanocomposite hydrogels for monitoring human motion with fast, stable, and reproducible signal collection, which has some potential in stretchable self-healing sensors.

8.3.3.2 Tannic acid

TA, which is rich in natural resources, widely available, inexpensive, and carries natural structural units of polyphenols compared to PDA, has received extensive attention in the preparation of biomimetic mussel hydrogels (Fu et al., 2021). Yan et al. prepared conductive nanocomposite hydrogels with stretchable, self-adhesive, and self-healing properties by introducing water-soluble TA/PANI@CNC into a polymer network with different functional acrylic monomers inspired by the adhesion mechanism of mussels (Fig. 8.10b) (Yan et al., 2020). The TA large number of catechol groups in the polymer network can form a metal phenol network, which further improves the mechanical properties of the nanocomposite hydrogels (stretchability over 900% and tensile strength over 700 kPa). Furthermore, the abundant supramolecular interactions and the presence of borate bonds in the polymer network give the hydrogels long-lasting and reproducible adhesion to the skin without the need for additional immobilization, as well as self-healing properties within 30 min in the absence of external stimuli. Flexible sensors assembled based on tough, self-adhesive, and electrically conductive composite hydrogels are also able to monitor minute movements including pulses and small droplets and can exhibit excellent strain sensitivity. Moreover, Liu et al. innovatively prepared a dual network hydrogel based on gelatin methacrylate (GelMA) using TA as a multifunctional hydrogen bonding provider (Liu et al., 2018b). They also investigated the mechanical properties and morphological changes of the hybridized hydrogel at different TA concentrations and treatment times. Compared with the original GelMA hydrogel, after the introduction of TA, the elongation of the hybrid hydrogel is six times higher than the original one, the ultimate stress and compression modulus are substantially increased, and water repellency is demonstrated. Catechols and lignans in the molecular structure of TA have also been exploited to confer adhesion to different substrate surfaces and tissue-like self-healing properties to hybrid hydrogels. To test the feasibility of hybridized hydrogels in vivo, they were first applied to skin wounds and gastric incisions, which demonstrated the ability of the hydrogels to promote wound healing. In addition, the integration of multi-walled CNTs into GelMA-TA hydrogels can be assembled into flexible strain-free electronic skins that monitor finger movements and capture signals. Inspired by the rapid development of skin-inspired multilayer structures, Qu et al. fabricated bilayer conductive composite hydrogels with high stretchability, toughness, and adhesion through a layer-by-layer polymerization process to obtain multilayered hydrogels satisfying mechanical resistance (Qu et al., 2021). This benefits from the synergistic effect of the bilayer structure (a tough layer formed by the integration of CNFs into the PAM network through hydrophobic linking, and the introduction of TA-coated CNFs into the PAA network as an adhesive layer). The hydrophobic linking and the presence of hydrogen bonds in the system give the conductive nanocomposite hydrogels excellent tensile properties (up to 3600%), toughness (up to 200 kPa), and good adhesion performance. As the catechol group in TA binds to the hydroxyl group in CNF and forms a large number of hydrogen bonds, this gives the adhesive layer high adhesion strength and high compliance to various substrates. In addition, biomolecules such as tannins make the composite hydrogels biocompatible to promote wound healing and inhibit bacterial growth. The flexible strain sensors based on conductive nanocomposite hydrogels have stable, immediate, and fast response time (134 ms), which can be prepared for the monitoring of human physiological activities and body movements.

The catechol moiety has a unique structure of hydroxyl and phenolic groups and thus can participate in various interfacial interactions, such as π-π interactions, metal coordination, hydrogen bonding, and thiol reduction (Zhang et al., 2020b). These interactions can confer adhesion properties between conductive nanocomposite hydrogels and substrates, and the adhesion properties between hydrogels and substrates are of great significance for flexible wearable sensors and electronics applications (Guo et al., 2023). However, the adhesion properties of hydrogels are severely weakened by the interaction of large amounts of water and adhesive groups, and although some strategies have been proposed to confer good skin adhesion properties to hydrogels, it is still a challenge to combine strong adhesion properties under humid conditions. Moreover, the water retention, conductivity, and stability of hydrogels in humid environments require simultaneous attention. Therefore, the development of more stable and effective interfacial adhesion will be the focus of future research, which is important for the practical application of flexible wearable sensors and other electronic devices.

8.3.4 Freeze resistance and water retention

To maintain a stable ionic conductivity and complete 3D structure, hydrogels typically have high water content. However, water inevitably freezes into ice at freezing temperatures (Ding et al., 2022). In high-temperature environments, the water in the hydrogels also evaporates rapidly, which can cause the hydrogels to become hard and brittle, losing mechanical properties and electrical conductivity, among others, seriously hindering their practical applications (Yang et al., 2021; Ma et al., 2023). In hydrogel systems, water usually exists in three states of bound, intermediate, and free water. Free water freezes at the normal freezing point, and most of it can be rapidly exchanged with different organic reagents. Intermediate water binds to chemicals in the polymer network and freezes at temperatures below the normal freezing point. Bound water, on the other hand, does not freeze at low temperatures because it is bound to multiple hydrogen bonds in the polymer network. Therefore, the key to improving the freeze-out resistance of hydrogels is to optimize their composition and molecular structure design. Most scholars would address this issue by using organic solvent substitutions or introducing ionic salts to lower the freezing point of the hydrogels.

8.3.4.1 Organic solvent substitution

The composition of the hydrogel, as well as the optimal design of the structure, is crucial for antifreeze properties, and the suppression of ice crystal growth in subzero external environments is key to the development of hydrogel flexible devices that work properly in extreme environments. In general, the evaporation of water is caused by the vapor pressure of the external environment being less than that of the hydrogel, and organic solvents with higher boiling points can reduce the vapor pressure when mixed with water, further enabling water retention and antifreeze functions. EG has been widely used in the development of hydrogel-based flexible wearable electronics due to its advantages of good hygroscopicity and low price. For example, Wen et al. fabricated conductive hydrogels with UV filtration, water retention, antifreezing, and biocompatibility properties by uniformly dispersing TA-coated hydroxyapatite nanowires in a PVA solution with a binary solvent of water and EG (Wen et al., 2021). The binary reagent of EG and water reduces the evaporation of water at high temperatures, as well as the freezing of water at subzero temperatures, and the free motion of aluminium ions in molecular-level water channels also gives the hydrogels a higher temperature. The free movement of aluminium ions in molecular-level water channels also endows the hydrogel with higher electrical conductivity (1.00×10^{-2} S/cm) and sensing properties. Flexible sensors assembled from such conductive hydrogels can also be used for strain, pressure, and temperature sensing. When used as a strain sensor, it has a GF value of 2.84 over a wider strain range of 350%, high linearity (R2 of 0.99003), excellent cyclic stability, and fast response time

(50 ms). Moreover, Zheng et al. prepared conductive nanocomposite hydrogels with excellent antifreeze properties and moisture retention by introducing CNTs into TA and PVA borate polymer networks with water/EG binary solvent (Zheng et al., 2021). The abundant hydrogen bonding and the presence of dynamic covalent disulfide bonds in the binary system lead to composite hydrogels with excellent plasticity and self-healing properties. Moreover, the formation of abundant molecular clusters in the water/EG binary solvent, as well as the formation of hydrogen bonds with the PVA chains, can reduce the water vapor pressure to retain water, which lowers the solidification point of the hydrogel, resulting in the organohydrogel's freezing resistance up to −60°C and its ability to be moisturized for a long period of time within 30 days. The flexible pressure sensor based on this organic hydrogel has good sensitivity (S = 0.625 kPa) and stability (400 cycles). To integrate a conductive hydrogel with high strength, antifreezing, and moisturizing and antimicrobial properties, Bao et al. introduced one-dimensional (1D) silk nanofibers and 2D graphite carbon nitride nanosheets into a PVA polymer network and used a water/EG binary solvent system to inhibit the volatilization and freezing of the free water, which further endowed the organohydrogel with antifreezing properties, flexibility, and conductivity even at −18°C (Bao et al., 2021). The addition of aluminium ions also gives the organohydrogel excellent bactericidal properties, with a more than 99% bactericidal rate against *Escherichia coli* and *Staphylococcus aureus*. Besides, the sandwich flexible sensor based on the conductive composite hydrogel also has a fast response (276 ms) and excellent fatigue resistance (1000 cycles), which can be used to monitor human movement and microexpressions of the face, maintaining the stability of the operation and the accuracy of the output signals even after 20 days of use.

Gly inhibits the growth of ice crystals in subzero temperatures by virtue of its higher viscosity, lower vapor pressure, and higher boiling point compared with similarly hygroscopic EGs, which will therefore confer higher water retention and frost resistance to the material. To combine strong mechanical properties and antifreezing properties, Wen et al. prepared nanocomposite reinforced hydrogels with a sandwich structure by introducing Gly (Wen et al., 2023). The synergy of different sizes and the toughening mechanism resulted in composite hydrogels with excellent mechanical properties (tensile strength of 7.38 MPa, strain at break of 941%, toughness of 31.59 MJ/m^3, and energy at break of 5.41 ± 0.63 kJ/m^2). Furthermore, in the presence of Gly, the composite hydrogel also showed good stability and long-term stability as a sensor. Li et al. introduced MXenes into a chitosan/PAM-PAA covalent network by a one-pot method using a Gly-water binary solvent and prepared a conductive nanocomposite hydrogel with good mechanical properties (tensile strain of ~1000%), self-healing properties, adhesive properties, and temperature resistance in 10 min (Fig. 8.11a) (Li et al., 2021a). The incorporation of MXenes nanosheets not only improves the mechanical properties and conductivity of the conducting hydrogel but also imparts certain self-healing and

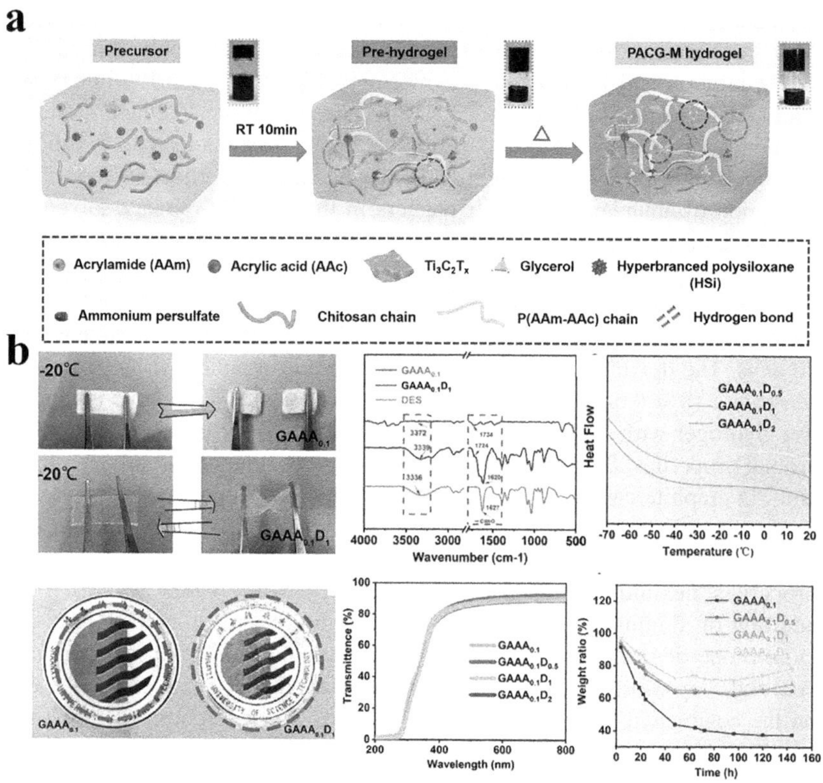

FIGURE 8.11 Antifreezing and water retention properties of nanocomposite conductive hydrogels. (a) Antifreezing, antidrying, and mechanical properties based on Gly-P(AAm-AAc) hydrogel. (b) Network structure of conductive hydrogels based on the DES–water binary system, mechanisms of DES–water interactions, and applications as sensors. *(a) Reproduced from Li et al. (2021a) with permission from Elsevier. (b) Reproduced from Zou et al. (2023) with permission from Elsevier.*

adhesion properties to the hydrogel through the dynamical interactions present in the system. The use of Gly-water binary reagent also makes the conductive hydrogel maintain good elasticity and structural stability in a wide temperature range of −20–80°C. In addition, flexible pressure/strain sensors can also be assembled based on this conductive composite hydrogel, which exhibit stable and reproducible performance over a range of different strains (1%–600%) and stresses (80–3200 Pa), even at low temperatures or in dry environments. Han et al. prepared hydrogels with good mechanical properties, frost and heat resistance, electrical conductivity, environmental stability, and tissue adhesion by using PDA-modified CNTs as fillers in a Gly-water binary system (Fig. 8.11e) (Han et al., 2017). Using quantum mechanical calculations to study the interaction between Gly-water and monomers, the binary system in the polymer network provides more noncovalent bonds that can efficiently dissipate energy, giving the composite hydrogel good mechanical properties and long-term stability, with a 700% tensile property and a toughness of 2300 J/m^2. The use of PDA modifications to promote the dispersion of CNTs on the one hand, and to improve the poor dispersion of the original CNTs in polymer substrates on the other hand, has been shown to give excellent adhesion up to 60 kPa to conductive nanocomposite hydrogels. Composite hydrogel-based electrodes can be used to collect biological signals in extreme environments and also as wearable dressings.

8.3.4.2 Introduction of ionic salts

The introduction of organic solvents not only imparts a certain degree of frost resistance and water retention to the hydrogels but also results in a reduction in electrical conductivity, requiring the introduction of additional conductive components to maintain their conductivity. Inorganic salts, amphoteric ions or ionic liquids, inspired by the fact that they are employed in antiicing applications, are used to confer cryogenic resistance and water retention to hydrogels. Some common inorganic salts, such as $ZnCl_2$, LiCl, $CaCl_2$, etc., have been favored by scholars for the construction of ionically conducting hydrogels with freeze-out resistance and moisture retention. For example, Cai et al. prepared ionic conductive hydrogels with superstretchability, self-adhesion, and antifreezing and moisturizing properties by adding alginate and LiCl to a polymer network copolymerized with acrylic and acrylamide (Cai et al., 2023). The covalent-like hydrogen bonding interactions present in the polymer network and the strong hydration of LiCl endowed the hydrogels with excellent mechanical properties (tensile strains up to 4500%) and ionic conductivity, as well as good self-adhesion to various substrates. Strain sensors based on ionic conductive hydrogels also possess a wide detection range (0%–4200%) and sensitivity (GF = 3.59 in the strain range of 2200%–4200%) for accurate monitoring of muscle movements. Moreover, Peng et al. prepared conductive nanocomposite hydrogels with ultralow-temperature antifreezing, moisture regeneration, and self-adhesion properties by introducing $CaCl_2$ and MXene into a polymer network of PVA and PAM (Peng et al., 2023). The introduction of $CaCl_2$ and MXene nanosheets also endows conducting hydrogels with electronic and ionic conduction channels to compensate for the lack of a single conduction channel. The presence of $CaCl_2$ enables hydrogels with hydrodynamic regeneration and self-adhesion properties, and the excellent water retention can maintain the normal state of the composite hydrogel at room temperature for 70 consecutive days, which makes the flexible wearable sensors based on conductive hydrogel have the potential of long-term stability. To obtain conductive nanocomposite hydrogels with durability and versatility, our group fabricated conductive hydrogels with antifreezing moisture retention and conformational adhesion by introducing $ZnCl_2$, 2D MXenes nanosheets, and TA-modified CNFs into a polymer network copolymerized with betaine and acrylamide in less than 1 min (Guo et al., 2023). The presence of amphiphilic ionic groups, catechol groups, Zn^{2+}, as well as intermolecular interactions and hydrogen bonding resulted in conductive hydrogels with wide environmental adaptability (−60 to 40°C), good electrical conductivity (20°C, 30 mS/cm), and certain mechanical properties (0.311 MPa, 970% strain). Conductive hydrogels can also be assembled based on conductive hydrogels into flexible wearable sensors (for human motion detection, speech and handwriting recognition) and 3D arrays for detecting the magnitude and spatial distribution of strain or force.

High salt concentrations can fulfill the requirements of frost protection and moisture retention, as well as electrical conductivity, while somewhat impairing the mechanical properties of the hydrogels. Inspired by some plants evolving a unique cold resistance free from frost damage, amphipathic proline (ZP), a highly hydration-capable amino acid, can disrupt the original hydrogen-bonding structure between water molecules by forming hydrogen bonds with water molecules. Yue et al. fabricated conductive nanocomposite hydrogels with antifreeze and self-healing properties by chemically modifying CNCs@PDA by polymerizing functional polydopamine on the surface of CNCs, and then introducing ZP and CNCs@PDA into a polymer network of gellan gum/PAA (Yue et al., 2023b). The mechanical properties, antifreeze properties, water retention properties, and stability of the conductive composite hydrogels were greatly enhanced by the introduction of $CaCl_2$, which shielded the gellan gum from electrostatic repulsion and formed strong ionic bonding, combined with the noncovalent bonding formed by functional groups, such as catechol groups and imines on the surface of PDA, the synergistic effect of ZP and nanomaterials, and the hydrogen bonding in the polymer substrate. Even at −30°C, it showed excellent mechanical properties and good self-healing, and its tensile strength also reached 2.7 MPa, tensile strain over 600%, and self-healing efficiency was 82.9%. Li et al. prepared conductive nanocomposite hydrogels with good mechanical properties and antifreeze properties by physically cross-linking GO dispersed into starch/PVA cross-linked network and introducing a green solvent, IL, to further improve the compatibility of the network (Li et al., 2023a). The interplay between different components is also confirmed by combining theoretical simulations and multiple characterization. The synergistic effect of GO and IL containing abundant oxygen-containing groups resulted in composite hydrogels with good tensile properties (>650%), antifreezing properties (<−20°C), and electrical conductivity (1.98 S/m). Based on these advantages, multimodal flexible sensors can be constructed for sensing strain (GF of 6.04), pressure (strain factor: 3.27), and temperature (sensitivity: 0.71%/°C). Similar to ILs in nature, low-eutectic solvents, which are mainly binary and ternary systems consisting of hydrogen bond donors and acceptors, can also be used to construct antifreeze moisturizing hydrogels. Zou et al. prepared conductive ionic hydrogels with good antifreeze, water retention, and self-healing properties using an integrated gelation and solvent exchange strategy with solvent substitution via DES (Fig. 8.11b). This enables the hydrogel to have good adhesion properties and integrated sensing properties for multiple stimuli despite the extreme environment of the DES−water binary system (Zou et al., 2023). Even when frozen at −20°C for 48 h, it still shows good environmental stability and can be used for human sleep monitoring.

8.4 Chapter summary

In recent decades, flexible composites have emerged as promising alternatives to traditional rigid electronic materials due to their soft and stretchable properties. Among these, conductive hydrogels have garnered significant attention in the field of flexible electronics owing to their high electrical conductivity, adjustable mechanical properties, exceptional biocompatibility, and ability to minimize impedance when conforming to curved human bodies. Notably, nanocomposite conductive hydrogels constructed with nanoconductive fillers and polymer networks effectively leverage the advantages of each component, resulting in conductive materials with diverse mechanical and chemical properties that cater to the requirements of next-generation wearable electronics.

Nanocomposite conductive hydrogels, prepared using diverse nanofillers, exhibit distinct mechanical and chemical properties. The incorporation of these fillers, along with innovative preparation strategies, effectively enhances the performance of conductive hydrogels for their successful application in flexible wearable electronic devices. This chapter provides an in-depth discussion on the utilization of hydrogels in the realm of flexible wearable sensors based on nanofiller classification and conductive hydrogel properties. Currently, significant advancements have been made in both fundamental research and practical applications of conductive hydrogels employing various nanocomposite fillers. Furthermore, apart from electrical conductivity improvements, they have also achieved multifunctionality, including exceptional mechanical properties, self-healing capabilities, adhesion characteristics, and antifreeze attributes. This progress represents a significant advancement in the development of fields such as flexible wearable sensors and electronic devices. Moreover, there has been a substantial increase in the number of publications and papers within this domain over recent decades. This surge indicates that the electronics industry recognizes the potential of flexible, self-healing, adhesive, and frost-resistant sensing technologies. It also highlights the motivation among numerous academics worldwide to propel further advancements. Despite these notable achievements, it is important to note that the development of nanocomposite conductive hydrogels for flexible wearable sensors is still ongoing with considerable strides yet to be made before large-scale production can be realized.

Firstly, hydrogels face the challenge of inevitable water evaporation, which poses a practical obstacle for flexible electronics and necessitates the development of environmentally resistant hydrogel sensors. Although certain strategies have successfully mitigated this issue to some extent for a limited duration, optimizing both properties simultaneously remains challenging. The high and stable electrical conductivity of hydrogels typically arises from either ionic or electronic conduction mechanisms. In the case of ionic hydrogels, their electrical conductivity relies on the unhindered movement of ions within the gel matrix; however, extreme cold and heat significantly impact conductivity. As a result, frost resistance and water retention are crucial factors enabling hydrogel sensors to operate effectively in harsh environments. In the case of electron-conducting hydrogels, conductive nanofillers or polymer nanofillers are typically incorporated in conjunction with conductive components. However, excessive addition of these fillers tends to compromise the mechanical properties of hydrogels, necessitating a delicate balance between mechanical strength and electrical conductivity. Hence, there is an imperative need for the development of uncomplicated yet effective strategies to fabricate nanocomposite conductive hydrogels that exhibit excellent mechanical properties, electrical conductivity, water retention capacity, and temperature resistance. Continuous efforts should be dedicated toward achieving this goal.

Secondly, despite the successful enhancement of hydrogel performance and its application in sensor devices through the integration of conductive nanocomposite fillers and innovative preparation strategies, there still exist certain challenges pertaining to processability and chemical stability. In terms of processability, one-dimensional metal nanowires possess a high aspect ratio and exceptional electrical conductivity, enabling them to effectively respond to significant external stimuli. Nonetheless, their large size and weight render them susceptible to precipitation and nonuniform dispersion, necessitating chemical or physical modifications for optimal utilization in sensors or other electronic domains. With respect to the following chemical stability, hydrogels based on two-dimensional material MXenes exhibit favorable performance in the short term. Nevertheless, over time, the conductive filler is susceptible to oxidation, thereby compromising the stability and reliability of sensors in practical applications. Accordingly, it is crucial to enhance the stability of MXenes and develop scalable synthesis methods with high yields. Despite extensive efforts by researchers to improve the oxidation resistance of MXenes through approaches such as polyanionic group adsorption and silylation treatment, achieving large-scale production of nonoxidized MXenes remains a long-standing challenge that necessitates further exploration by scholars.

Finally, during physical exercise, the human skin undergoes deformation. To ensure precise and real-time capture and monitoring of external stimuli, it is imperative to maintain close contact between the hydrogel and the human skin. However, a significant challenge that necessitates further exploration lies in ensuring sustained adhesion and comfort of the hydrogel with the skin despite sweat accumulation postmovement. This challenge must be addressed to meet the demands for accuracy and immediacy in practical applications of flexible sensors. Furthermore, in this era dominated by Internet of things and artificial intelligence, there is an increasing emphasis on pursuing intelligent, humanized, and multifunctional electronic products as a mainstream direction. Nevertheless, the development of lightweight and multifunctional nanocomposite conductive hydrogel sensors holds great promise. Nonetheless, for practical applications, it is imperative to develop effective performance integration and technological additives such as excellent biocompatibility and biodegradability, real-time wireless sensing capabilities, etc., to steer these sensors toward microscopic, integrated, intelligent, and systematic advancements.

Acknowledgments

The financial support from the Beijing Nova Program (20230484431) and the Opening Project of State Silica-based Materials Laboratory of Anhui Province (2022KF12) is gratefully acknowledged.

References

Ahmed, A., Sharma, S., Adak, B., Hossain, M.M., LaChance, A.M., Mukhopadhyay, S., Sun, L., 2022. InfoMat 4, e12295.

Bai, Y., Yan, S., Wang, Y., Wang, Q., Duan, X., 2023. Facile preparation of a self-adhesive conductive hydrogel with long-term usability. ACS Appl. Mater. Interfaces 15, 48744–48753.

Bao, S., Gao, J., Xu, T., Li, N., Chen, W., Lu, W., 2021. Chem. Eng. J. 411, 128470.

Cai, G., Wang, J., Qian, K., Chen, J., Li, S., Lee, P.S., 2017. Adv. Sci. 4, 1600190.

Cai, H., Zhang, D., Zhang, H., Tang, M., Xu, Z., Xia, H., Li, K., Wang, J., 2023. Chem. Eng. J. 472, 144849.

Cao, L., Zhao, Z., Li, J., Yi, Y., Wei, Y., 2022. Biomacromolecules 23, 1278–1290.

Chae, A., Murali, G., Lee, S.Y., Gwak, J., Kim, S.J., Jeong, Y.J., Kang, H., Park, S., Lee, A.S., Koh, D.Y., In, I., Park, S.J., 2023. Highly oxidation-resistant and self-healable MXene-based hydrogels for wearable strain sensor. Adv. Funct. Mater. 33, 2213382.

Chai, X., Tang, J., Li, Y., Cao, Y., Chen, X., Chen, T., Zhang, Z., 2023. ACS Appl. Mater. Interfaces 15, 18262–18271.

Chen, J., Peng, Q., Thundat, T., Zeng, H., 2019. Chem. Mater. 31, 4553–4563.

Chen, Q., Tang, H., Liu, J., Wang, R., Sun, J., Yao, J., Shao, Z., Chen, X., 2021a. Chem. Eng. J. 422, 130091.

Chen, H., Huang, J., Liu, J., Gu, J., Zhu, J., Huang, B., Bai, J., Guo, J., Yang, X., Guan, L., 2021b. J. Mater. Chem. A 9, 23243–23255.

Chen, J., Xu, X., Liu, M., Li, Y., Yu, D., Lu, Y., Xiong, M., Wyman, I., Xu, X., Wu, X., 2021c. Carbohydr. Polym. 264, 117978.

Chen, C., Feng, J., Li, J., Guo, Y., Shi, X., Peng, H., 2022. Chem. Rev. 123, 613–662.

Cheng, Y., Zang, J., Zhao, X., Wang, H., Hu, Y., 2022. Carbohydr. Polym. 277, 118872.

Cui, S., Zhang, S., Zhang, F., Lin, R., Tang, C., Jing, X., 2024. Tannic acid-coated cellulose nanocrystal-reinforced transparent multifunctional hydrogels with UV-filtering for wearable flexible sensors. Carbohydr. Polym. 323, 121385.

Deng, Z., Guo, Y., Zhao, X., Ma, P.X., Guo, B., 2018. Multifunctional stimuli-responsive hydrogels with self-healing, high conductivity, and rapid recovery through host-guest interactions. Chem. Mater. 30, 1729–1742.

Ding, Q., Zhou, Z., Wang, H., Wu, Z., Tao, K., Yang, B.R., Xie, X., Fu, J., Wu, J., 2022. Self-healable, recyclable, ultrastretchable, and high-performance NO2 sensors based on an organohydrogel for room and sub-zero temperature and wireless operation. SmartMat 4, e1141.

Fang, H., Zheng, P., Ma, R., Xu, C., Yang, G., Wang, Q., Wang, H., 2018. Mater. Horiz. 5, 1000–1007.

Fang, L., Zhang, J., Wang, W., Zhang, Y., Chen, F., Zhou, J., Chen, F., Li, R., Zhou, X., Xie, Z., 2020. ACS Appl. Mater. Interfaces 12, 56393–56402.

Fang, Y., Xu, J., Gao, F., Du, X., Du, Z., Cheng, X., Wang, H., 2021. Compos. B Eng. 219, 108965.

Feng, Y., Liu, H., Zhu, W., Guan, L., Yang, X., Zvyagin, A.V., Zhao, Y., Shen, C., Yang, B., Lin, Q., 2021. Adv. Funct. Mater. 31, 2105264.

Firdous, I., Fahim, M., Mushtaq, F., Daoud, W.A., 2023. Nano Energy 116, 108817.

Fu, Q., Hao, S., Meng, L., Xu, F., Yang, J., 2021. Engineering self-adhesive polyzwitterionic hydrogel electrolytes for flexible zinc-ion hybrid capacitors with superior low-temperature adaptability. ACS Nano 15, 18469–18482.

Fu, H., Wang, B., Li, J., Xu, J., Li, J., Zeng, J., Gao, W., Chen, K., 2022. Mater. Horiz. 9, 1412–1421.

Ge, G., Lu, Y., Qu, X., Zhao, W., Ren, Y., Wang, W., Wang, Q., Huang, W., Dong, X., 2020. Muscle-inspired self-healing hydrogels for strain and temperature sensor. ACS Nano 14, 218–228.

Geng, L., Liu, W., Fan, B., Wu, J., Shi, S., Huang, A., Hu, J., Peng, X., 2023. Chem. Eng. J. 462, 142226.

Gu, J., Huang, J., Chen, G., Hou, L., Zhang, J., Zhang, X., Yang, X., Guan, L., Jiang, X., Liu, H., 2020. ACS Appl. Mater. Interfaces 12, 40815–40827.

Guo, M., Yang, X., Yan, J., An, Z., Wang, L., Wu, Y., Zhao, C., Xiang, D., Li, H., Li, Z., Zhou, H., 2022. J. Mater. Chem. A 10, 16095–16105.

Guo, W.-Y., Mai, T., Huang, L.-Z., Zhang, W., Qi, M.-Y., Yao, C., Ma, M.-G., 2023. ACS Appl. Mater. Interfaces 15, 24933–24947.

Han, L., Liu, K., Wang, M., Wang, K., Fang, L., Chen, H., Zhou, J., Lu, X., 2017. Adv. Funct. Mater. 28, 1704195.

Han, L., Yan, L., Wang, M., Wang, K., Fang, L., Zhou, J., Fang, J., Ren, F., Lu, X., 2018. Chem. Mater. 30, 5561–5572.

Han, X., Xiao, G., Wang, Y., Chen, X., Duan, G., Wu, Y., Gong, X., Wang, H., 2020. J. Mater. Chem. A 8, 23059–23095.

Han, X., Lv, Z., Ran, F., Dai, L., Li, C., Si, C., 2021. Int. J. Biol. Macromol. 176, 78–86.

Hardman, D., Hughes, J., Thuruthel, T.G., Gilday, K., Iida, F., 2021. IEEE Rob. Autom. Lett. 6, 5269–5275.

He, X., Shi, J., Hao, Y., He, M., Cai, J., Qin, X., Wang, L., Yu, J., 2022a. Carbon Energy 4, 621–632.

He, S., Guo, B., Sun, X., Shi, M., Zhang, H., Yao, F., Sun, H., Li, J., 2022b. ACS Appl. Mater. Interfaces 14, 45869–45879.

He, X.-F., Zeng, Z.-F., Ni, Q.-Y., Xu, Z.-C., Mao, P.-F., Jiang, B., Wu, Q., Wang, B., Gong, L.-X., Tang, L.-C., Li, S.-N., 2023. Mechanical robust and highly conductive composite hydrogel reinforced by a combination of cellulose nanofibrils/polypyrrole toward high-performance strain sensor. Compos. B Eng. 266, 111022.

Heidarian, P., Gharaie, S., Yousefi, H., Paulino, M., Kaynak, A., Varley, R., Kouzani, A.Z., 2022. Carbohydr. Polym. 291, 119545.

Huang, J., Zhao, M., Cai, Y., Zimniewska, M., Li, D., Wei, Q., 2019. Adv. Electron. Mater. 6, 1900934.

Huang, X., Wang, L., Shen, Z., Ren, J., Chen, G., Li, Q., Zhou, Z., 2022. Super-Stretchable and Self-Healing hydrogel with a Three-Dimensional silver nanowires network structure for wearable sensor and electromagnetic interference shielding. Chem. Eng. J. 446, 137136.

Ionov, L., 2014. Mater. Today 17, 494–503.

Jamatia, T., Matyas, J., Olejnik, R., Danova, R., Maloch, J., Skoda, D., Slobodian, P., Kuritka, I., 2023. Polymers 15, 1618.

Jiang, L., He, S., Liu, A., Zhang, J., Liu, J., He, S., Shao, W., 2022. Preparation and characterization of self-healable and wearable hydrogels with ultrasensitive sensing performances. Compos. B Eng. 239, 109982.

Jiao, Y., Lu, Y., Lu, K., Yue, Y., Xu, X., Xiao, H., Li, J., Han, J., 2021. J. Colloid Interface Sci. 597, 171–181.

Jing, X., Mi, H.-Y., Peng, X.-F., Turng, L.-S., 2018. Carbon 136, 63–72.

Karimzadeh, Z., Mahmoudpour, M., Rahimpour, E., Jouyban, A., 2022. Adv. Colloid Interface Sci. 305, 102705.

Lee, K.H., Zhang, Y.Z., Kim, H., Lei, Y., Hong, S., Wustoni, S., Hama, A., Inal, S., Alshareef, H.N., 2021. Small Methods 5, e2100819.

Li, S.-N., Yu, Z.-R., Guo, B.-F., Guo, K.-Y., Li, Y., Gong, L.-X., Zhao, L., Bae, J., Tang, L.-C., 2021a. Nano Energy 90, 106502.

Li, X., He, L., Li, Y., Chao, M., Li, M., Wan, P., Zhang, L., 2021b. Constructing heterogeneous structure in metal-organic framework-derived hierarchical sulfur hosts for capturing polysulfides and promoting conversion kinetics. ACS Nano 15, 7765–7773.

Li, J., Yang, P., Li, X., Jiang, C., Yun, J., Yan, W., Liu, K., Fan, H.J., Lee, S.W., 2022a. ACS Energy Lett. 8, 1–8.

Li, M., Chen, D., Sun, X., Xu, Z., Yang, Y., Song, Y., Jiang, F., 2022b. Carbohydr. Polym. 284, 119199.

Li, X., Zhang, S., Li, X., Lu, L., Cui, B., Yuan, C., Guo, L., Yu, B., Chai, Q., 2023a. Starch/polyvinyl alcohol with ionic liquid/graphene oxide enabled highly tough, conductive and freezing-resistance hydrogels for multimodal wearable sensors. Carbohydr. Polym. 320, 121262.

Li, B., Chen, Y., Wu, W., Cao, X., Luo, Z., 2023b. Carbohydr. Polym. 317, 121092.

Li, P., Ling, Z., Liu, X., Bai, L., Wang, W., Chen, H., Yang, H., Yang, L., Wei, D., 2023c. Chem. Eng. J. 466, 143306.

Li, Y., Yang, D., Wu, Z., Gao, F.-L., Gao, X.-Z., Zhao, H.-Y., Li, X., Yu, Z.-Z., 2023d. Nano Energy 109, 108324.

Liao, H., Guo, X., Wan, P., Yu, G., 2019. Adv. Funct. Mater. 29, 1904507.

Lin, F., Wang, Z., Shen, Y., Tang, L., Zhang, P., Wang, Y., Chen, Y., Huang, B., Lu, B., 2019. J. Mater. Chem. A 7, 26442–26455.

Ling, Q., Ke, T., Liu, W., Ren, Z., Zhao, L., Gu, H., 2021. Ind. Eng. Chem. Res. 60, 18373–18383.

Liu, Y.J., Cao, W.T., Ma, M.G., Wan, P., 2017. ACS Appl. Mater. Interfaces 9, 25559–25570.

Liu, H., Li, M., Ouyang, C., Lu, T.J., Li, F., Xu, F., 2018a. Small 14, 1801711.

Liu, B., Wang, Y., Miao, Y., Zhang, X., Fan, Z., Singh, G., Zhang, X., Xu, K., Li, B., Hu, Z., Xing, M., 2018b. Biomaterials 171, 83–96.

Liu, C., Wang, X., Zhang, H.J., You, X., Yue, O., 2021. ACS Appl. Mater. Interfaces 13, 36240–36252.

Liu, L., Tanguy, N.R., Yan, N., Wu, Y., Liu, X., Qing, Y., 2022a. Carbohydr. Polym. 280, 119005.

Liu, D., Zhou, H., Zhao, Y., Huyan, C., Wang, Z., Torun, H., Guo, Z., Dai, S., Xu, B.B., Chen, F., 2022b. Small 18, 2203258.

Liu, C., Zhang, R., Li, P., Qu, J., Chao, P., Mo, Z., Yang, T., Qing, N., Tang, L., 2022c. ACS Appl. Mater. Interfaces 14, 26088–26098.

Liu, D., Huyan, C., Wang, Z., Guo, Z., Zhang, X., Torun, H., Mulvihill, D., Xu, B.B., Chen, F., 2023. Mater. Horiz. 10, 2800–2823.

Long, Y., Zhang, Z., Sun, K., Wang, C., Zeng, N., Gao, B., Tang, X., Qi, X., Fan, R., 2023. Adv. Compos. Hybrid Mater. 6, 173.

Lu, J., Han, X., Dai, L., Li, C., Wang, J., Zhong, Y., Yu, F., Si, C., 2020. Carbohydr. Polym. 250, 117010.

Lu, J., Hu, S., Li, W., Wang, X., Mo, X., Gong, X., Liu, H., Luo, W., Dong, W., Sima, C., Wang, Y., Yang, G., Luo, J.-T., Jiang, S., Shi, Z., Zhang, G., 2022. ACS Nano 16, 3744–3755.

Ma, Y., Gao, Y., Liu, L., Ren, X., Gao, G., 2020. Chem. Mater. 32, 8938–8946.

Ma, Y., Zhang, D., Wang, Z., Zhang, H., Xia, H., Mao, R., Cai, H., Luan, H., 2023. ACS Appl. Mater. Interfaces 15, 29413–29424.

Naahidi, S., Jafari, M., Logan, M., Wang, Y., Yuan, Y., Bae, H., Dixon, B., Chen, P., 2017. Biotechnol. Adv. 35, 530–544.

Nie, Y., Yue, D., Xiao, W., Wang, W., Chen, H., Bai, L., Yang, L., Yang, H., Wei, D., 2022. Chem. Eng. J. 436, 135243.

Pan, X., Wang, Q., He, P., Liu, K., Ni, Y., Chen, L., Ouyang, X., Huang, L., Wang, H., Xu, S., 2020. Chem. Eng. J. 379, 122271.

Pang, J., Wang, L., Xu, Y., Wu, M., Wang, M., Liu, Y., Yu, S., Li, L., 2020. Carbohydr. Polym. 240, 116360.

Park, C., Kim, M.S., Kim, H.H., Sunwoo, S.-H., Jung, D.J., Choi, M.K., Kim, D.-H., 2022a. Appl. Phys. Rev. 9, 021312.

Park, J., Jeon, N., Lee, S., Choe, G., Lee, E., Lee, J.Y., 2022b. Chem. Eng. J. 446, 137344.

Pei, X., Zhang, H., Zhou, Y., Zhou, L., Fu, J., 2020. Mater. Horiz. 7, 1872–1882.

Peng, W., Han, L., Huang, H., Xuan, X., Pan, G., Wan, L., Lu, T., Xu, M., Pan, L., 2020. J. Mater. Chem. A 8, 26109–26118.

Peng, W., Pan, X., Liu, X., Gao, Y., Lu, T., Li, J., Xu, M., Pan, L., 2023. J. Colloid Interface Sci. 634, 782–792.

Qian, C., Li, Y., Chen, C., Han, L., Han, Q., Liu, L., Lu, Z., 2023. Chem. Eng. J. 454, 140263.

Qin, Z., Zhao, G., Zhang, Y., Gu, Z., Tang, Y., Aladejana, J.T., Ren, J., Jiang, Y., Guo, Z., Peng, X., Zhang, X., Xu, B.B., Chen, T., 2023. Small 19, 2303038.

Qu, X., Wang, S., Zhao, Y., Huang, H., Wang, Q., Shao, J., Wang, W., Dong, X., 2021. Skin-inspired highly stretchable, tough and adhesive hydrogels for tissue-attached sensor. Chem. Eng. J. 425, 131523.

Rahmani, P., Shojaei, A., 2021. Adv. Colloid Interface Sci. 298, 102553.

Ren, Z., Ke, T., Ling, Q., Zhao, L., Gu, H., 2021. Carbohydr. Polym. 273, 118533.

Salimiyan, N., Gholami, M., Sedghi, R., 2023. Chem. Eng. J. 471, 144648.

Shen, Z., Zhang, Z., Zhang, N., Li, J., Zhou, P., Hu, F., Rong, Y., Lu, B., Gu, G., 2022. Adv. Mater. 34, e2203650.

Song, M., Yu, H., Zhu, J., Ouyang, Z., Abdalkarim, S.Y.H., Tam, K.C., Li, Y., 2020. Constructing stimuli-free self-healing, robust and ultrasensitive biocompatible hydrogel sensors with conductive cellulose nanocrystals. Chem. Eng. J. 398, 125547.

Su, G., Yin, S., Guo, Y., Zhao, F., Guo, Q., Zhang, X., Zhou, T., Yu, G., 2021. Mater. Horiz. 8, 1795–1804.

Sun, Z., Dong, C., Chen, B., Li, W., Hu, H., Zhou, J., Li, C., Huang, Z., 2023. Small 19, 2303612.

Tang, J., Javaid, M.U., Pan, C., Yu, G., Berry, R.M., Tam, K.C., 2020. Carbohydr. Polym. 229, 115486.

Wang, Y., Song, C., Yu, X., Liu, L., Han, Y., Chen, J., Fu, J., 2017. Compos. Sci. Technol. 151, 139–146.

Wang, J., Tang, F., Wang, Y., Lu, Q., Liu, S., Li, L., 2020. ACS Appl. Mater. Interfaces 12, 1558–1566.

Wang, S., Xiang, J., Sun, Y., Wang, H., Du, X., Cheng, X., Du, Z., Wang, H., 2021. Skin-inspired nanofibrillated cellulose-reinforced hydrogels with high mechanical strength, long-term antibacterial and self-recovery ability for wearable strain/pressure sensors. Carbohydr. Polym. 261, 117894.

Wang, D., Zhang, D., Tang, M., Zhang, H., Sun, T., Yang, C., Mao, R., Li, K., Wang, J., 2022a. Nano Energy 100, 107509.

Wang, M., Chen, H., Li, X., Wang, G., Peng, C., Wang, W., Zhang, F., Wang, J., Liu, H., Yan, G., Qin, H., 2022b. J. Mater. Chem. A 10, 24096–24105.

Wang, Z., Bu, M., Xiu, K., Sun, J., Hu, N., Zhao, L., Gao, L., Kong, F., Zhu, H., Song, J., Lau, D., 2022c. A flexible, stretchable and triboelectric smart sensor based on graphene oxide and polyacrylamide hydrogel for high precision gait recognition in Parkinsonian and hemiplegic patients. Nano Energy 104, 107978.

Wang, C., Liu, Y., Qu, X., Shi, B., Zheng, Q., Lin, X., Chao, S., Wang, C., Zhou, J., Sun, Y., Mao, G., Li, Z., 2022d. Adv. Mater. 34, e2105416.

Wang, S., Bi, S., Zhang, L., Liu, R., Wang, H., Gu, J., 2023a. J. Mater. Chem. A 11, 14096–14107.

Wang, J., Cheng, H., Chen, W., Han, P., Yao, X., Tang, B., Duan, W., Li, P., Wei, X., Chu, P.K., Zhang, X., 2023b. Chem. Eng. J. 452, 139474.

Wang, Y., Qin, W., Yang, M., Tian, Z., Guo, W., Sun, J., Zhou, X., Fei, B., An, B., Sun, R., Yin, S., Liu, Z., 2023c. Adv. Funct. Mater. 33, 2301587.

Wen, J., Tang, J., Ning, H., Hu, N., Zhu, Y., Gong, Y., Xu, C., Zhao, Q., Jiang, X., Hu, X., Lei, L., Wu, D., Huang, T., 2021. Adv. Funct. Mater. 31, 2011176.

Wen, J., Wu, Y., Gao, Y., Su, Q., Liu, Y., Wu, H., Zhang, H., Liu, Z., Yao, H., Huang, X., Tang, L., Shi, Y., Song, P., Xue, H., Gao, J., 2023. Nano-Micro Lett. 15, 174.

Xia, W., Yu, Y., Zhou, C., Wang, W., Wu, Z., Chen, H., 2023. J. Mater. Chem. C. https://doi.org/10.1039/d3tc03708h.

Xu, C., Lu, H., Liu, Z., Luo, N., Wei, A., 2023a. J. Mater. Sci. Mater. Electron. 34, 906.

Xu, L., Chen, Y., Yu, M., Hou, M., Gong, G., Tan, H., Li, N., Xu, J., 2023b. Nano Energy 107, 108119.

Xu, J., Zhang, H., Guo, Z., Zhang, C., Tan, H., Gong, G., Yu, M., Xu, L., 2023c. Int. J. Biol. Macromol. 230, 123195.

Xu, Q., Wu, Z., Zhao, W., He, M., Guo, N., Weng, L., Lin, Z., Taleb, M.F.A., Ibrahim, M.M., Singh, M.V., Ren, J., El-Bahy, Z.M., 2023d. Adv. Compos. Hybrid Mater. 6, 203.

Yan, Q., Zhou, M., Fu, H., 2020. Study on mussel-inspired tough TA/PANI@CNCs nanocomposite hydrogels with superior self-healing and self-adhesive properties for strain sensors. Compos. B Eng. 201, 108356.

Yang, Y., Yang, Y., Cao, Y., Wang, X., Chen, Y., Liu, H., Gao, Y., Wang, J., Liu, C., Wang, W., Yu, J.-K., Wu, D., 2021. Chem. Eng. J. 403, 126431.

Yang, X., Zhang, B., Li, J., Shen, M., Liu, H., Xu, X., Shang, S., 2023. Carbohydr. Polym. 313, 120813.

Yin, J., Liu, N., Jia, P., Ren, Z., Zhang, Q., Lu, W., Yao, Q., Deng, M., Gao, Y., 2023a. SusMat 3, 859–876.

Yin, R., Zhang, C., Chen, Y., Wang, Y., Feng, Q., Liu, Y., Yu, M., Yuan, Y., Xu, C.-Y., Liu, F., Zhao, W., 2023b. Chem. Eng. J. 475, 145794.

You, L., Shi, X., Cheng, J., Yang, J., Xiong, C., Ding, Z., Zheng, Z., Wang, S., Wang, J., 2022. J. Colloid Interface Sci. 625, 197–209.

Yu, K.J., Yan, Z., Han, M., Rogers, J.A., 2017. npj Flexible Electronics 1, 4.

Yu, J., Feng, Y., Sun, D., Ren, W., Shao, C., Sun, R., 2022a. ACS Appl. Mater. Interfaces 14, 10886–10897.

Yu, X., Qin, W., Li, X., Wang, Y., Gu, C., Chen, J., Yin, S., 2022b. J. Mater. Chem. A 10, 15000–15011.

Yue, J., Li, C., Ji, X., Tao, Y., Lu, J., Cheng, Y., Du, J., Wang, H., 2023a. Chem. Eng. J. 466, 143358.

Yue, D., Chen, Y., Wu, Y., Chen, H., Bai, L., Wang, W., Yang, H., Yang, L., Wei, D., 2023b. Sustain. Mater. Technol. 37, e00653.

Zhang, Q., Liu, X., Duan, L., Gao, G., 2019. Chem. Eng. J. 365, 10–19.

Zhang, H., Ren, P., Yang, F., Chen, J., Wang, C., Zhou, Y., Fu, J., 2020a. J. Mater. Chem. B 8, 10549–10558.

Zhang, W., Wang, R., Sun, Z., Zhu, X., Zhao, Q., Zhang, T., Cholewinski, A., Yang, F., Zhao, B., Pinnaratip, R., Forooshani, P.K., Lee, B.P., 2020b. Chem. Soc. Rev. 49, 433–464.

Zhang, Y., Ren, E., Li, A., Cui, C., Guo, R., Tang, H., Xiao, H., Zhou, M., Qin, W., Wang, X., Liu, L., 2021a. J. Mater. Chem. B 9, 719–730.

Zhang, Z., Tang, L., Chen, C., Yu, H., Bai, H., Wang, L., Qin, M., Feng, Y., Feng, W., 2021b. J. Mater. Chem. A 9, 875–883.

Zhang, Q., Wang, Q., Wang, G., Zhang, Z., Xia, S., Gao, G., 2021c. ACS Appl. Mater. Interfaces 13, 50411–50421.

Zhang, X., Zhang, Y., Zhang, W., Dai, Y., Xia, F., 2021d. Chem. Eng. J. 420, 130447.

Zhang, H., Zhang, D., Zhang, B., Wang, D., Tang, M., 2022a. ACS Appl. Mater. Interfaces 14, 48907–48916.

Zhang, L., Wang, J., Wang, S., Wang, L., Wu, M., 2022b. J. Mater. Chem. C 10, 4327–4335.

Zhang, D., Jian, J., Xie, Y., Gao, S., Ling, Z., Lai, C., Wang, J., Wang, C., Chu, F., Dumont, M.-J., 2022c. Chem. Eng. J. 427, 130921.

Zhang, C., Wang, M., Jiang, C., Zhu, P., Sun, B., Gao, Q., Gao, C., Liu, R., 2022d. Nano Energy 95, 106991.

Zhang, W., Ji, X.X., Ma, M.G., 2023a. Emerging MXene/cellulose composites: design strategies and diverse applications. Chem. Eng. J. 458, 141402.

Zhang, C., Nayeem, M.O.G., Wang, Z., Pu, X., Dagdeviren, C., Wang, Z.L., Zhang, X., Liu, R., 2023b. Conductive hydrogels for bioenergy harvesting and self-powered application. Prog. Mater. Sci. 138, 101156.

Zhang, D., Tang, Y., Gong, X., Chang, Y., Zheng, J., 2023c. SmartMat, e1160. https://doi.org/10.1002/smm2.1160.

Zhang, H., Zhang, D., Wang, Z., Xi, G., Mao, R., Ma, Y., Wang, D., Tang, M., Xu, Z., Luan, H., 2023d. ACS Appl. Mater. Interfaces 15, 5128–5138.

Zhang, Z., Chen, G., Xue, Y., Duan, Q., Liang, X., Lin, T., Wu, Z., Tan, Y., Zhao, Q., Zheng, W., Wang, L., Wang, F., Luo, X., Xu, J., Liu, J., Lu, B., 2023e. Adv. Funct. Mater. 33, 2305705.

Zhao, D., Zhu, Y., Cheng, W., Chen, W., Wu, Y., Yu, H., 2021. Adv. Mater. 33, e2000619.

Zhao, L., Xu, H., Liu, L., Zheng, Y., Han, W., Wang, L., 2023. Adv. Sci. 10, 2303922.

Zheng, W., Xu, L., Li, Y., Huang, Y., Li, B., Jiang, Z., Gao, G., 2021. J. Colloid Interface Sci. 594, 584–592.

Zheng, H., Chen, M., Sun, Y., Zuo, B., 2022. Chem. Eng. J. 446, 136931.

Zheng, Y., Baidya, A., Annabi, N., 2023. Bioact. Mater. 29, 214–229.

Zhou, J., Zhuo, F., Long, X., Liu, Y., Lu, H., Luo, J., Chen, L., Dong, S., Fu, Y., Duan, H., 2022. Chem. Eng. J. 447, 137259.

Zhu, J., Tao, J., Yan, W., Song, W., 2023. Natl. Sci. Rev. 10, nwad180.

Zhuo, F., Zhou, J., Liu, Y., Xie, J., Chen, H., Wang, X., Luo, J., Fu, Y., Elmarakbi, A., Duan, H., 2023. Adv. Funct. Mater. 33, 2308487.

Zong, S., Lv, H., Liu, C., Zhu, L., Duan, J., Jiang, J., 2023. Chem. Eng. J. 465, 142831.

Zou, X., Wang, X., Bai, Z., Yue, O., Wei, C., Xie, L., Zhang, H., Liu, X., 2023. Environment-tolerant versatile ion-conductive eutectic hydrogels for self-powered wearable flexible sensors. Chem. Eng. J. 463, 142349.

Chapter 9

Promising cellulose-based aerogel composites: Preparation methods and advanced applications

Tian Mai,[1,#] Peilin Wang[1,#] and Mingguo Ma[1,2,*]
[1]*MOE Engineering Research Center of Forestry Biomass Materials and Bioenergy, Research Center of Biomass Clean Utilization, Beijing Key Laboratory of Lignocellulosic Chemistry, College of Materials Science and Technology, Beijing Forestry University, Beijing, PR China;*
[2]*State Silica-based Materials Laboratory of Anhui Province, Bengbu, PR China*
[*]*Corresponding author: E-mail: mg_ma@bjfu.edu.cn.*

9.1 Introduction

With the advancement of society, the integration of high-strength and lightweight functional materials has attracted significant attention in the field of materials science, emerging as a prominent research focus. The pursuit of lightweight high-strength materials has become a crucial area across various sectors of production within society (Nepal et al., 2023). For instance, in the automotive industry, the development of such materials can enhance driving efficiency, braking performance, and vehicle crashworthiness (Zhang and Xu, 2022). In military applications, such materials are essential for optimizing soldier mobility when designing body armor (Yin et al., 2022). Aerogel stands out as an extraordinary material with its three-dimensional (3D) (see Table 9.1) solid–interconnected network comprising gas-filled pores, often referred to as "miracle material" in the 21st century. To be specific, aerogel is defined as a material with characteristics of low density and high porosity (Rahmanian et al., 2021). As early as 1931, Kistler transformed the liquid in

[#]Equal contribution to this work.

Surface Science and Advanced Functional Applications of Cellulose Composites
https://doi.org/10.1016/B978-0-443-13675-7.00009-9

TABLE 9.1 List of abbreviations.

Abbreviation	Meaning
2D	Two-dimensional
3D	Three-dimensional
AgNWs	Silver nanowires
AH NPs	Aluminum hydroxide nanoparticles
APTES	3-Aminpropyltriethoxysilane
BC	Bacterial cellulose
BGPC/BT	Biguanide phosphonate Cellulose/$BaTiO_3$ composite aerogel
CFs	Cellulose fibers
CGPA	CNFs/3-(glycidoxypropyl)-trimethoxysilane/polyethyleneneimine
CH	Chitosan
CMC	Carboxymethyl cellulose
CMFs	Cellulose microfibers
CNCs	Cellulose nanocrystals
CNFs	Cellulose nanofibers
CNFAs	Carbon nanofiber aerogels
CNTs	Carbon nanotubes
CR	Congo red
d-DCPG	dialdehyde-CNFs/betaine hydrazide chloride/polyvinyl alcohol
DMOS	Dimethyl sulfoxide
ECH	Epichlorohydrin
PVA	Polyvinyl alcohol
PVB	Polyvinyl butyral
RC	Regenerated cellulose
SHB-CAC	Superhydrophobic waste paper-based cellulose aerogel
TA	Tannic acid
TOC	TEMPO-oxidized cellulose
TOCNF	Tetracycline hydrochloride2,2,6,6-tetramethylpiperidine-1-oxyl radical-oxidized cellulose nanofibers
UiO	Universitetet i Oslo
WSM	Wood sponge/MXene
WPU	Waterborne polyurethane
ZIFs	Zeolitic imidazolate frameworks
EPTMAC	2,3-epoxypropyl trimethyl ammonium chloride
GO	Graphene oxide
HNF	Hierarchically structured cellulose aerogel composites
ImClO4	Imidazole perchlorate
I-CNF	Ion-imprinted functional aerogels
IL	Ionic liquid
KFs	Kapok fibers
LACM	Lignocellulose aerogel membrane
MB	Methylene blue
MIL	Materials of Institute Lavoisier frameworks
MTMS	Methyltrimethoxysilane
MOFs	Metal–organic frameworks
MG	Malachite green
M-MXene/Gr	CMC aerogel comodified by MXene and graphene
CANI-CNF	Cellulose aerogels without ion-imprinted functional site
PA	Phytic acid
PAN	Polyacrylonitrile
PANI	Polyaniline
PDMS	Polydimethylsiloxane
PS	Polystyrene

jelly into gas by exceeding the critical temperature and obtained the world's first aerogel (Kistler, 1931). Various materials can be used to construct aerogels, such as silica, alumina, ovalbumin, and cellulose, notably distinguished by their exceptionally large surface area and unique porous structure alongside low density and high porosity attributes. Hence, aerogel-based materials have experienced rapid development over subsequent decades while finding extensive applications across wearable devices, medical equipment, aerospace military, and energy storage (Liao et al., 2024; Sonu et al., 2023).

With people's in-depth understanding of the concept of sustainable development and the continuous implementation of energy conservation and emission reduction measures, the preparation of green biobased aerogel materials (also referred to as third-generation aerogels) has been put on the agenda (Nita et al., 2020). As one of the richest biomass energy sources on the earth, cellulose has also been used to prepare environmentally friendly aerogel composites (Prasad et al., 2024). Different from nonrenewable petrochemical products such as plastics, cellulose can be extracted from renewable natural resources, such as wood, cotton, wheat grass, straw, reed, hemp, mulberry bark, and other plant-based materials. Cellulose materials have their unique advantages such as renewable, nontoxic, and cheap (Mai et al., 2023b). More importantly, due to its adjustable chemical properties, cellulose can be chemically modified to render it functional (Deng et al., 2022b). Cellulose can be further processed into smaller-diameter nanocellulose with specific physical or chemical treatments (Delgado-Aguilar et al., 2015). Besides, the hydroxyl group and aldehyde group in cellulose can be modified into a variety of other functional groups, which are easy to compound with other substances (Fig. 9.1) (Voisin et al., 2017; Lin and Dufresne, 2014). As early as 1993,

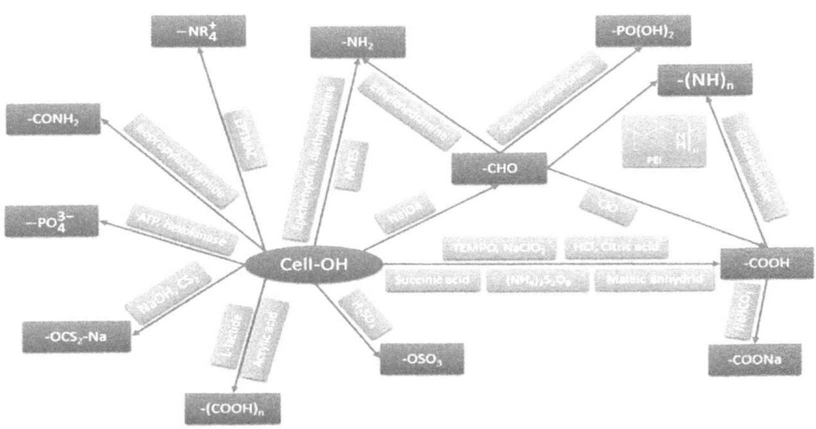

FIGURE 9.1 Relationship of modification methods of cellulose (Voisin et al., 2017).

cellulose aerogels were obtained from viscose by Ookuna's group and named "porous cellulose" (Ookuna et al., n.d.). Subsequently, cellulose aerogels have been developed rapidly and are widely used in many fields in the next few decades (Figs. 9.2 and 9.3). The advantages of cellulose aerogel are

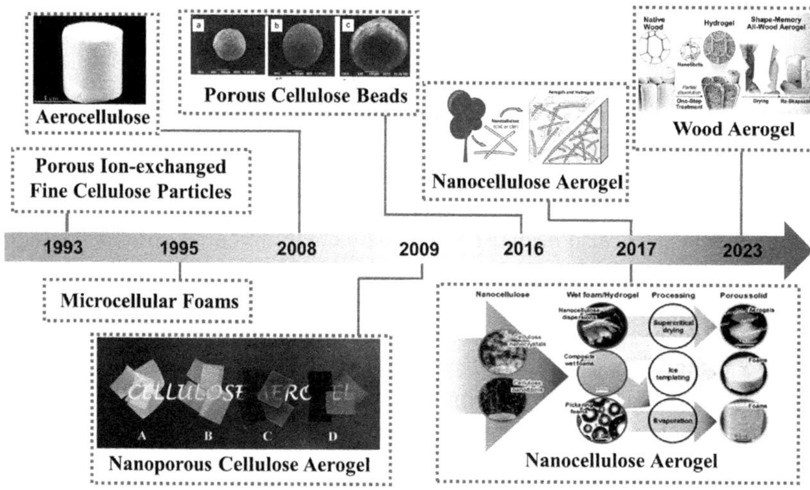

FIGURE 9.2 The development of cellulose aerogels (Gavillon and Budtova, 2008; Cai et al., 2009; Voon et al., 2016; Lavoine and Bergström, 2017; De France et al., 2017; Garemark et al., 2023).

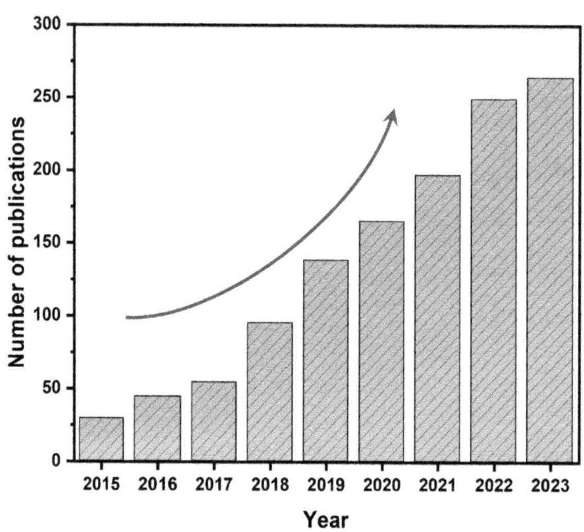

FIGURE 9.3 The number of published papers based on cellulose, aerogel, and composite from the Web of Science.

renewability, biocompatibility and biodegradability, huge reserves, and low cost, which make it an ideal candidate for green materials. In addition, good compressibility and high mechanical strength, hydrophilicity, and other characteristics also give cellulose functional applications. These properties make cellulose an excellent raw material for the preparation of functional composite aerogels for a variety of applications, such as implantable medical devices, wearable devices attached to the skin, degradable and environmentally friendly water purification materials, edible drug carriers, etc. (Liu et al., 2021; Jayan et al., 2023; Jabbari and Babaeipour, 2024).

Although cellulose aerogels have many advantages, the inherent flammability, structural randomness, and lower mechanical strength of pure cellulose aerogels limit their applications (Rostamitabar et al., 2021; Liu et al., 2023a). To expand the range of applications for cellulose aerogels, the combination of cellulose aerogels with functional components to prepare cellulose aerogel composites that have complementary strengths can effectively compensate for the disadvantages of each (Wang et al., 2022). The large specific surface of cellulose aerogel can be used to introduce a large number of active sites into the rich pore structure, so that cellulose aerogel and functional components can be combined, which can greatly increase the loading of functional components. In addition, cellulose aerogel has an interlaced 3D network structure and high mechanical strength, which can be used to strengthen the phase to increase the mechanical properties of composite aerogel. Aghvami-Panah et al. (2024) summarized the fabrication methods and shielding performance of cellulose/MXene aerogel composites, revealing the relationships between microstructural design and shielding performance. However, the systematic preparation methods and wider application scope of cellulose aerogel composites with different functional components need to be further developed. In the following chapters, the different preparation methods of cellulose aerogel composites are reviewed, and the latest research and application progress are summarized, looking forward to the cellulose aerogel composites development prospect. The structural modification and performance improvement of cellulose aerogel by functional components in cellulose aerogel composites are emphasized.

9.2 Fabrication approach of cellulose aerogel composites

The fabrication approach of cellulose aerogel composites consists of two main steps. The first step is the preparation of cellulose gel. To construct a strong skeleton of the cellulose aerogel composites, the cellulose suspension obtained usually needs to undergo a gelation process (Jiang et al., 2018). The gelation process of cellulose suspension can be divided into two main categories: physical cross-linking gelation and chemical cross-linking gelation (Long et al., 2018; Chen et al., 2021b). In physical cross-linking gelation, individual gel components are incorporated into cellulose aerogel precursors through physical entanglement, hydrogen bonding, or van der Waals interactions. This method focuses on the formation of gels by mixing functional presynthesized materials with cellulose in a solvent. The stability of the functional material in the cellulose aerogel structure prepared by physical cross-linking gelation is caused by physical entanglement or interaction between the functional material and cellulose, which forms a weak interaction force (Shuai et al., 2020). Namata et al. (2023) prepared the high water content hybrid gel based on dendrimer cross-linked cellulose nanofibers (CNFs) through physical interactions. The positive charge on the surface of the dendrimer forms a hybrid gel through electrostatic interaction with anion CNFs. The chemical cross-linking gelation method involves the interaction of the chemical bonds between the functional components of the gel and cellulose. This method involves chemically cross-linking cellulose or a large amount of surface-modified cellulose functional groups (e.g., hydroxyl, carboxyl, and methoxy) with the functional component to form a strong interaction force (Liu et al., 2023a). Liu et al. (2023d) fabricated chemically cross-linked cellulose fibers with epichlorohydrin (ECH) in a polyvinyl chloride hose. The tensile strength of cellulose fiber can reach 368 MPa, which is due to the stronger chemical bond between ECH and the cellulose chain than the hydrogen bond in the pure cellulose chain. Physical and chemical cross-linking has its advantages. The bond energy produced by chemical cross-linking is stronger, which can give the aerogel skeleton stronger mechanical properties (Yang and Cranston, 2014). However, the chemically cross-linked gel has the problem of difficult recovery (Huang et al., 2022). Recently, the focus has been on the preparation of integrated physical and chemical cross-linking cellulose aerogel composites (Sun et al., 2021).

The second step is the removal of the solvent from the gel. Under ideal conditions, the initial liquid must be replaced with air so that the formed network is unaffected by the process. When conventional drying methods are used, the capillary pressure caused by the bending of the gas−liquid interface will cause the gel pore structure to collapse and rupture (Qin et al., 2019). Therefore, to obtain cellulose aerogel materials with porous properties, special solvents or special drying methods must be used for further treatment (Fattakhova-Rohlfing et al., 2014). Based on the pressure−temperature curve proposed by Zhao et al. (2018), the preparation methods of cellulose aerogels can be divided into three types, including supercritical carbon dioxide drying, freeze-drying, and ambient pressure drying methods (Peng et al., 2022) (Fig. 9.4). Supercritical carbon dioxide drying uses pressure and temperature to reach the supercritical point and exchange solvent in the aerogel frame (Tabernero et al., 2020). Shi et al. (2021) manufactured the mesoporous nanoscale nitrocellulose/graphene oxide (GO) nanocomposite aerogels by a sol−gel method and supercritical carbon dioxide drying. The prepared composite aerogels show low density (about 0.14 g/cm^3), mesoporous/large pores structure (several nanometers to 140 nm), and high specific surface area

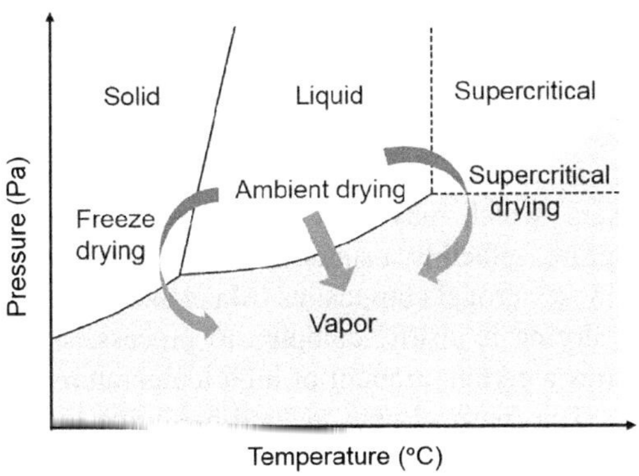

FIGURE 9.4 Phase diagram of pure compounds for different drying routes (Peng et al., 2022).

(around 140 m^2/g). The advantage of the supercritical carbon dioxide drying method is that it can be used to dissolve some organic solvents or nonvolatile components. However, the supercritical carbon dioxide drying method has some disadvantages, such as a long processing cycle, complex process, and high cost of solvent replacement, which makes it unsuitable for some low-quality products (Long et al., 2018). The freeze-drying method is another used method for preparing aerogel materials. The process of freeze-drying is a process of sublimation drying. First, the cellulose wet gel is frozen into a solid state, and then at a certain low pressure and low temperature, the frozen solvent inside the ice gel is sublimed directly from the solid state to the gas state (Liao et al., 2019). Illa et al. (2019) compared the morphological changes of bacterial cellulose (BC) aerogels before and after the wet gel was directly dried at atmospheric pressure and freeze-dried. BC after freeze-drying can effectively retain the porous microstructure due to the sublimation of ice. One of the great advantages of freeze-drying is control of the nanoscale structure of the aerogel. Isotropic or anisotropic structures can be obtained by different freezing treatments of prewet gels. The precursor solution was poured into the mold in the heat insulation device. The bottom of the mold is in contact with the copper plate frozen in liquid nitrogen, so as to achieve the directional growth of ice crystals (Liu et al., 2023c). However, recent studies have found that the pores of aerogels obtained by freeze-drying often show a lamellar structure, which has certain defects in some aerogels requiring specific structures (Chen et al., 2021b). Although supercritical carbon dioxide and freeze-drying methods can effectively prepare aerogel materials, the change in temperature and pressure is energy-consuming (Tang et al., 2022). The atmospheric drying method is the simplest and most cost-effective. By introducing a solvent with low surface tension into the hydrogel, and then drying under the condition of ambient pressure (Li et al., 2023b), Wu et al. (2023) reported the solvent exchange strategy to dissolve ice crystals in ethanol by adding ethanol with a lower freezing point than water after forming a hybrid phase of ice crystals between MXene and CNFs. The hydrogen bond between the aerogels is strengthened, and the surface tension of ethanol is low, which will not cause the aerogel to collapse. However, it should be noted that aerogel materials obtained by atmospheric drying methods do not have great property advantages. And the atmospheric drying method depends on the subsequent addition of chemical cross-linkers, which may affect the properties of the aerogels (Li et al., 2023a). Therefore, selecting a suitable solvent is important to maintain the shape of the cellulose aerogel composites (Ma et al., 2022). As mentioned before, supercritical drying is costly, complex to process, safety issues, and equipment that requires a certain amount of high temperature and high pressure (Zhu et al., 2015). The freeze-drying method has some limitations, such as it is only suitable for low-temperature resistant materials, and the internal network may be destroyed due to the formation of ice crystals, forming cracks and reducing mechanical durability. In addition, the properties of aerogels produced by freeze-drying and supercritical drying are also different. Aerogel samples obtained by the supercritical carbon dioxide drying method have higher specific surface area and lower density. However, the aerogel samples obtained by freeze-drying have better structural controllability and larger pore structure due to the ice crystal expansion (Ciftci et al., 2017; Plazzotta et al., 2019; Wang et al., 2020). In contrast, ambient pressure drying is a more economical and safer method for producing aerogels. However, to avoid irreversible network collapse caused by capillary force, additional modification of aerogels is needed (Syeda and Yap, 2022). The advantages and disadvantages of various drying methods for cellulose aerogel are listed in Table 9.2. Overall, the most commonly used methods of drying cellulose aerogel composites in laboratories are supercritical carbon dioxide drying and freeze-drying, but are limited by the scale of preparation and equipment. The advantages of atmospheric drying are gradually being highlighted, especially in large-scale preparation, but it is still in the exploratory stage. Researchers need to consider their research equipment conditions as well as the properties and applications of the prepared materials to select a suitable method for operation.

9.3 Performance and applications of cellulose aerogel composites

9.3.1 Water pollutant treatment

The problem of water pollution has become increasingly serious in the development of agricultural and industrial practices in today's world (Liu et al., 2023b). Sewage with high pollutant concentrations contains heavy metal ions, azo carcinogens, and other toxic components (Niu et al., 2023). At a certain concentration, all heavy metal ions and dyes are almost toxic (Wang et al., 2022). The ideal water pollution treatment materials must have the

TABLE 9.2 The advantages and disadvantages of different aerogel drying methods (Uscanga et al., 2021; Liu et al., 2022; Xiao et al., 2022; Lee et al., 2017; Guastaferro et al., 2021).

Methods	Advantages	Disadvantages
Supercritical carbon dioxide drying	Minimal damage to the microstructure; low shrinkage; high porosity and surface area	Energy-consuming; high cost; dependent on solvation properties; device requirement
Freeze drying	Low shrinkage; controllable microstructure; larger pore structure	Time-consuming; high cost; pre-freezing required; Ice crystal expansion; device requirement
Ambient pressure drying	Energy saving, safety; economy; low equipment requirements; simple operation process	High shrinkage; low porosity; chemical cross-linkers are required

following characteristics: excellent water purification ability, faster water purification process, long recycling ability, no pollution, environmental friendliness, and good mechanical properties (Wu et al., 2023; Liu et al., 2022). Cellulose aerogels are porous with a large specific surface area and are a suitable material for wastewater treatment, which increases the contact area between pollutants and materials (Paul and Ahankari, 2023). In addition, the porous structure serves as an anchor point for various functional substances, further enhancing the treatment capabilities of cellulose aerogels (Kong et al., 2022). While cellulose aerogels have good adsorption capabilities, there is potential for further performance optimization. The effective approach involves incorporating functional materials to prepare cellulose aerogel composites with enhanced overall performance (Vu et al., 2022).

Functional sites designed by ion imprinting were a promising method to enhance the binding selectivity of adsorbents to target ions (Cao et al., 2023). Oxygen and nitrogen-rich compounds were added to nanocellulose aerogels, and then ion-imprinted uranyl ions was used to induce them to produce a special spatial ion-imprinted functional site structure in the aerogels (I-CNF). The ordered layered structure, high porosity (96.65%), and hydrophilicity (water contact angle of 22.81 degrees) of the nanocellulose aerogel complex can promote the entry of seawater into the material, and the functional ion-imprinted functional sites can endow the material with high uranium adsorption capacity (9.46 mg/g). The adsorption capacity of I-CNF is 2.57 times larger, and the adsorption speed is 2.5 times faster than the cellulose aerogels without ion-imprinted functional site (NI-CNF). In addition to the ion-imprinted method, anchoring metal—organic frameworks (MOFs) functionalized nanoparticles on cellulose aerogels to prepare cellulose aerogels composites can also improve water pollutant treatment performance (Olorunnisola et al., 2023). Ma et al. (2019) used bacterial cellulose (BC) as the template of aerogel composites and immersed them into the metal ion solution and organic ligands in two steps. The abundant hydroxyl groups of BC promote the adsorption of metal ions through weak interactions, followed by the nucleation of MOFs by organic ligands on the BC surface (Fig. 9.5a). The introduction of MOFs greatly improved the adsorption capacity of cellulose aerogel composites for Pb^{2+} (390 mg/g) and Cd^{2+} (220 mg/g), while bacterial cellulose aerogels as a skeleton gave the composites good compressibility and water absorption. However, no obvious removal ability was detected in bacterial cellulose aerogels without MOFs functionalization. The preparation of cellulose aerogel composites can not only enhance the water treatment ability, but also improve the mechanical properties of cellulose aerogels. Hu et al. (2023) used 3-aminpropyltriethoxysilane (APTES) and glutaraldehyde as cross-linking agents via Schiff base reaction to construct the double cross-linked network. The modified cellulose aerogel composites have better water resistance in the Congo red dye solution and compression modulus than the unmodified cellulose aerogels. In addition to water-soluble pollutants, oil spills during extraction have become a ubiquitous source of pollution in water bodies, posing a great threat to the marine ecosystem and national economy (Sui et al., 2024). Traditional adsorption methods have been limited in effectively addressing the issue due to the high viscosity characteristic of oil spills, necessitating a reduction in viscosity for effective adsorption. Thus, incorporating functionalized substances with photothermal or electrothermal conversion ability onto cellulose aerogel complex can increase the temperature of the crude oil, reducing the viscosity (Fan et al., 2024). Bi et al. (2023) prepared the superhydrophobic carboxymethyl cellulose (CMC) aerogel comodified by MXene and graphene (M-MXene/Gr CA) with an aligned channel structure to realize all-weather crude oil recovery. The introduction of MXene and graphene functional substances can effectively improve the photothermal conversion performance, while cellulose aerogel with vertical pore structure can effectively reduce the resistance of crude oil adsorption. The average crude oil adsorption rate of M-MXene/Gr CA can reach 1.3×10^7 $g/m^3/s$. However, unmodified CMC aerogels are difficult to realize effective photothermal conversion to reduce the viscosity of crude oil (Chen et al., 2024). Utilizing natural wood for the direct production of cellulose aerogel composites is a more convenient method. Because natural wood itself is anisotropic, there is no need to construct vertically by a specific method (Wang et al., 2022). Wang et al. (2022) compounded 2D MXene with the wood sponge (the wood block aerogels that have been removed lignin and

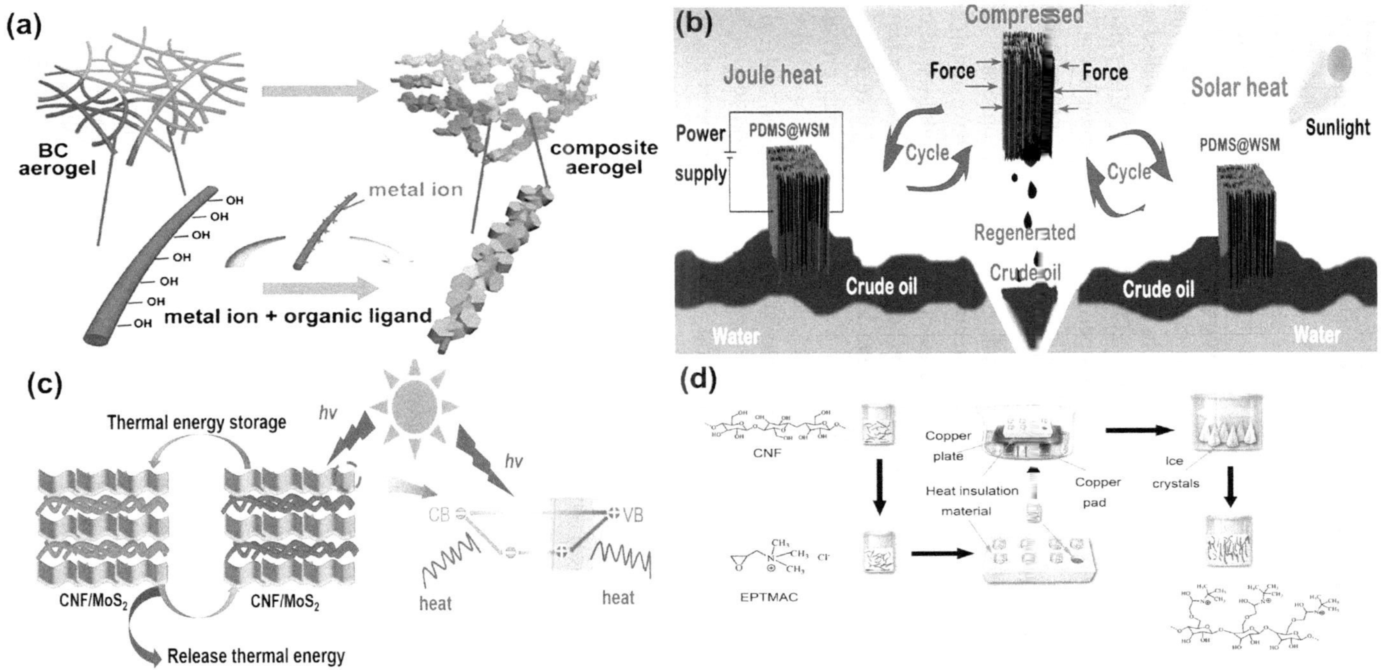

FIGURE 9.5 (a) The fabrication process of BC@MOFs composite aerogels (Ma et al., 2019). (b) Schematic diagram of all-weather adsorption of crude oil (Wang et al., 2022). (c) Photothermal conversion schematic diagram of CNFs/MoS_2 aerogels (Yuan et al., 2022). (d) Directional structure and reaction mechanism of modified cellulose aerogel composites (Zhuang et al., 2022).

hemicellulose and are mainly composed of cellulose) and then coated with polydimethylsiloxane (PDMS) to prepare the cellulose aerogel composites (Fig. 9.5b). The wood sponge composites have a 3D framework with high porosity and vertical channels for effective absorption of crude oil, the MXene layer has sunlight absorption and metal conductivity for rapid in situ heat generation, and the PDMS layer has hydrophobicity for enhanced selective absorption properties. The oil absorption rate (64s) of wood sponge composites modified by MXene coating is much higher than that of unmodified wood sponge composites (150s). In addition to 2D carbon materials, MoS_2 also has extremely high photothermal conversion performance. As shown in Fig. 9.5c, Yuan et al. (2022) fabricated the CNFs/MoS_2 aerogels that generate a high temperature of up to 75°C under 1 sun, which can completely absorb the 1 mL crude oil in 120 s. However, due to the lack of photothermal conversion ability of pure CNFs aerogels, the surface temperature can only be maintained at room temperature, which is difficult to adsorb crude oil. In addition to crude oil, the application of oil/organic pollution adsorption, and oil/water separation are also important. These pollutants have led to severe negative effects on aquatic life and the environment (Gu et al., 2020). Guo's group prepared a hierarchical framework adsorbent nanocellulose aerogel foam by high-speed mechanical foaming and solvent-free method by adding sodium dodecyl sulfate, which has the adsorption capacity of cyclohexane (206.79 g/g), ethyl acetate (194.75 g/g), and vacuum pump oil (145.20 g/g) (Zhang et al., 2019). Furthermore, by optimizing the preparation process, a solvent-free method for preparing alumina/cellulose aerogel composites was proposed (Zhou et al., 2021). The thiophene adsorption capacity of cellulose aerogel composites obtained by mechanical stirring and liquid nitrogen freezing casting can reach 108.07 g/g, which is much higher than that of pure nanocellulose aerogel (74.07 g/g). In addition, to improve the hydrophobicity and adsorption performance of cellulose aerogel composites, nanochitosan and reduced graphene oxide were introduced into cellulose aerogels to construct nanocomposite aerogels. The prepared nanocomposite aerogel exhibited a high hydrophobicity (115.26 degrees) and the high adsorption capacities high adsorption capacities to nine kinds of oil or organic pollution (>120 g/g) (Gu et al., 2021). Besides, the influence of cellulose aerogel composites prepared from different types of cellulose materials on the adsorption capacity of pollutants was explored. Hoviatdoost et al. (2023) prepared three types of aerogel graphene composites (wood-based cellulose nanofiber, paper-based cellulose nanofiber, and cellulose microfiber). It is found that graphene/wood-based cellulose nanofiber aerogel has higher adsorption capacity, because this type of cellulose aerogel composites has higher crystallinity and more functional groups in the structure, while increasing the oil absorption capacity. Besides, the environmental impact of widespread plastic use in our daily lives is significant, particularly due to the continuous accumulation of microplastics (plastic fragments smaller than 5 mm). Each year, nearly 100,000 tons of plastic waste are deposited into the ocean, posing a significant threat to both the marine ecosystem and the health of humanity (Peng et al., 2020). As such, it is important to carefully consider the long-term implications of our plastic usage and take proactive steps to reduce the amount of waste (Zhou et al., 2022). As shown in Fig. 9.5d, Zhuang et al. (Zhou et al., 2022; Zhuang et al., 2022) used 2,3-epoxypropyl trimethyl ammonium chloride (EPTMAC) as a modifier to modify CNFs aerogel, and then cellulose aerogel composites with oriented structures were prepared by liquid nitrogen freezing method. The adsorption capacity of fluorescent polystyrene microspheres (PS) of cellulose aerogel composites can reach 146.38 mg/g. Conversely, the adsorption capacity of unmodified CNFs aerogels is only 15.58 mg/g, which is due to the different electrostatic interactions between microplastics and cellulose aerogels. Further, to address the slow adsorption speed of microplastics, researchers have developed CNFs/3-glycidoxypropyltrimethoxysilane/polyethyleneneimine (CGPA) aerogels. The CGPA aerogels have a large number of active sites, which allows for the rapid adsorption of microplastic particles. Important, the cellulose aerogel can reach saturation within just 20 min with the adsorption capacity of 117 mg/g. In contrast, the adsorption capacity of the unmodified cellulose aerogels was only 31 mg/g (Zhuang et al., 2023). Liu et al. (2024b) reported the cationic-modified directional dialdehyde-CNFs/betaine hydrazide chloride/polyvinyl alcohol (d-DCPG) aerogel using dialdehyde-modified CNFs as the matrix, betaine chloride hydrazide provided cation modification, and polyvinyl alcohol (PVA) provided cross-linking function. Functional modification of betaine chloride hydrazide can greatly improve the surface charge of cellulose aerogel and enhance electrostatic adsorption. The adsorption of PS by d-DCPG gels needs 60 min to reach equilibrium. The adsorption capacity of d-DCPG aerogels can reach 145.05 mg/g (4.58 times that of CNFs/PVA aerogel). PVA can physically cross-link with cellulose, thus enhancing the mechanical strength of cellulose aerogel composites.

The summary of the purification performance of representative cellulose aerogel composites for water pollutant treatment is listed in Table 9.3. Although cellulose aerogel composites have been widely researched for water pollutant treatment, there are still some issues that need to be solved. The development of cellulose aerogel composites that can remove multiple pollutants at the same time is an aspect worthy of attention. For example, in the adsorption of dyes and microplastics procession, the electrostatic interaction between cellulose aerogel composites and dyes or microplastics is one of the most important mechanisms, but in fact, dyes or microplastics that pollute the water body may be positively charged, negatively charged, or electrically neutral (Mai et al., 2021; Zhu et al., 2023). This requires different modifications for cellulose aerogel composites, which makes it difficult to effectively purify a variety of dyes or microplastics at the same time. The reasonable control of the addition of functional substances and the improvement of

TABLE 9.3 Purification performance of the representative cellulose aerogel composites for water pollutant treatment.

Composites	Pollutants	Purification performance	Refs.
I-CNF	Uranium	9.46 mg/g	Cao et al. (2023)
Pineapple peel waste aerogel	MB	29.7 mg/g	Le et al. (2023)
	CR	101.4 mg/g	
	MG	48.6 mg/g	
CNC-CMC-ZIF-8	Cr (VI)	39.06 mg/g	Zhu et al. (2016)
CNC-CMC-UiO-66		5.33 mg/g	
CNC-CMC-MIL-100 (Fe)		2.32 mg/g	
TOC/CNFs	MB	248.27 mg/g	Zhang et al. (2023)
	MG	227.30 mg/g	
PDMS@WSM	Crude oil	11.2×10^5 g/m^3	Wang et al. (2022)
M-MXene/Gr CA	Crude oil	18 g/g	Bi et al. (2023)
MTMS/paper waste cellulose	Ruby crude oil	18.4 g/g	Nguyen et al. (2013)
	Te Giac Trang crude oil	18.5 g/g	
	Rang Dong crude oil	20.5 g/g	
CNF/EPTMAC	PS	146.38 mg/g	Zhuang et al. (2022)
CGPA	PS	117.04 mg/g	Zhuang et al. (2023)
d-DCPG-1	PS	145.05 mg/g	Liu et al. (2024b)

environmental weather ability require the expected control of the temperature change process, while photothermal conversion improves the oil spill adsorption capacity of cellulose aerogel composites. Because the current oil spill adsorption process is greatly influenced by the environment, too low photothermal conversion temperature will cause crude oil to block pores, and too high temperature will have the risk of burning crude oil (Yuan et al., 2022; Zhou et al., 2023). This may need to be further improved by adding additional flame-retardant substances or optimizing the structure.

9.3.2 Piezoresistive sensor

As a recently emerging technology, wearable devices have emerged as a promising technology with broad applications in human health monitoring, virtual reality, biometrics, and artificial intelligence (Zhang et al., 2024). Of particular importance is the role that pressure sensors play in detecting human movement, monitoring physiological signals, and preventing cardiovascular diseases (Gackowski et al., 2023; Zhang et al., 2024). The excellent piezoresistive sensor usually needs high sensitivity, strong stretchability, long durability, high linearity, high selectivity, outstanding detection limit, and short response time (Ma et al., 2021). The utilization of sustainable biomass cellulose resources for producing the aerogel-based piezoresistive sensor can confer the sensor with eco-friendly, low-density, non-hazardous, biodegradable, recyclable, and sustainable characteristics (Zhang et al., 2024). After reasonable structural design, cellulose aerogels can have good compressibility and durability (Xiang et al., 2024; Hu et al., 2023). However, cellulose is usually insulated, limiting pressure information from being converted into an electrical signal for detection (Wei et al., 2022). One of the ways is to carry out carbonization pyrolysis of cellulose aerogel. However, high-temperature pyrolysis leads to extremely fragile biological materials, which greatly limits their applications as sensing materials (Wang et al., 2022). Another method is to add conductive fillers, but this method is limited by the uniformity and binding force of the fillers and matrix. Therefore, structural design or further modification is needed to improve the defects (Chen et al., 2021d).

The modified cellulose aerogel composites can effectively solve the problem of poor mechanical properties of pressure sensors prepared by directly carbonized cellulose aerogels. Cellulose aerogel composites with a 3D interconnected honeycomb structure have been produced modifying by $(NH_4)_2SO_4$. The inclusion of $(NH_4)_2SO_4$ has been used to significantly increase the carbon yield while also effectively preventing shrinkage and deformation of the cellulose aerogel composites during carbonization. Cellulose aerogels that have not been modified by $(NH_4)_2SO_4$ have a disordered microstructure and are easy to deform and collapse during compression. However, the volume shrinkage of cellulose aerogel composites modified by $(NH_4)_2SO_4$ was restrained during compression. The prepared cellulose aerogel composites exhibit 99% strain compressibility, superelasticity (97.4% height retention after 500 cycles at 90% compression), and 10,000 cycles of fatigue resistance (Wang et al., 2022). Besides, Wang et al. (2023) designed the cellulose aerogel composites by sequential microfibrillation, periodate oxidization, and tannic acid (TA) cross-linking of cellulose fibers (CFs). The periodate oxidation and TA cross-linking effectively improved the mechanical properties of cellulose aerogel composites. The prepared sensor shows a sensitivity of 4 kPa^{-1}, response/recovery time of 32/30 ms, detection limit of 0.8 Pa, and reproducibility (3000 cycles at 0.5 kPa). The addition of conductive fillers is another effective way to improve the conductivity and pressure sensing ability of cellulose aerogel composites. The conductivity of conductive fillers varies with the distance between layers or contact areas (Wang et al., 2021). As shown in Fig. 9.6a, Cheng et al. (2021) designed the silver nanowires (AgNWs)/CNFs aerogel multifunctional strain sensor via directional freeze-casting and thermal welding. Direct thermal welding can effectively reduce the interface resistance of AgNWs junction. AgNWs form a stable, tough, and complete network as an electron transfer channel to endow the sensor with fast sensing performance. The assembled sensor displayed 3.86 kPa^{-1} sensitivity and over 10,000 cycles of durability. However, the larger aspect ratio of cellulose will endow the cellulose aerogel with higher rigidity and greater fragility. To prevent the failure of the cellulose aerogel under a strong mechanical load, chitosan (CH) is added to the bacterial cellulose as a reinforcing phase. A lightweight conductive polyaniline (PANI)/BC/CH aerogel composites were obtained by in situ polymerization of PANI on the surface. A piezoresistive sensor with high strength and high sensitivity could be obtained by further immersion of PDMS (Huang et al., 2019) (Fig. 9.6b). To achieve the purpose of total component cellulose, a strategy has been developed to combine carbonized BC with CNFs solution, which was then prepared using the ice template method to produce the BC/CNFs composite aerogel (CBCC) film with a complete cellulose component (Fig. 9.6c) (Chen et al., 2021c). These cellulose aerogel composites with full cellulose components provide new insights into the development of green pressure sensors and the utilization of sustainable natural biomaterials in high-tech fields. Inspired by tracheid, Wei et al. (2023) used CH as the "bridge junction" to tightly connect the CNFs and MXene. The cellulose aerogels modified by CH show an elastic modulus about 62.90 kPa, while that of CNFs/MXene aerogels without CH modification is only about 6.64 kPa (Fig. 9.6d). The prepared CNFs/MXene/CH cellulose aerogel composites exhibit fast response time (6.3 ms) and recovery times (8.1 ms). Thanks to the excellent compression performance, the sensitivity retention rate can reach 94.2% after 10,000 cycles. In addition to component regulation, structural design is also an effective method to improve the performance of cellulose carbonized aerogel composites. By adjusting the freezing rate, cellulose aerogel composites with different surface structures ("regular gully" morphology, "hair" morphology, and "branch-channeled" morphology) can be obtained. The functional functionalized cellulose nanocrystals (f-CNCs)/waterborne polyurethane/MXene aerogel composites after structural optimization show good compressibility (Cai et al., 2021). Besides, Zhuo et al. (2019) skillfully demonstrated by

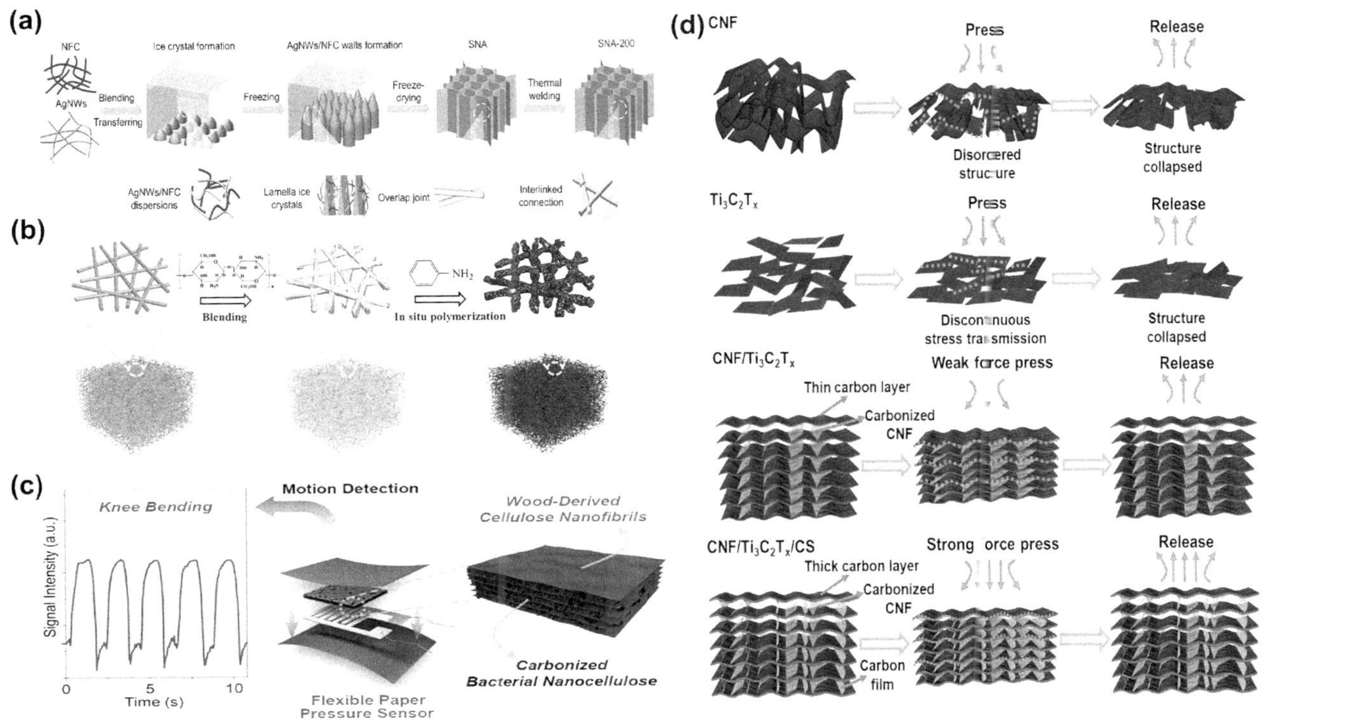

FIGURE 9.6 (a) Fabrication process of AgNWs/CNFs aerogel (Cheng et al., 2021). (b) Fabrication process of lightweight conductive PANI/BC/CH aerogel composites (Huang et al., 2019). (c) The structure of CBC-C cellulose aerogel composites (Chen et al., 2021c). (d) Compression elasticity mechanisms of the different aerogel samples (Wei et al., 2023).

adopting a strategy that involved the combination of carbonization and conductive polymer. The combination of the oxygen group followed CNCs carbonization with Ti in MXene. The electron pairs from the hydroxyl group, which were dissociated by CNCs, were utilized to occupy the vacant orbital of the titanium atom and form TiO_2, creating a robust interaction between MXene and CNCs. The CNCs/MXene cellulose aerogel composites had a distinctive undulating shape, while C—CNC cellulose aerogels without MXene modification possessed discontinuous fragment shapes not suitable for the piezoresistive sensor application. All-cellulose aerogel composites can also be optimized by structural design. Dong et al. (2024) reported shape memory bamboo-derived cellulose microfibers (CMFs)/CNFs/regenerated cellulose (RC) all-cellulose aerogel composites. The CNFs/RCs act as aggregated layers, and the CMFs act as bridges. CMFs/CNFs/RCall-cellulose aerogel composites can be soaked in water for 24 h without any structural deformation, while CNF cellulose aerogels have suffered from severe peeling on the surface layer. The piezoresistive sensor with low response time (150—200 ms), high sensitivity (0.45—19.8 kPa^{-1}), and high fatigue resistance can be obtained by further cross-linking the CMFs/CNFs/RC all-cellulose aerogel composites with Li ions.

The summary of the main features of representative cellulose aerogel composites for piezoresistive sensor is listed in Table 9.4. Although cellulose aerogel composites have been widely researched for piezoresistive sensor, there are still some issues that need to be solved. High sensitivity, low detection limit, and wide detection range are usually requirements for piezoresistive sensors, but in most cellulose aerogel composites sensors reported at present, the sensitivity and detection limits in different pressure ranges are different. In addition, too high sensitivity also has some problems, such as noise interference, which may distort data acquisition in the actual monitoring process (Ji et al., 2022). In addition, environmental pressure drying methods have recently received a lot of attention. However, reports of piezoresistive sensors based on ambient pressure drying are rare, which may be due to the difficulty in the structural design of ambient pressure drying cellulose aerogels. Therefore, it is necessary to seek more optimized technical means to improve the way of environmental pressure drying.

9.3.3 Thermal insulation

In the construction industry, how to effectively improve the energy efficiency of buildings is an important aspect of energy conservation and emission reduction measures (Adamczyk and Dylewski, 2017). Because of the diffusivity of thermal energy, thermal insulation is important for energy management to reduce energy waste and improve energy efficiency (Hu et al., 2019). Thermal conductivity (λ) affects the thermal insulation performance of materials, and low thermal conductivity is preferred for insulation. Due to the low

TABLE 9.4 Main features of the representative cellulose aerogel composites for piezoresistive sensor.

Composites	Sensitivity (kPa^{-1})	Pressure range (kPa)	Response/ recovery time (ms)	Fatigue resistance	Refs.
CNF–C	6.63	0.002–7	400	50,000	Lai et al. (2021)
CNFAs	5.66	0–28	100	10,000	Wang et al. (2022)
CBC-C	0.003–0.358	0–100	50/110	10,000	Chen et al. (2021a)
Ag/PA@BC	6.92	18.87	200	1000	Huang et al. (2022)
rGO@BC	13.89 2.71	0.16–2.54 2.54–22.2	120/540	1000	Wei et al. (2021)
AgNWs/CNFs	3.86	0.5–1.5	180	10,000	Cheng et al. (2021)
CNT/WPU/ CNC	0.25	1–10	120	800	Zhai et al. (2021)
	114.55	21.76	150/120	24,000	Qin et al. (2023)
C-MX/CNC	114.6	10	189	10,000	Zhuo et al. (2019)
CMFs/CNFs/RC	0.45–19.8	0–100	150/200	500	Ma et al. (2021)

bulk density, small pore size, and radiation-absorbing elements of cellulose aerogel composites, cellulose aerogel composites can significantly reduce solid conduction, gas conduction/convection, and thermal radiation. Hence, the thermal insulation properties of cellulose aerogel composites which is mainly affected by surface functional groups, pore size, pore size distribution, density, and microstructure (Jiménez-Saelices et al., 2018; Wu et al., 2023). The control of the structure of the cellulose aerogel composites has a great influence on thermal conductivity. From the macroscopic aspect, the design of macroscopic 3D porous cellulose aerogel structure or the cellulose fiber arrangement has a great influence on the solid-phase thermal conduction (Du et al., 2018; Dong et al., 2023). From the microscopic aspect, the distribution and arrangement of holes will also affect the thermal insulation performance, and the different air volumes contained in different holes will also affect the thermal conductivity of air (Dong et al., 2023). The structure and composition of the aerogel can be effectively regulated through early design or later treatment, and the excellent properties of cellulose and the rich functional groups on the surface also make the development of cellulose aerogel composites thermal insulation materials possible (Liu et al., 2021). The thermal insulation sector is one of the segments occupied by the aerogel market according to the report by Grand View Research. Therefore, cellulose aerogel composites have great application prospects in the field of heat insulation.

As displayed in Fig. 9.7a, Jimenez-Saelices et al. (2018) used TOCNFs-stabilized Pickering emulsions as ice templates to design the hierarchical morphology with three levels of porosity. Ice crystal induction forms macropores, emulsion template forms spherical smaller macropores, and mesoporous domains localized at the pore walls level. The resistance of the Pickering emulsions template can avoid shrinkage of cellulose aerogel composites. The prepared cellulose aerogel composites show a low thermal conductivity of 0.018 W/(m K). Additionally, the natural wood sponge aerogels with the unique pore structure prepared by the top-down method have been widely concerned. However, due to the anisotropic characteristics of wood sponge aerogel, it is challenging to prepare it as an isotropic thermal insulation material. Garemark et al. (2022) proposed a method to fill the lumen with an ionic liquid (dimethyl sulfoxide [DMSO] and a guanidinium phosphorus-based ionic liquid [MTBD][MMP]), so that a portion of the cell wall was dissolved and the mean free path of air was introduced to increase the heat insulation capacity. The cellulose aerogel composites show low thermal conductivity in both radial direction (0.037 W/mK) and axial direction (0.057 W/mK) (Fig. 9.7b). Inspired by the natural hollow insulation animal hair, Xu et al. (2023) prepared the hollow kapok fibers/polyvinyl alcohol/biguanide phosphonate (KF-PVA-BGP) cellulose aerogel composites. The thermal conductivity of KF-PVA-BGP aerogels was reduced to 0.0531 W/m/K. The thermal insulation effect of KF-PVA-BGP aerogels is obviously better than that of KF aerogel, and the surface temperature (70.3°C) of KF-PVA-BGP aerogels is lower than that of KF aerogel (82.6°C) when the heating platform temperature is about 200°C. With the introduction of PVA, the volume retention rate of cellulose aerogel composites and the construction of 3D network are improved, which is helpful to reduce the thermal conductivity of the materials.

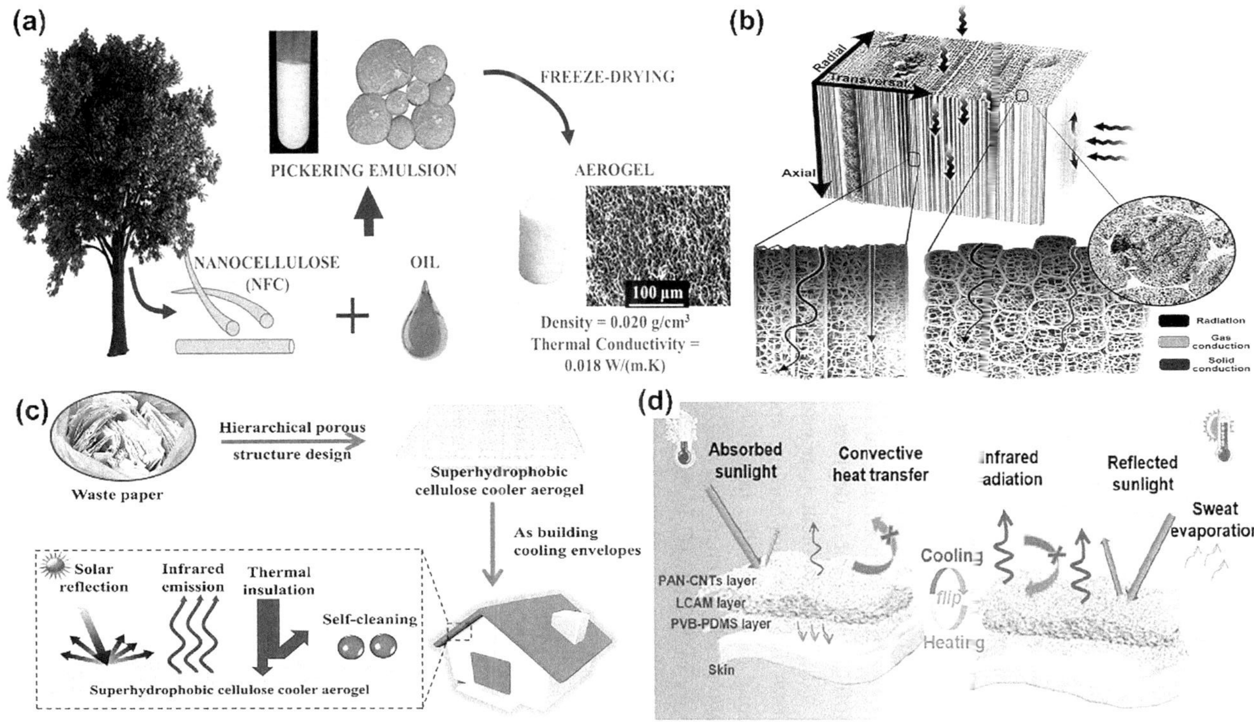

FIGURE 9.7 (a) The preparation of Pickering emulsions-CNFs aerogel composites (Jiménez-Saelices et al., 2018). (b) Schematic illustrations of the thermal conduction in the wood aerogel (Garemark et al., 2022). (c) Schematic illustrations of preparation of SHB-CAC (Yue et al., 2022). (d) Schematic illustrations of radiation cooling/heating mechanism of HNF textiles (Gu et al., 2023).

Besides, the Radiant cooling materials are one of the most concentrated practical application fields of thermal insulation materials (Wu et al., 2023). Cai et al. (2022) prepared the CNCs/methyltrimethoxysilane (CNCs/MTMS) aerogel composites to apply to the thermal insulator and daytime passive radiation cooler. A large number of chemical bonds, including CH, CO, SiOSi, and COC bonds, favor high radiation cooling performance using the cross-linking reaction between CNC and silane agents. The excellent infrared absorption ability of CNC, the layered structure of the absorber, and the uniform and orderly porous nano/micropore structure reduce the heat conduction ability in the solid phase. On the contrary, unmodified cellulose aerogels show greater anisotropy in thermal conductivity. CNCs/MTMS aerogel composites have excellent weather resistance and radiation cooling performance, which can be widely used in extreme environments. Besides, Yue et al. (2022) designed the superhydrophobic waste paper–based cellulose aerogel composites (SHB-CAC) with a high-water contact angle about 152 degrees and a low roll-off angle about 5 degrees. And the cellulose aerogel composites achieved a high subambient temperature drop of ~8.5°C on the daytime, which shows that 43.4% of cooling energy could be saved. Superhydrophobic self-cleaning radiant cooling materials are expected to play an important role in the future field of energy-saving buildings (Fig. 9.7c). The low thermal conductivity of SHB-CAC is mainly due to the low solid phase conduction caused by high porosity and porous structure, thus forming a tortuous path of phonon conduction to realize scattering and diffusion heat transfer. In addition, the existence of many nanopores also leads to a significant decrease in heat transfer in the air phase. Moreover, Gu et al. (2023) prepared a hierarchically structure cellulose aerogel composites (HNF) textile to meet the demand of wearable radiation cooling/heating materials. The porous lignocellulose aerogel membrane (LACM) is the medial layer with low thermal conductivity (0.0366 W/m/K), which significantly enhances the thermal insulation performance of textile. The top layer of textiles is a radiation cooling layer (polyvinyl butyral (PVB)–PDMS), and the bottom layer is a radiation heating layer (polyacrylonitrile–CNTs) (Fig. 9.7d). The diameter of PVB nanofibers increases after PDMS is doped, which enhances the scattering of sunlight by nanofibers. At the same time, the addition of PDMS gives PVB–PDMS excellent thermal radiation. The thermal emissivity (90.4%) of the PVB–PDMS side of HNF textiles is much higher than that of pure PVB (69.9%). By combining efficient heat insulation and regulating radiation heat transfer, the cooling and heat preservation performance in extremely hot and cold environments are improved, respectively.

The summary of the thermal insulation/radiative cooling performance of representative cellulose aerogel composites is listed in Table 9.5. Although cellulose aerogel composites have been extensively researched for thermal insulation/radiative cooling, there are still some issues that need to be addressed. High temperature is often accompanied by the potential possibility

TABLE 9.5 Main features of the representative cellulose aerogel composites for thermal insulation/Radiative cooling.

Composites	Thermal conductivity	Radiative cooling	Refs.
Pickering emulsions-CNFs	0.018 W/m/K	–	Jimenez-Saelices et al. (2018)
Wood aerogels/IL	0.037 W/m/K (radial)	–	Garemark et al. (2022)
	0.057 W/m/K (axial)		
CNF@Al-MIL-53	40 mW/m/K	–	Zhou et al. (2019)
CNFs/starch/clay	41.5–48.3 mW/m/K	–	Zhao et al. (2021)
CNCs/MTMS	0.029–0.056 W/m/K	9.2°C (direct sunlight)	Cai et al. (2022)
		7.4°C (extreme conditions)	
SHB-CAC	28 mW/m/K	8.5°C (800 W/m^2)	Yue et al. (2022)
BC/PVA bilayer gel	0.03302 W/m/K	~18°C (direct sunlight)	Zhao et al. (2022)

of combustion, and cellulose aerogel composites with dual functions of flame retardant and heat insulation are more reliable in practical application. However, cellulose is flammable, so it is necessary to add flame-retardant substances to avoid the harm caused by the combustion of cellulose aerogel composites, but adding additional flame retardant substances will increase the density of aerogel (Han et al., 2023; Yang et al., 2022). Therefore, it is necessary to find a balance between the two or use a material that can be both heat insulation and flame retardant for modification. In addition, one of the most promising applications of radiation cooling is in the preparation of cellulose aerogel composites for textile clothing. However, most radiation cooling materials are currently white or translucent because this reduces the absorption of incident light and increases reflection (Sun et al., 2023; Liu et al., 2022). This cannot meet people's demand for diversified colors of clothing fabrics, and the exploration of radiation cooling materials with various colors is still very scarce. Therefore, the development of cellulose aerogel composites in the future should meet aesthetic and diversified needs while ensuring the thermal insulation and radiation cooling properties of materials.

Table 9.4 lists the main features of the representative cellulose aerogel composites for thermal insulation/radiative cooling.

9.3.4 Other applications

In addition to water purification, piezoresistive sensors, and thermal insulation applications, cellulose aerogel composites also show potential for other applications.

Food package. To enhance food safety and reduce environmental pollution, natural cellulose-based biodegradable packaging materials have received increasing attention in recent years. An ideal food packaging material should not only have the basic characteristics of light weight, biodegradability, and nontoxicity but also have the function of showing food freshness and prolonging food shelf life (Deng et al., 2022a). Mirmoeini et al. (2023) used freshly cut grape stalks cellulose/Salep/anthocyanins to develop intelligent freshness cellulose aerogel composites. The aerogels show a rapid response to meat deterioration and distinguish fresh and spoiled minced beef according to changes in volatiles produced by anthocyanins in the protein dispersion of meat. As a modifier, Salep can maintain the structural integrity of cellulose aerogel composites and fine-tune the porous structure by enhancing hydrogen bonding so as to improve the sensitivity of indicators to spoilage metabolites. Besides, Fontes-Candia et al. extracted cellulose suspension and water-based bioactive components from *Arundo donax* waste biomass, thus preparing cellulose aerogel composites for the preservation of red meat. Due to the porous structure and hydrophilic ability of cellulose aerogels, the extracted polyphenols can be released more easily, which can effectively reduce discoloration and lipid oxidation of red meat during cold storage. However, the meat preserved in cellulose aerogels without modification of active polyphenol substances showed more serious oxidation.

Triboelectric nanogenerator (TENG). TENG is a new type of energy harvesting device, which can convert mechanical energy (Zhang et al., 2023). There are many negative charges on the surface of cellulose materials, and the rich surface area of aerogels can provide many sites for charge capture, so TENG devices with cellulose aerogel composites have great potential (Huang et al., 2023).Cheng et al. (2019) used CMC as a cross-linking agent to interact with MXene to form a porous 3D network aerogel structure, while MXene provided a conductive path. The peak-to-peak open-circuit voltage and short-circuit current can reach 54.37 V and 1.22 μA. The prepared cellulose aerogel composites can be used as motion sensors to detect various human movements. The introduction of MXene can effectively improve the triboelectric properties about open-circuit voltage and short-circuit current of cellulose aerogel composites. Besides, Shi et al. (2019) loaded tetragonal $BaTiO_3$ with a high piezoelectric coefficient and high dielectric constant into cellulose aerogel to prepare cellulose/$BaTiO_3$ composite aerogel (C/BT). Then the C/BT aerogel is compressed into flexible paper and then coated with PDMS to realize the preparation of C/BT/PDMS. The high loading of tetragonal BT nanoparticles can significantly enhance the piezoelectric property of the C/BT/PDMS aerogel composites. The C/BT/PDMS can reach the maximum output voltage, current, and maximum power of 15.5 V, 3.3 μA, and 11.8 μW.

Supercapacitor. Supercapacitor is a new type of energy storage device between traditional capacitor and rechargeable battery (Liu et al., 2024a). Flexible electrodes, electrolytes, and separators composed of cellulose aerogels have become potential flexible supercapacitor materials because of their rich pore structure, good thermal and chemical stability, and easy biodegradation (Sun et al., 2021; Park et al., 2023). However, pure cellulose aerogels have defects in properties and need to be further modified to be more suitable for supercapacitor materials (Li et al., 2023a). Liu et al. (2022) proposed a novel cellulose aerogel composites composed of TOCNFs and GO. The role of TOCNFs is to act as an elastic network structure, and GO acts as the skeleton in cellulose aerogel composites to prevent shrinkage of TOCNFs during carbonization. The prepared cellulose aerogel composites showed high compressibility under 80% strain and good elasticity about 96% stress retention after 2000 compressive cycles at 30% strain. Interestingly, inspired by a specialized network of rigidity-flexible integration in *aestivum,* Cai et al. (2023) proposed the half-dry membrane aerogel. The prepared cellulose aerogel composites showed low-tortuosity topology, 870 S/cm electron-conductive network and 215 F/g electrochemical property. Assembled supercapacitors can harvest energy from sunlight and result in a 196% enhancement in capacitance due to the synergistic photothermal capability.

Electromagnetic shielding. Radiation can disrupt the normal function of equipment and even induce cancerous changes in human cells (Mai et al., 2024). Aerogel-based composites exhibit low density, significant porosity, and elevated specific surface area, making them an ideal candidate for EMI shielding applications (Chen et al., 2021a). However, pure cellulose is nonconductive and needs to be modified to improve its conductivity (Mai et al., 2023a). Xin et al. (2021) prepared the silicone-coated MXene/CNFs aerogel membranes with the EMI shielding of 39.5 dB. Moreover, due to silicone coating, cellulose aerogel composites exhibit outstanding water-repellent properties. In addition, to achieve the designability and diversity of miniature electromagnetic shielding materials. Wu et al. (2023) prepared the anisotropic cellulose/polyacrylamide/graphene nanosheet/silver nanowire hybrid aerogels via the directional freeze-drying method. The novel anisotropic cellulose aerogel composites exhibited the EMI SE of 71.08 dB along the transverse direction and 35.21 dB along the longitudinal direction. Anisotropic electromagnetic shielding performance makes this kind of material play an important role in electromagnetic shielding applications with special needs. Chen et al. (2023) reported CNFs/MXene cellulose aerogel composites electromagnetic shielding devices with adjustable 2D/3D macroscopic shapes by direct ink write 3D printing technology, which provides meaningful insights into the shielding of diverse electronic devices (such as Bluetooth module) in the future.

Sound absorption. Noise has become one of the main pollution in modern life. The World Health Organization (WHO) emphasizes that noise pollution is a public health problem, because direct exposure to noise for a long time will lead to physical or mental illness. The complex internal structure, high porosity, and large specific surface area of cellulose aerogel composites can increase the friction loss between sound waves and materials. Ordered structures can also induce the propagation direction of sound waves, thus achieving sound absorption. However, the frequency of noise in daily life is widely distributed, and it is difficult for pure cellulose aerogel materials to absorb sound at higher frequencies and need further functionalization (Basic Properties of Building Decorative Materials, 2011), optimized performance, and wider applications. The combination of the two can be used to construct cellulose aerogel composites with many advantages. In this paper, the latest progress in preparation methods and multifunctional applications of cellulose-based aerogels is reviewed, and the improvement of the properties of cellulose aerogels by composite functional components is emphasized. Over the past decades, cellulose aerogel composites have been explored in many fields. However, to achieve industrial-scale production, there are still several practical challenges to overcome.

First, whether it is cellulose aerogel composites or other types of cellulose products, one of the primary purposes of considering the use of cellulose and its derivatives is to reduce or replace traditional nonrenewable petrochemical materials, so as to achieve an environmentally friendly production process (Calvino et al., 2020). This is one way to comply with the generally accepted concept of minimizing the impact of human production on the environment. However, no matter what kind of biomass material, its components are very complex, and lignin and hemicellulose are closely cross-linked with cellulose. In the current cellulose extraction process, to extract high purity cellulose raw materials, a lot of energy and chemical agents need to be consumed in the process of pretreatment, bleaching, extraction, and separation (Lu et al., 2021). Although cellulose itself is a green material, its production process is not completely clean and environmentally friendly. Therefore, although the use of cellulosic materials can effectively reduce the environmental pollution caused by oil extraction and refining processes, the pollution of the cellulose extraction process also needs to be paid attention to (Tahir et al., 2022). A more environmentally friendly method for extracting cellulose is worth further exploration. In addition, there is still a lack of research on the environmental impact of cellulose raw material preparation (Kong et al., 2022). If the actual cost and environmental impact are much higher than traditional nondegradable raw materials, it will be difficult to achieve the original goal of using cellulose materials to achieve green and friendly chemical production processes.

Therefore, in the follow-up research and development process, researchers should also pay attention to the waste produced in the cellulose extraction process and the environmental pollution caused by the chemical reagents used. Optimize the cellulose extraction method and production process, or recycle the resulting wastewater.

Second, sustainability is another concern for cellulose aerogel composites. There have been many reports exploring the degradability of cellulose (Luz et al., 2020; Frank et al., 2021). However, the sustainability of functional components in cellulose aerogel composites is often overlooked in most reports. For example, some cellulose aerogel composites are loaded with metal nanoparticles. Once cellulose as a supporting structure framework is degraded, metal nanoparticles will diffuse into the environment, but may be more difficult to recycle, resulting in secondary pollution. And sometimes to achieve higher functional performance, a large number of functional components will be filled into cellulose aerogel composites. Although magnetic cellulose aerogel composites can be recycled through magnets, reusability still needs to be improved. Lu et al. (2022) reported a biodegradable and recyclable piezoelectric sensor with imidazole perchlorate ($ImClO_4$) molecular ferroelectric as the functional element and BC as the substrate. Interestingly, BC hydrogel can be completely degraded into glucose and oligosaccharides, while $ImClO_4$ can be recycled and reused in the same equipment, without leaving electronic waste harmful to the environment. However, similar reports on cellulose aerogel composites are rarely reported. Future development can focus on cellulose aerogel composites consisting of completely degradable components.

Third, future manufacturing methods need to adapt to the efficient and rapid preparation of cellulose aerogels without damaging their complex structure. The process of preparing gel precursors for cellulose aerogels is complex and time-consuming. In addition, a large amount of energy is required in the drying process. For example, freeze-drying or supercritical CO_2 drying usually takes 24–96 h. Even if the method of ambient pressure drying is used, a large number of organic solvents need to be consumed as substitute solvents in the repeated solvent exchange process (such as ethanol). However, the three commonly used methods of drying of cellulose aerogel composites make it difficult to achieve rapid and effective mass preparation. Finding a fast and efficient method to prepare aerogels with complex structures is also a way worth exploring in the future.

Fourth, the research and development of multifunctional integrated cellulose aerogels is also an important direction to promote the development of cellulose aerogels in the future. Although cellulose aerogels have been extensively studied, most of the current research is focused on a single functional or the same type of application. This greatly limits the development of cellulose aerogels. For example, in the field of sensing, we can develop the self-powered cellulose aerogel composites, which can monitor temperature, humidity, pressure, and physiological signals at the same time, and has ultralight, flexible, flame-retardant, thermal insulation, and breathable properties. The integration of multiple functions can endow cellulose aerogel composites with high value.

Fifth, the cost problem is one of the main constraints limiting the development of emerging materials. Although cellulose aerogel composites have developed rapidly at present, the cost problem in the extraction and processing process still needs to be further improved. Ai et al. (2022) summarized the reports on the life cycle assessment of nanocellulose and suggested several ways to improve the industrialization of cellulose materials, starting with the selection of raw materials with high cellulose content and requiring as little pretreatment as possible. In addition, sustainable and recyclable reagents must be selected during processing, processing must be expanded to minimize average energy consumption, and mature and proven industrial technologies must be selected. These methods have effectively promoted industrial cellulose.

All in all, the achievement of these goals will greatly promote the development of cellulose-based aerogel composites in various fields, whether at the laboratory level or the industrial level.

Acknowledgments

The financial support from Beijing Nova Program (20230484431) and Opening Project of State Silica-Based Materials Laboratory of Anhui Province (2022KF12) is gratefully acknowledged.

References

Adamczyk, J., Dylewski, R., 2017. The impact of thermal insulation investments on sustainability in the construction sector. Renew. Sustain. Energy Rev. 80, 421–429. https://doi.org/10.1016/j.rser.2017.05.173.

Aghvami-Panah, M., Ameli, A., 2024. MXene/cellulose composites as electromagnetic interference shields: relationships between microstructural design and shielding performance. Compos. Appl. Sci. Manuf. 176, 107879. https://doi.org/10.1016/j.compositesa.2023.107879.

Ai, Y., Zhang, L., Cui, M., Huang, R., Qi, W., He, Z., Klemeš, J.J., Su, R., 2022. Toward cleaner production of nanocellulose: a review and evaluation. Green Chem. 24 (17), 6406–6434. https://doi.org/10.1039/d2gc01669a.

Basic Properties of Building Decorative Materials, 2011. Elsevier BV, pp. 10–24. https://doi.org/10.1533/9780857092588.10.

Bi, B., Guan, Y., Qiao, D., Chen, X., Bao, M., Wang, Z., Li, Y., 2023. MXene/Graphene modified cellulose aerogel for photo-electro-assisted all-weather cleanup of high-viscous crude oil from spill. J. Hazard Mater. 460, 132353. https://doi.org/10.1016/j.jhazmat.2023.132353.

Cai, C., Wei, Z., Deng, L., Fu, Y., 2021. Temperature-invariant superelastic multifunctional MXene aerogels for high-performance photoresponsive supercapacitors and wearable strain sensors. ACS Appl. Mater. Interfaces 13 (45), 54170–54184. https://doi.org/10.1021/acsami.1c16318.

Cai, C., Wei, Z., Ding, C., Sun, B., Chen, W., Gerhard, C., Nimerovsky, E., Fu, Y., Zhang, K., 2022. Dynamically tunable all-weather daytime cellulose aerogel radiative supercooler for energy-saving building. Nano Lett. 22 (10), 4106−4114. https://doi.org/10.1021/acs.nanolett.2c00844.

Cai, C., Liu, J., Wei, Z., Huang, Y., Fu, Y., 2023. Aestivum-inspired photothermal 3D-MXene-unity membrane-aerogel for flexible supercapacitor with solar enhanced performance. Compos. B Eng. 260, 110748. https://doi.org/10.1016/j.compositesb.2023.110748.

Cai, J., Kimura, S., Wada, M., Kuga, S., 2009. Nanoporous cellulose as metal nanoparticles support. Biomacromolecules 10 (1), 87−94. https://doi.org/10.1021/bm800919e.

Calvino, C., Macke, N., Kato, R., Rowan, S.J., 2020. Development, processing and applications of bio-sourced cellulose nanocrystal composites. Prog. Polym. Sci. 103, 101221. https://doi.org/10.1016/j.progpolymsci.2020.101221.

Cao, M., Wang, Y., Feng, L., Zhao, S., Wang, H., Liu, T., Yuan, Y., Wang, N., 2023. Ion imprinted nanocellulose aerogel with comprehensive optimized performance for uranium extraction from seawater. Chem. Eng. J. 475, 146048. https://doi.org/10.1016/j.cej.2023.146048.

Chen, Y., Yang, Y., Xiong, Y., Zhang, L., Xu, W., Duan, G., Mei, C., Jiang, S., Rui, Z., Zhang, K., 2021a. Porous aerogel and sponge composites: assisted by novel nanomaterials for electromagnetic interference shielding. Nano Today 38. https://doi.org/10.1016/j.nantod.2021.101204.

Chen, Y., Zhang, L., Yang, Y., Pang, B., Xu, W., Duan, G., Jiang, S., Zhang, K., 2021b. Recent progress on nanocellulose aerogels: preparation, modification, composite fabrication, applications. Adv. Mater. 33 (11). https://doi.org/10.1002/adma.202005569.

Chen, L., Mai, T., Ji, X.-X., Wang, P.-L., Qi, M.-Y., Liu, Q., Ding, Y., Ma, M.-G., 2023. 3D printing of customizable and lightweight multilayer MXene/nanocellulose architectures for tunable electromagnetic interference shielding via direct ink writing. Chem. Eng. J. 476, 146652. https://doi.org/10.1016/j.cej.2023.146652.

Chen, S., Chen, Y., Li, D., Xu, Y., Xu, F., 2021c. Flexible and sensitivity-adjustable pressure sensors based on carbonized bacterial nanocellulose/wood-derived cellulose nanofibril composite aerogels. ACS Appl. Mater. Interfaces 13 (7), 8754−8763. https://doi.org/10.1021/acsami.0c21392.

Chen, Z., Yan, T., Pan, Z., 2021d. Review of flexible strain sensors based on cellulose composites for multi-faceted applications. Cellulose 28 (2), 615−645. https://doi.org/10.1007/s10570-020-03543-6.

Chen, Z., Weng, P., Song, Y., Zheng, L., Tan, Y., Yin, X., 2024. Loofah-inspired sodium alginate/carboxymethyl cellulose sodium-based porous frame for all-weather super-viscous crude oil adsorption and wastewater treatment in harsh environment. Carbohydr. Polym. 323, 121450. https://doi.org/10.1016/j.carbpol.2023.121450.

Cheng, Y., Zhu, W.D., Lu, X.F., Wang, C., 2019. Lightweight and flexible MXene/carboxymethyl cellulose aerogel for electromagnetic shielding, energy harvest and self-powered sensing. Nano Energy 57, 450−458.

Cheng, R., Zeng, J., Wang, B., Li, J., Cheng, Z., Xu, J., Gao, W., Chen, K., 2021. Ultralight, flexible and conductive silver nanowire/nanofibrillated cellulose aerogel for multifunctional strain sensor. Chem. Eng. J. 424, 130565. https://doi.org/10.1016/j.cej.2021.130565.

Ciftci, D., Ubeyitogullari, A., Huerta, R.R., Ciftci, O.N., Flores, R.A., Saldaña, M.D.A., 2017. Lupin hull cellulose nanofiber aerogel preparation by supercritical CO2 and freeze drying. J. Supercrit. Fluids 127, 137−145. https://doi.org/10.1016/j.supflu.2017.04.002.

De France, K.J., Hoare, T., Cranston, E.D., 2017. Review of hydrogels and aerogels containing nanocellulose. Chem. Mater. 29 (11), 4609–4631. https://doi.org/10.1021/acs.chemmater.7b00531.

Delgado-Aguilar, M., Tarrés, Q., Pèlach, M.À., Mutjé, P., Fullana-i-Palmer, P., 2015. Are cellulose nanofibers a solution for a more circular economy of paper products? Environ. Sci. Technol. 49 (20), 12206–12213. https://doi.org/10.1021/acs.est.5b02676.

Deng, J., Zhu, E.Q., Xu, G.F., Naik, N., Murugadoss, V., Ma, M.G., Guo, Z., Shi, Z.J., 2022a. Overview of renewable polysaccharide-based composites for biodegradable food packaging applications. Green Chem. 24 (2), 480–492. https://doi.org/10.1039/d1gc03898b.

Deng, J., Song, Q., Liu, S., Pei, W., Wang, P., Zheng, L., Huang, C., Ma, M., Jiang, Q., Zhang, K., 2022b. Advanced applications of cellulose-based composites in fighting bone diseases. Compos. B Eng. 245, 110221. https://doi.org/10.1016/j.compositesb.2022.110221.

Dong, H., Li, X., Cai, Z., Wei, S., Fan, S., Ge, Y., et al., 2024. Lightweight, and shape-memory bamboo-derived all-cellulose aerogels for versatile Scaffolds of sustainable multifunctional materials. Small 20 (4).

Dong, S., Maciejewska, B., Millar, R., Grobert, N., 2023. 3D Electrospinning of Al2O3/ZrO2 fibrous aerogels for multipurpose thermal insulation. Adv. Compos. Hybrid Mater. 6 (5). https://doi.org/10.1007/s42114-023-00760-y.

Du, A., Wang, H., Zhou, B., Zhang, C., Wu, X., Ge, Y., Niu, T., Ji, X., Zhang, T., Zhang, Z., Wu, G., Shen, J., 2018. Multifunctional silica nanotube aerogels inspired by polar bear hair for light management and thermal insulation. Chem. Mater. 30 (19), 6849–6857. https://doi.org/10.1021/acs.chemmater.8b02926.

Fan, B., Bao, X., Pan, S., Liu, Y., Yu, Y., Cui, L., Wang, Q., Wang, P., 2024. High capillary effect and solar dual-drive nanofibrillated cellulose aerogels for efficient crude oil spill cleanup. Chem. Eng. J. 480, 148149. https://doi.org/10.1016/j.cej.2023.148149.

Fattakhova-Rohlfing, D., Zaleska, A., Bein, T., 2014. Three-dimensional titanium dioxide nanomaterials. Chem. Rev. 114 (19), 9487–9558. https://doi.org/10.1021/cr500201c.

Frank, B.P., Smith, C., Caudill, E.R., Lankone, R.S., Carlin, K., Benware, S., Pedersen, J.A., Fairbrother, D.H., 2021. Biodegradation of functionalized nanocellulose. Environ. Sci. Technol. 55 (15), 10744–10757. https://doi.org/10.1021/acs.est.0c07253.

Gackowski, B.M., Goh, G.D., Sharma, M., Idapalapati, S., 2023. Additive manufacturing of nylon composites with embedded multi-material piezoresistive strain sensors for structural health monitoring. Compos. B Eng. 261, 110796. https://doi.org/10.1016/j.compositesb.2023.110796.

Garemark, J., Perea-Buceta, J.E., Felhofer, M., Chen, B., Cortes Ruiz, M.F., Sapouna, I., Gierlinger, N., Kilpeläinen, I.A., Berglund, L.A., Li, Y., 2023. Strong, shape-memory lignocellulosic aerogel via wood cell wall nanoscale reassembly. ACS Nano 17 (5), 4775–4789. https://doi.org/10.1021/acsnano.2c11220.

Garemark, J., Perea-Buceta, J.E., Rico Del Cerro, D., Hall, S., Berke, B., Kilpeläinen, I., Berglund, L.A., Li, Y., 2022. Nanostructurally controllable strong wood aerogel toward efficient thermal insulation. ACS Appl. Mater. Interfaces 14 (21), 24697–24707. https://doi.org/10.1021/acsami.2c04584.

Gavillon, R., Budtova, T., 2008. Aerocellulose: new highly porous cellulose prepared from cellulose-NaOH aqueous solutions. Biomacromolecules 9 (1), 269–277. https://doi.org/10.1021/bm700972k.

Gu, H., Gao, C., Zhou, X., Du, A., Naik, N., Guo, Z., 2021. Nanocellulose nanocomposite aerogel towards efficient oil and organic solvent adsorption. Adv. Compos. Hybrid Mater. 4 (3), 459–468. https://doi.org/10.1007/s42114-021-00289-y.

Gu, B., Xu, Q., Wang, H., Pan, H., Zhao, D., 2023. A hierarchically nanofibrous self-cleaning textile for efficient personal thermal management in severe hot and cold environments. ACS Nano 17 (18), 18308–18317. https://doi.org/10.1021/acsnano.3c05460.

Gu, H., Zhou, X., Lyu, S., Pan, D., Dong, M., Wu, S., Ding, T., Wei, X., Seok, I., Wei, S., Guo, Z., 2020. Magnetic nanocellulose-magnetite aerogel for easy oil adsorption. J. Colloid Interface Sci. 560, 849–856. https://doi.org/10.1016/j.jcis.2019.10.084.

Guastaferro, M., Reverchon, E., Baldino, L., Agarose, 2021. Alginate and chitosan nanostructured aerogels for pharmaceutical applications: a short review. Front. Bioeng. Biotechnol. 9.

Han, X., Ding, S., Zhu, L., Wang, S., 2023. Preparation and characterization of flame-retardant and thermal insulating bio-based composite aerogels. Energy Build. 278, 112656. https://doi.org/10.1016/j.enbuild.2022.112656.

Hoviatdoost, A., Naderi, M., Ghazitabar, A., Gholami, F., 2023. Fabrication of high-performance ultralight and reusable graphene aerogel/cellulose fibers nanocomposite to remove organic pollutants. Mater. Today Commun. 34, 105077. https://doi.org/10.1016/j.mtcomm.2022.105077.

Hu, F., Wu, S., Sun, Y., 2019. Hollow-structured materials for thermal insulation. Adv. Mater. 31 (38). https://doi.org/10.1002/adma.201801001.

Hu, X., Zhang, S., Yang, B., Hao, M., Chen, Z., Liu, Y., Wang, X., Yao, J., 2023. Preparation of ambient-dried multifunctional cellulose aerogel by freeze-linking technique. Chem. Eng. J. 477, 147044. https://doi.org/10.1016/j.cej.2023.147044.

Hu, Z., Zhang, X., Sun, Q., Gu, P., Liang, X., Yang, X., Liu, M., Huang, J., Wu, G., Zu, G., 2023. Biomimetic transparent layered tough aerogels for thermal superinsulation and triboelectric nanogenerator. Small. https://doi.org/10.1002/smll.202307602.

Huang, J., Li, D., Zhao, M., Ke, H., Mensah, A., Lv, P., Tian, X., Wei, Q., 2019. Flexible electrically conductive biomass-based aerogels for piezoresistive pressure/strain sensors. Chem. Eng. J. 373, 1357–1366. https://doi.org/10.1016/j.cej.2019.05.136.

Huang, Y., Zhou, P., Zhang, X., 2022. Green synthesis of Ag-doped cellulose aerogel for highly sensitive, flame retardant strain sensors. Cellulose 29 (16), 8719–8731. https://doi.org/10.1007/s10570-022-04802-4.

Huang, L.-Z., Qi, M.-Y., Zhang, W., Guo, W.-Y., Ma, M.-G., 2023. Multifunctional composite fabric with outstanding self-cleaning and high photothermal effect for biomechanical energy harvester. Compos. Appl. Sci. Manuf. 172, 107623. https://doi.org/10.1016/j.compositesa.2023.107623.

Huang, Z., Zhang, H., Guo, M., Zhao, M., Liu, Y., Zhang, D., Terrones, M., Wang, Y., 2022. Large-scale preparation of electrically conducting cellulose nanofiber/carbon nanotube aerogels: ambient-dried, recyclable, and 3D-Printable. Carbon 194, 23–33. https://doi.org/10.1016/j.carbon.2022.03.056.

Illa, M.P., Sharma, C.S., Khandelwal, M., 2019. Tuning the physiochemical properties of bacterial cellulose: effect of drying conditions. J. Mater. Sci. 54 (18), 12024–12035. https://doi.org/10.1007/s10853-019-03737-9.

Jabbari, F., Babaeipour, V., 2024. Bacterial cellulose as an ideal potential treatment for burn wounds: a comprehensive review. Wound Repair Regen. 32 (3), 323–339. https://doi.org/10.1111/wrr.13163.

Jayan, S.S., Jayan, J.S., Saritha, A., 2023. A review on recent advances towards sustainable development of bio-inspired agri-waste based cellulose aerogels. Int. J. Biol. Macromol. 248, 125928. https://doi.org/10.1016/j.ijbiomac.2023.125928.

Ji, F., Sun, Z., Hang, T., Zheng, J., Li, X., Duan, G., Zhang, C., Chen, Y., 2022. Flexible piezoresistive pressure sensors based on nanocellulose aerogels for human motion monitoring: a review. Compos. Commun. 35. https://doi.org/10.1016/j.coco.2022.101351.

Jiang, F., Hu, S., Hsieh, Y.L., 2018. Aqueous synthesis of compressible and thermally stable cellulose nanofibril-silica aerogel for CO2 adsorption. ACS Appl. Nano Mater. 1 (12), 6701–6710. https://doi.org/10.1021/acsanm.8b01515.

Jiménez-Saelices, C., Seantier, B., Grohens, Y., Capron, I., 2018. Thermal superinsulating materials made from nanofibrillated cellulose-stabilized pickering emulsions. ACS Appl. Mater. Interfaces 10 (18), 16193–16202. https://doi.org/10.1021/acsami.8b02418.

Kistler, S.S., 1931. Coherent expanded aerogels and jellies. Nature 127 (3211), 741. https://doi.org/10.1038/127741a0.

Kong, H., Chen, Y., Yang, G., Liu, B., Guo, L., Wang, Y., Zhou, X., Wei, G., 2022. Two-dimensional material-based functional aerogels for treating hazards in the environment: synthesis, functional tailoring, applications, and sustainability analysis. Nanoscale Horizons 7 (2), 112–140. https://doi.org/10.1039/d1nh00633a.

Kong, X., Zhou, S., Strømme, M., Xu, C., 2022. All-cellulose-based freestanding porous carbon nanocomposites and their versatile applications. Compos. B Eng. 232, 109602. https://doi.org/10.1016/j.compositesb.2021.109602.

Lai, H., Zhuo, H., Hu, Y., Shi, G., Chen, Z., Zhong, L., Zhang, M., 2021. Anisotropic carbon aerogel from cellulose nanofibers featuring highly effective compression stress transfer and pressure sensing. ACS Sustain. Chem. Eng. 9 (29), 9761–9769. https://doi.org/10.1021/acssuschemeng.1c02051.

Lavoine, N., Bergström, L., 2017. Nanocellulose-based foams and aerogels: processing, properties, and applications. J. Mater. Chem. A 5 (31), 16105–16117. https://doi.org/10.1039/c7ta02807e.

Le, H.V., Dao, N.T., Bui, H.T., Kim Le, P.T., Le, K.A., Tuong Tran, A.T., Nguyen, K.D., Mai Nguyen, H.H., Ho, P.H., 2023. Bacterial cellulose aerogels derived from pineapple Peel waste for the adsorption of dyes. ACS Omega 8 (37), 33412–33425. https://doi.org/10.1021/acsomega.3c03130.

Lee, S., Kang, K.Y., Jeong, M.J., Potthast, A., Liebner, F., 2017. Evaluation of supercritical CO2 dried cellulose aerogels as nano-biomaterials. J. Kor. Phys. Soc. 71 (8), 483–486. https://doi.org/10.3938/jkps.71.483.

Li, J., Chen, S., Li, X., Zhang, J., Nawaz, H., Xu, Y., Kong, F., Xu, F., 2023a. Anisotropic cellulose nanofibril aerogels fabricated by directional stabilization and ambient drying for efficient solar evaporation. Chem. Eng. J. 453, 139844. https://doi.org/10.1016/j.cej.2022.139844.

Li, Q., Tang, R., Zhou, H., Hu, X., Zhang, S., 2023b. A high-performance and flexible electrode film based on bacterial cellulose/polypyrrole/nitrogen-doped graphene for supercapacitors. Carbohydr. Polym. 311, 120754. https://doi.org/10.1016/j.carbpol.2023.120754.

Liao, W., Xiao, K., Tian, T., Pan, M., 2024. Carbon aerogel monoliths from polymers: a review. J. Clean. Prod. 437. https://doi.org/10.1016/j.jclepro.2024.140736.

Liao, W., Zhao, H.B., Liu, Z., Xu, S., Wang, Y.Z., 2019. On controlling aerogel microstructure by freeze casting. Compos. B Eng. 173. https://doi.org/10.1016/j.compositesb.2019.107036.

Lin, N., Dufresne, A., 2014. Nanocellulose in biomedicine: current status and future prospect. Eur. Polym. J. 59, 302–325. https://doi.org/10.1016/j.eurpolymj.2014.07.025.

Liu, Q., Yan, K., Chen, J., Xia, M., Li, M., Liu, K., Wang, D., Wu, C., Xie, Y., 2021. Recent advances in novel aerogels through the hybrid aggregation of inorganic nanomaterials and polymeric fibers for thermal insulation. Aggregate 2 (2). https://doi.org/10.1002/agt2.30.

Liu, W., Li, H., Tay, R.Y., 2024a. Recent progress of high-performance in-plane zinc ion hybrid micro-supercapacitors: design, achievements, and challenges. Nanoscale 16 (9), 4542–4562. https://doi.org/10.1039/d3nr06120e.

Liu, Z., Zhang, S., He, B., Wang, S., Kong, F., 2021. Synthesis of cellulose aerogels as promising carriers for drug delivery: a review. Cellulose 28 (5), 2697–2714. https://doi.org/10.1007/s10570-021-03734-9.

Liu, C., Huang, C., Li, Y., Liu, Y., Bian, H., Xiang, Z., Wang, H., Wang, H., Xiao, H., 2023a. Freeze-casting production of thermal insulating and fire-retardant lightweight aerogels based on nanocellulose and boron nitride. Int. J. Biol. Macromol. 252, 126370. https://doi.org/10.1016/j.ijbiomac.2023.126370.

Liu, F., Zhang, Y., Zhong, L., Feng, Q., Dong, Z., Xu, Z., 2024b. Dialdehyde modified and cationic aerogel for efficient microplastics adsorption from environmental waters. Int. J. Biol. Macromol. 256, 128326. https://doi.org/10.1016/j.ijbiomac.2023.128326.

Liu, H., Xu, T., Liang, Q., Zhao, Q., Zhao, D., Si, C., 2022. Compressible cellulose nanofibrils/reduced graphene oxide composite carbon aerogel for solid-state supercapacitor. Adv. Compos. Hybrid Mater. 5 (2), 1168–1179. https://doi.org/10.1007/s42114-022-00427-0.

Liu, H., Sun, F., Li, X., Ma, Q., Liu, G., Yu, H., Yu, W., Dong, X., Su, Z., 2023b. g-C3N4/TiO2/ZnIn2S4 graphene aerogel photocatalysts with double S-scheme heterostructure for improving photocatalytic multifunctional performances. Compos. B Eng. 259, 110746. https://doi.org/10.1016/j.compositesb.2023.110746.

Liu, Q., Liu, Y., Feng, Q., Chen, C., Xu, Z., 2023c. Preparation of antifouling and highly hydrophobic cellulose nanofibers/alginate aerogels by bidirectional freeze-drying for water-oil separation in the ocean environment. J. Hazard Mater. 441, 129965. https://doi.org/10.1016/j.jhazmat.2022.129965.

Liu, X., Zhu, X., Peng, J., Song, D., Xu, W., Zhu, K., 2023d. Fabrication and characterization of chemical crosslinked cellulose fiber by dissolving cellulose in alkali urea aqueous solvent. J. Appl. Poly. Sci. 140. https://doi.org/10.1002/app.5353310974628.

Liu, W., Liu, K., Du, H., Zheng, T., Zhang, N., Xu, T., Pang, B., Zhang, X., Si, C., Zhang, K., 2022. Cellulose nanopaper: fabrication, functionalization, and applications. Nano-Micro Lett. 14 (1). https://doi.org/10.1007/s40820-022-00849-x.

Liu, Y., Zhang, Z., Hu, L., 2022. High efficient freeze-drying technology in food industry. Crit. Rev. Food Sci. Nutr. 62 (12), 3370–3388. https://doi.org/10.1080/10408398.2020.1865261.

Liu, Y., Zhang, H., Zhang, Y., Liang, C., An, Q., 2022. Rendering passive radiative cooling capability to cotton textile by an alginate/CaCO3 coating via synergistic light manipulation and high water permeation. Compos. B Eng. 240, 109988. https://doi.org/10.1016/j.compositesb.2022.109988.

Long, L.Y., Weng, Y.X., Wang, Y.Z., 2018. Cellulose aerogels: synthesis, applications, and prospects. Polymers 8 (6). https://doi.org/10.3390/polym10060623.

Lu, Y., He, Q., Fan, G., Cheng, Q., Song, G., 2021. Extraction and modification of hemicellulose from lignocellulosic biomass: a review. Green Process. Synth. 10 (1), 779–804. https://doi.org/10.1515/gps-2021-0065.

Lu, J., Hu, S., Li, W., Wang, X., Mo, X., Gong, X., Liu, H., Luo, W., Dong, W., Sima, C., Wang, Y., Yang, G., Luo, J.T., Jiang, S., Shi, Z., Zhang, G., 2022. A biodegradable and recyclable piezoelectric sensor based on a molecular ferroelectric embedded in a bacterial cellulose hydrogel. ACS Nano 16 (3), 3744–3755. https://doi.org/10.1021/acsnano.1c07614.

Luz, E.P.C.G., Chaves, P.H.S., Vieira, L.D.A.P., Ribeiro, S.F., Borges, M.D.F., Andrade, F.K., Muniz, C.R., Infantes-Molina, A., Rodríguez-Castellón, E., Rosa, M.D.F., Vieira, R.S., 2020. In vitro degradability and bioactivity of oxidized bacterial cellulose-hydroxyapatite composites. Carbohydr. Polym. 237. https://doi.org/10.1016/j.carbpol.2020.116174.

Ma, C., Ma, M.G., Si, C., Ji, X.X., Wan, P., 2021. Flexible MXene-based composites for wearable devices. Adv. Funct. Mater. 31 (22). https://doi.org/10.1002/adfm.202009524.

Ma, X., Lou, Y., Chen, X.B., Shi, Z., Xu, Y., 2019. Multifunctional flexible composite aerogels constructed through in-situ growth of metal-organic framework nanoparticles on bacterial cellulose. Chem. Eng. J. 356, 227–235. https://doi.org/10.1016/j.cej.2018.09.034.

Ma, B., Hou, X., Li, Y., 2022. Mechanically strong and thermal insulating polyimide/SiO2 aerogel prepared by ambient drying. Mater. Lett. 327, 133039. https://doi.org/10.1016/j.matlet.2022.133039.

Mai, T., Wang, P.L., Yuan, Q., Ma, C., Ma, M.G., 2021. In situ anchoring Zn-doped ZIF-67 on carboxymethylated bacterial cellulose for effective indigo carmine capture. Nanoscale 13 (43), 18210–18217. https://doi.org/10.1039/d1nr05388d.

Mai, T., Chen, L., Wang, L., Ma, M.G., 2024. Hollow metal–organic framework/MXene/nanocellulose composite films for giga/terahertz electromagnetic shielding and photothermal conversion. Nano-Micro Lett. 16 (1).

Mai, T., Guo, W.-Y., Wang, P.-L., Chen, L., Qi, M.-Y., Liu, Q., Ding, Y., Ma, M.-G., 2023a. Bilayer metal-organic frameworks/MXene/nanocellulose paper with electromagnetic double loss for absorption-dominated electromagnetic interference shielding. Chem. Eng. J. 464, 142517. https://doi.org/10.1016/j.cej.2023.142517.

Mai, T., Li, D.-D., Chen, L., Ma, M.-G., 2023b. Collaboration of two-star nanomaterials: the applications of nanocellulose-based metal organic frameworks composites. Carbohydr. Polym. 302, 120359. https://doi.org/10.1016/j.carbpol.2022.120359.

Mirmoeini, S.S., Moradi, M., Tajik, H., Almasi, H., Gama, F.M., 2023. Cellulose/Salep-based intelligent aerogel with red grape anthocyanins: preparation, characterization and application in beef packaging. Food Chem. 425, 136493. https://doi.org/10.1016/j.foodchem.2023.136493.

Namata, F., Sanz del Olmo, N., Wågberg, L., Malkoch, M., 2023. High water content physically cross-linked hybrid hydrogels based on polyester dendrimers and cellulose nanofibrils: a comprehensive Study. Chem. Mater. 35 (20), 8561–8573. https://doi.org/10.1021/acs.chemmater.3c01476.

Nepal, D., Kang, S., Adstedt, K.M., Kanhaiya, K., Bockstaller, M.R., Brinson, L.C., Buehler, M.J., Coveney, P.V., Dayal, K., El-Awady, J.A., Henderson, L.C., Kaplan, D.L., Keten, S., Kotov, N.A., Schatz, G.C., Vignolini, S., Vollrath, F., Wang, Y., Yakobson, B.I., Heinz, H., 2023. Hierarchically structured bioinspired nanocomposites. Nat. Mater. 22 (1), 18–35. https://doi.org/10.1038/s41563-022-01384-1.

Nguyen, S.T., Feng, J., Le, N.T., Le, A.T.T., Hoang, N., Tan, V.B.C., Duong, H.M., 2013. Cellulose aerogel from paper waste for crude oil spill cleaning. Ind. Eng. Chem. Res. 52 (51), 18386–18391. https://doi.org/10.1021/ie4032567.

Nita, L.E., Ghilan, A., Rusu, A.G., Neamtu, I., Chiriac, A.P., 2020. New trends in bio-based aerogels. Pharmaceutics 12 (5). https://doi.org/10.3390/pharmaceutics12050449.

Niu, L., Zhang, B., Sun, J., Wang, J., Qin, C., Dai, L., 2023. Fabrication of photocatalytic PAN nanofiber membrane loading with TiO2@RGO by electro-spinning & electro-spraying. Compos. B Eng. 266, 111046. https://doi.org/10.1016/j.compositesb.2023.111046.

Olorunnisola, D., Olorunnisola, C.G., Otitoju, O.B., Okoli, C.P., Rawel, H.M., Taubert, A., Easun, T.L., Unuabonah, E.I., 2023. Cellulose-based adsorbents for solid phase extraction and recovery of pharmaceutical residues from water. Carbohydr. Polym. 318. https://doi.org/10.1016/j.carbpol.2023.121097.

Ookuna, I.K., Hara, M., Aso, K., Yoshidone, H., Nakayama, H., Suzuki, Nakajima, n.d. Porousion-Exchanged Fine Cellulose Particles, Method for Production Thereof, and Affinity Carrier.

Park, S., Song, J., Lee, W.C., Jang, S., Lee, J., Kim, J., Kim, H.-K., Min, K., 2023. Advances in biomass-derived electrode materials for energy storage and circular carbon economy. Chem. Eng. J. 470, 144234. https://doi.org/10.1016/j.cej.2023.144234.

Paul, J., Ahankari, S.S., 2023. Nanocellulose-based aerogels for water purification: a review. Carbohydr. Polym. 309, 120677. https://doi.org/10.1016/j.carbpol.2023.120677.

Peng, H., Xiong, W., Yang, Z., Xu, Z., Cao, J., Jia, M., Xiang, Y., 2022. Advanced MOFs@aerogel composites: construction and application towards environmental remediation. J. Hazard Mater. 432, 128684. https://doi.org/10.1016/j.jhazmat.2022.128684.

Peng, L., Fu, D., Qi, H., Lan, C.Q., Yu, H., Ge, C., 2020. Micro- and nano-plastics in marine environment: source, distribution and threats — a review. Sci. Total Environ. 698, 134254. https://doi.org/10.1016/j.scitotenv.2019.134254.

Plazzotta, S., Calligaris, S., Manzocco, L., 2019. Structure of oleogels from κ-carrageenan templates as affected by supercritical-CO2-drying, freeze-drying and lettuce-filler addition. Food Hydrocolloids 96, 1–10. https://doi.org/10.1016/j.foodhyd.2019.05.008.

Prasad, C., Jeong, S.-G., Won, J.S., Ramanjaneyulu, S., Sangaraju, S., Kerru, N., Choi, H.Y., 2024. Review on recent advances in cellulose nanofibril based hybrid aerogels: synthesis, properties and their applications. Int. J. Biol. Macromol. 261, 129460. https://doi.org/10.1016/j.ijbiomac.2024.129460.

Qin, F., Zhang, K., Zhang, Z., Fang, J., Li, J., Lai, Y., Huang, H., 2019. Graphene/carbon aerogel for high areal capacity sulfur cathode of Li-S batteries. Ionics 25 (10), 4615–4624. https://doi.org/10.1007/s11581-019-03046-w.

Qin, Z., Chen, X., Lv, Y., Zhao, B., Fang, X., Pan, K., 2023. Wearable and high-performance piezoresistive sensor based on nanofiber/sodium alginate synergistically enhanced MXene composite aerogel. Chem. Eng. J. 451, 138586. https://doi.org/10.1016/j.cej.2022.138586.

Rahmanian, V., Pirzada, T., Wang, S., Khan, S.A., 2021. Cellulose-based hybrid aerogels: Strategies toward design and functionality. Adv. Mater. 33 (51). https://doi.org/10.1002/adma.202102892.

Rostamitabar, M., Subrahmanyam, R., Gurikov, P., Seide, G., Jockenhoevel, S., Ghazanfari, S., 2021. Cellulose aerogel micro fibers for drug delivery applications. Mater. Sci. Eng. C 127, 112196. https://doi.org/10.1016/j.msec.2021.112196.

Shi, K., Huang, X., Sun, B., Wu, Z., He, J., Jiang, P., 2019. Cellulose/BaTiO 3 aerogel paper based flexible piezoelectric nanogenerators and the electric coupling with triboelectricity. Nano Energy 57, 450–458. https://doi.org/10.1016/j.nanoen.2018.12.076.

Shi, X.R., Yang, L.F., Tian, L., Shen, J.P., Li, Z.Q., Pei, C.H., 2021. Laser ignited combustion of mesoporous nitrocellulose/graphene oxide composite aerogel with strong photo-thermal sensitivity and high flame propagation rate. J. Appl. Phys. 130 (22). https://doi.org/10.1063/5.0069788.

Shuai, L., Guo, Z.H., Zhang, P., Wan, J., Pu, X., Wang, Z.L., 2020. Stretchable, self-healing, conductive hydrogel fibers for strain sensing and triboelectric energy-harvesting smart textiles. Nano Energy 78, 105389. https://doi.org/10.1016/j.nanoen.2020.105389.

Sonu, S.S., Rai, N., Chauhan, I., 2023. Multifunctional Aerogels: a comprehensive review on types, synthesis and applications of aerogels. J. Sol. Gel Sci. Technol. 105 (2), 324–336. https://doi.org/10.1007/s10971-022-06026-1.

Sui, S., Quan, H., Yang, X., Dong, X., Ji, Y., Liu, C., Xu, G., Guo, S., Zhang, Y., 2024. Multifunctional modified polyurethane sponge for recovery of oil spills and photocatalytic degradation. Compos. B Eng. 271, 111176. https://doi.org/10.1016/j.compositesb.2023.111176.

Sun, H., Hou, C., Ji, T., Zhou, X., Ren, Z., Song, Y., 2023. Processing bulk wood into a light-permeable passive radiative cooling material for energy-efficient building. Compos. B Eng. 250, 110426. https://doi.org/10.1016/j.compositesb.2022.110426.

Sun, Y., Chu, Y., Wu, W., Xiao, H., 2021. Nanocellulose-based lightweight porous materials: a review. Carbohydr. Polym. 255, 117489. https://doi.org/10.1016/j.carbpol.2020.117489.

Sun, Z., Qu, K., You, Y., Huang, Z., Liu, S., Li, J., Hu, Q., Guo, Z., 2021. Overview of cellulose-based flexible materials for supercapacitors. J. Mater. Chem. A 9 (12), 7278–7300. https://doi.org/10.1039/d0ta10504j.

Syeda, H.I., Yap, P.-S., 2022. A review on three-dimensional cellulose-based aerogels for the removal of heavy metals from water. Sci. Total Environ. 807, 150606. https://doi.org/10.1016/j.scitotenv.2021.150606.

Tabernero, A., Baldino, L., Misol, A., Cardea, S., del Valle, E.M.M., 2020. Role of rheological properties on physical chitosan aerogels obtained by supercritical drying. Carbohydr. Polym. 233, 115850. https://doi.org/10.1016/j.carbpol.2020.115850.

Tahir, D., Karim, Hu, H., Naseem, s, Rehan, M., Ahmad, M., et al., 2022. Chemical functionalization, and commercial applications of nanocellulose and nanocellulose-based composites: a review. Polymers 14 (21).

Tang, S., Ma, M., Zhang, X., Zhao, X., Fan, J., Zhu, P., Shi, K., Zhou, J., 2022. Covalent cross-links enable the Formation of ambient-dried biomass aerogels through the activation of a triazine derivative for energy storage and generation. Adv. Funct. Mater. 32 (36). https://doi.org/10.1002/adfm.202205417.

Uscanga, M.A., Salvador, A., Camacho, M.D.M., Martínez-Navarrete, N., 2021. Impact of freeze-drying shelf temperature on the bioactive compounds, physical properties and sensory evaluation of a product based on orange juice. Int. J. Food Sci. Technol. 56 (10), 5409–5416. https://doi.org/10.1111/ijfs.15086.

Voisin, H., Bergström, L., Liu, P., Mathew, A., 2017. Nanocellulose-based materials for water purification. Nanomaterials 7 (3), 57. https://doi.org/10.3390/nano7030057.

Voon, Ken, L., Pang, Cem, S., & Chin, Fun, S., 2016. Highly porous cellulose beads of controllable sizes derived from regenerated cellulose of printed paper wastes. Mater. Lett. 164, 264–266. https://doi.org/10.1016/j.matlet.2015.10.161.

Vu, P.V., Doan, T.D., Tu, G.C., Do, N.H.N., Le, K.A., Le, P.K., 2022. A novel application of cellulose aerogel composites from pineapple leaf fibers and cotton waste: removal of dyes and oil in wastewater. J. Porous Mater. 29 (4), 1137–1147. https://doi.org/10.1007/s10934-022-01238-w.

Wang, B., Yin, X., Cheng, R., Li, J., Ying, G., Chen, K., 2022. Compressible, superelastic and fatigue resistant carbon nanofiber aerogels derived from bacterial cellulose for multifunctional piezoresistive sensors. Carbon 199, 318–328. https://doi.org/10.1016/j.carbon.2022.08.006.

Wang, G., Liu, X., Song, Z., Yu, D., Li, G., Wang, H., Ge, S., Liu, W., 2023. Multifunctional flexible pressure sensor based on a cellulose fiber-derived hierarchical carbon aerogel. ACS Appl. Electron. Mater. 5 (3), 1581–1591. https://doi.org/10.1021/acsaelm.2c01628.

Wang, P.L., Ma, C., Yuan, Q., Mai, T., Ma, M.G., 2022. Novel Ti3C2Tx MXene wrapped wood sponges for fast cleanup of crude oil spills by outstanding Joule heating and photothermal effect. J. Colloid Interface Sci. 606, 971–982. https://doi.org/10.1016/j.jcis.2021.08.092.

Wang, P.L., Mai, T., Ji, X.X., Hu, P., Ma, M.G., 2022. Aerogel-based composites for electromagnetic interface shielding applications. Compos. Interfac. 29 (13), 1483–1503. https://doi.org/10.1080/09276440.2022.2069320.

Wang, Z., Zhou, Z., Wang, S., Yao, X., Han, X., Cao, W., Pu, J., 2022. An anti-freezing and strong wood-derived hydrogel for high-performance electronic skin and wearable sensing. Compos. B Eng. 239. https://doi.org/10.1016/j.compositesb.2022.109954.

Wang, X., Yu, J., Cui, Y., Li, W., 2021. Research progress of flexible wearable pressure sensors. Sensor Actuator Phys. 330, 112838. https://doi.org/10.1016/j.sna.2021.112838.

Wang, Y., Cheng, H., Min, Y., Li, X., You, L., Li, J., 2022. Fe3O4@m-ZrO2@Ag ternary magnetic nanocomposites for sensitive SERS sensing and photocatalytic removal of Cr(VI) and organic dyes. Compos. B Eng. 239, 109959. https://doi.org/10.1016/j.compositesb.2022.109959.

Wang, Z., Zhu, W., Huang, R., Zhang, Y., Jia, C., Zhao, H., Chen, W., Xue, Y., 2020. Fabrication and characterization of cellulose nanofiber aerogels prepared via two different drying techniques. Polymers 12 (11), 2583. https://doi.org/10.3390/polym12112583.

Wei, L., Wu, Z., Tang, S., Qin, X., Xiong, Y., Li, J., Ruiz-Hitzky, E., Wang, X., 2023. Tracheid-inspired nanoarchitectured carbon-based aerogels with ultra-compressibility for wearable piezoresistive sensors. Carbon 203, 386–396. https://doi.org/10.1016/j.carbon.2022.11.081.

Wei, S., Qiu, X., An, J., Chen, Z., Zhang, X., 2021. Highly sensitive, flexible, green synthesized graphene/biomass aerogels for pressure sensing application. Compos. Sci. Technol. 207, 108730. https://doi.org/10.1016/j.compscitech.2021.108730.

Wei, W., Yi, Y., Song, J., Chen, X., Li, J., Li, J., 2022. Tunable graphene/nitrocellulose temperature alarm sensors. ACS Appl. Mater. Interfaces 14 (11), 13790–13800. https://doi.org/10.1021/acsami.2c02340.

Wu, C., Zeng, L., Chang, G., Zhou, Y., Yan, K., Xie, L., Xue, B., Zheng, Q., 2023. Composite phase change materials embedded into cellulose/polyacrylamide/graphene nanosheets/silver nanowire hybrid aerogels simultaneously with effective thermal management and anisotropic electromagnetic interference shielding. Adv. Compos. Hybrid Mater. 6 (1). https://doi.org/10.1007/s42114-022-00618-9.

Wu, N., Yang, Y., Wang, C., Wu, Q., Pan, F., Zhang, R., Liu, J., Zeng, Z., 2023. Ultrathin cellulose nanofiber assisted ambient-pressure-dried, ultralight, mechanically robust, multifunctional MXene aerogels. Adv. Mater. 35 (1). https://doi.org/10.1002/adma.202207969.

Wu, S., Cao, Y., Li, Y., Sun, W., 2023. Recent advances in material engineering and applications for passive daytime radiative cooling. Adv. Opt. Mater. 11 (4). https://doi.org/10.1002/adom.202202163.

Wu, Y., Parandoust, A., Sheibani, R., Kargaran, F., Khorsandi, Z., Liang, Y., Xia, C., Van Le, Q., 2023. Advances in gum-based hydrogels and their environmental applications. Carbohydr. Polym. 318, 121102. https://doi.org/10.1016/j.carbpol.2023.121102.

Wu, Y., Wang, X., Yao, L., Chang, S., Wang, X., 2023. Thermal insulation mechanism, preparation, and modification of nanocellulose aerogels: a review. Molecules 28 (15), 5836. https://doi.org/10.3390/molecules28155836.

Xiang, Q., Zhang, H., Liu, Z., Zhao, Y., Tan, H., 2024. Engineered structural carbon aerogel based on bacterial cellulose/chitosan and graphene oxide/graphene for multifunctional piezoresistive sensor. Chem. Eng. J. 480, 147825. https://doi.org/10.1016/j.cej.2023.147825.

Xiao, H., Lv, J., Tan, W., He, X., Chen, M., Zeng, K., Hu, J., Yang, G., 2022. Ultrasound-assisted freeze-drying process for polyimide aerogels. Chem. Eng. J. 450, 138344. https://doi.org/10.1016/j.cej.2022.138344.

Xin, W., Ma, M.G., Chen, F., 2021. Silicone-coated MXene/cellulose nanofiber aerogel films with photothermal and Joule heating performances for electromagnetic interference shielding. ACS Appl. Nano Mater. 4 (7), 7234–7243. https://doi.org/10.1021/acsanm.1c01185.

Xu, Y., Yan, C., Du, C., Xu, K., Li, Y., Xu, M., Bourbigot, S., Fontaine, G., Li, B., Liu, L., 2023. High-strength, thermal-insulating, fire-safe bio-based organic lightweight aerogel based on 3D network construction of natural tubular fibers. Compos. B Eng. 261, 110809. https://doi.org/10.1016/j.compositesb.2023.110809.

Yang, X., Cranston, E.D., 2014. Chemically cross-linked cellulose nanocrystal aerogels with shape recovery and superabsorbent properties. Chem. Mater. 26 (20), 6016–6025. https://doi.org/10.1021/cm502873c.

Yang, W.J., Wei, C.X., Yuen, A.C.Y., Lin, B., Yeoh, G.H., Lu, H.D., Yang, W., 2022. Fire-retarded nanocomposite aerogels for multifunctional applications: a review. Compos. B Eng. 237. https://doi.org/10.1016/j.compositesb.2022.109866.

Yin, B.B., Sun, W.K., Zhang, X.Y., Liew, K.M., 2022. Deciphering structural biological materials: Viewing from the mechanics perspective and their prospects. Compos. B Eng. 245, 110213. https://doi.org/10.1016/j.compositesb.2022.110213.

Yuan, Q., Huang, L.Z., Wang, P.L., Mai, T., Ma, M.G., 2022. Cellulose nanofiber/molybdenum disulfide aerogels for ultrahigh photothermal effect. J. Colloid Interface Sci. 624, 70–78. https://doi.org/10.1016/j.jcis.2022.05.102.

Yue, X., Wu, H., Zhang, T., Yang, D., Qiu, F., 2022. Superhydrophobic waste paper-based aerogel as a thermal insulating cooler for building. Energy 245, 123287. https://doi.org/10.1016/j.energy.2022.123287.

Zhai, J., Zhang, Y., Cui, C., Li, A., Wang, W., Guo, R., Qin, W., Ren, E., Xiao, H., Zhou, M., 2021. Flexible waterborne polyurethane/cellulose nanocrystal composite aerogels by integrating graphene and carbon nanotubes for a highly sensitive pressure sensor. ACS Sustain. Chem. Eng. 9 (42), 14029–14039. https://doi.org/10.1021/acssuschemeng.1c03068.

Zhang, H., Lyu, S., Zhou, X., Gu, H., Ma, C., Wang, C., Ding, T., Shao, Q., Liu, H., Guo, Z., 2019. Super light 3D hierarchical nanocellulose aerogel foam with superior oil adsorption. J. Colloid Interface Sci. 536, 245–251. https://doi.org/10.1016/j.jcis.2018.10.038.

Zhang, S., Pan, Y., Wang, W., Lin, R., Liu, X., 2023. Carboxylated cellulose-based aerogel with cellular pores prepared by stir freezing for cationic dye adsorption. Process Saf. Environ. Protect. 177, 807–817. https://doi.org/10.1016/j.psep.2023.07.052.

Zhang, W., Wang, P.L., Huang, L.Z., Guo, W.Y., Zhao, J., Ma, M.G., 2023. A stretchable, environmentally tolerant, and photoactive liquid metal/MXene hydrogel for high performance temperature monitoring, human motion detection and self-powered application. Nano Energy 117. https://doi.org/10.1016/j.nanoen.2023.108875.

Zhang, W., Wang, P.L., Ji, X.X., Huang, L.Z., Cao, D.Q., Li, J., Ma, M.G., 2024. Ultrastretchable and adhesive MXene-based hydrogel for high-performance strain sensing and self-powered application. Compos. Appl. Sci. Manuf. 177. https://doi.org/10.1016/j.compositesa.2023.107957.

Zhang, H., Wang, S., Zhang, J., Zhou, G., Sun, X., Wang, Y., Wang, Y., Zhang, K., 2024. High-sensitivity piezoresistive sensors based on cellulose handsheets using origami-inspired corrugated structures. Carbohydr. Polym. 328, 121742. https://doi.org/10.1016/j.carbpol.2023.121742.

Zhang, W., Xu, J., 2022. Advanced lightweight materials for Automobiles: a review. Mater. Des. 221, 110994. https://doi.org/10.1016/j.matdes.2022.110994.

Zhao, S., Malfait, W.J., Guerrero-Alburquerque, N., Koebel, M.M., Nyström, G., 2018. Biopolymer aerogels and foams: chemistry, properties, and applications. Angew. Chem. Int. Ed. 57 (26), 7580–7608. https://doi.org/10.1002/anie.201709014.

Zhao, B.C., Yue, X.J., Tian, Q., Qiu, F.X., Li, Y.Q., Zhang, T., 2022. Bio-inspired BC aerogel/PVA hydrogel bilayer gel for enhanced daytime sub-ambient building cooling. Cellulose 29 (14), 7775–7787.

Zhou, S., Apostolopoulou-Kalkavoura, V., Tavares da Costa, M.V., Bergstrom, L., Stromme, M., Xu, C., 2019. Elastic aerogels of cellulose nanofibers@metal-organic frameworks for thermal insulation and fire retardancy. Nanomicro Lett. 12 (1), 9.

Zhou, T., Ge, L., Li, Q., Yang, L., Mai, L., Huang, J., Wang, Y., Xu, C., 2023. Combustion and gasification properties of petroleum coke and its pyrolytic semi-coke. Energy 266, 126414. https://doi.org/10.1016/j.energy.2022.126414.

Zhou, X., Fu, Q., Liu, H., Gu, H., Guo, Z., 2021. Solvent-free nanoalumina loaded nanocellulose aerogel for efficient oil and organic solvent adsorption. J. Colloid Interface Sci. 581, 299–306. https://doi.org/10.1016/j.jcis.2020.07.099.

Zhou, Z., Liu, T., Zhou, W., Chen, H., Bian, Y., Gong, S., Shi, S.Q., Li, J., 2022. A strong and biodegradable wood particles-based bioplastic modified by synergistically dynamic cross-linking with tannic acid and $Fe3+$. Compos. B Eng. 247, 110349. https://doi.org/10.1016/j.compositesb.2022.110349.

Zhu, G., Wang, J., Wang, X., Liu, Y., Wu, W., Xiao, H., Huang, L., Dai, H., Zhou, X., Bian, H., 2023. Aerogels fabricated from wood-derived functional cellulose nanofibrils for highly efficient separation of microplastics. ACS Sustain. Chem. Eng. 11 (38), 13928–13938. https://doi.org/10.1021/acssuschemeng.3c02582.

Zhu, H., Yang, X., Cranston, E.D., Zhu, S., 2016. Flexible and porous nanocellulose aerogels with high loadings of metal–organic-framework particles for separations applications. Adv. Mater. 28 (35), 7652–7657. https://doi.org/10.1002/adma.201601351.

Zhu, J., Guo, S., Li, X., 2015. Facile preparation of a SiO 2 $-Al_2O_3$ aerogel using coal gangue as a raw material via an ambient pressure drying method and its application in organic solvent adsorption. RSC Adv. 5 (125), 103656–103661. https://doi.org/10.1039/C5RA20392A.

Zhuang, J., Pan, M., Zhang, Y., Liu, F., Xu, Z., 2023. Rapid adsorption of directional cellulose nanofibers/3-glycidoxypropyltrimethoxysilane/polyethyleneimine aerogels on microplastics in water. Int. J. Biol. Macromol. 235, 123884. https://doi.org/10.1016/j.ijbiomac.2023.123884.

Zhuang, J., Rong, N., Wang, X., Chen, C., Xu, Z., 2022. Adsorption of small size microplastics based on cellulose nanofiber aerogel modified by quaternary ammonium salt in water. Separ. Purif. Technol. 293, 121133. https://doi.org/10.1016/j.seppur.2022.121133.

Zhuo, H., Hu, Y., Chen, Z., Peng, X., Liu, L., Luo, Q., Yi, J., Liu, C., Zhong, L., 2019. A carbon aerogel with super mechanical and sensing performances for wearable piezoresistive sensors. J. Mater. Chem. A 7 (14), 8092–8100. https://doi.org/10.1039/c9ta00596j.

Index

'*Note:* Page numbers followed by "f" indicate figures and "t" indicate tables.'

D

O

P

R

S

9780443136757